Legumes
Nutritional Quality, Processing and Potential Health Benefits

Food Chemistry, Function and Analysis

Series editors:
Gary Williamson, *University of Leeds, UK*
Alejandro G. Marangoni, *University of Guelph, Canada*
Juliet A. Gerrard, *University of Auckland, New Zealand*

Titles in the series:
1: Food Biosensors
2: Sensing Techniques for Food Safety and Quality Control
3: Edible Oil Structuring: Concepts, Methods and Applications
4: Food Irradiation Technologies: Concepts, Applications and Outcomes
5: Non-extractable Polyphenols and Carotenoids: Importance in Human Nutrition and Health
6: Cereal Grain-based Functional Foods: Carbohydrate and Phytochemical Components
7: Steviol Glycosides: Cultivation, Processing, Analysis and Applications in Food
8: Legumes: Nutritional Quality, Processing and Potential Health Benefits

How to obtain future titles on publication:
A standing order plan is available for this series. A standing order will bring delivery of each new volume immediately on publication.

For further information please contact:
Book Sales Department, Royal Society of Chemistry, Thomas Graham House, Science Park, Milton Road, Cambridge, CB4 0WF, UK
Telephone: +44 (0)1223 420066, Fax: +44 (0)1223 420247,
Email: booksales@rsc.org
Visit our website at www.rsc.org/books

Legumes
Nutritional Quality, Processing and Potential Health Benefits

Edited by

María Ángeles Martín-Cabrejas
Universidad Autónoma de Madrid, Spain
Email: maria.martin@uam.es

ROYAL SOCIETY OF CHEMISTRY

Food Chemistry, Function and Analysis No. 8

Print ISBN: 978-1-78801-161-7
PDF ISBN: 978-1-78801-572-1
EPUB ISBN: 978-1-78801-675-9
Print ISSN: 2398-0656
Electronic ISSN: 2398-0664

A catalogue record for this book is available from the British Library

The Royal Society of Chemistry is a charity, registered in England and Wales, Number 207890, and a company incorporated in England by Royal Charter (Registered No. RC000524), registered office: Burlington House, Piccadilly, London W1J 0BA, UK, Telephone: +44 (0) 20 7437 8656.

Visit our website at www.rsc.org/books

Printed in the United Kingdom by CPI Group (UK) Ltd, Croydon, CR0 4YY, UK

Preface

In recent years there has been a clear breakthrough in the discovery of new bioactive compounds and the development of functional food ingredients that have the potential to modulate specific functions for the prevention of diseases. Traditionally, legumes have been considered as foods with beneficial effects for human health. Legumes are plants of the family Fabaceae whose seeds represent a basic pillar of human nutrition since ancient times. In addition to a very adequate nutritional content in carbohydrates, proteins, dietary fiber, minerals and vitamins, legumes also have a variety of non-nutritional compounds considered bioactive compounds with antioxidant, hypoglycemic, hypolipidemic and anticarcinogenic properties. Nowadays, there is an increasing trend and research activity for incorporating legumes in innovative food products and formulations that will help towards increasing legume consumption and improving human diet throughout the world. Consumers as well as the food industry are increasingly demanding tasty foods, convenient quick-cooking and ready-to-eat products that contribute to their health and well-being. Likewise, they are aware of and search for food alternatives such as legumes that offer variety with a reduced carbon footprint.

This book intends to present the recent state-of-the art of (i) the main bioactive compounds of legumes such as phenolic compounds (flavonoids and non-flavonoids), dietary fibre, bioactive peptides, melatonin, non-nutritional factors with antioxidant, hypoglycemic, hypolipidemic and anticarcinogenic properties; (ii) the impact of different legume processing prior to consumption and the changes occurring in their bioactive components with the emphasis on the innovative legume foods; and (iii) the book emphasizes the role of legumes in the prevention and management of

Food Chemistry, Function and Analysis No. 8
Legumes: Nutritional Quality, Processing and Potential Health Benefits
Edited by María Ángeles Martín-Cabrejas

Published by the Royal Society of Chemistry, www.rsc.org

chronic diseases such as metabolic syndrome, obesity, cardiovascular disease, gastrointestinal health and cancer. With contributions from food and nutrition scientists around the world, this book provides a comprehensive overview of the nutritional quality, processing and potential health aspects of legumes.

The book would not have been possible without the admirable efforts of international renowned contributors, and I would like to greatly express my gratitude for the precious time they have taken to share their knowledge. I wish to express my appreciation to the editorial staff at the Royal Society of Chemistry for keeping me working at this project.

Finally, and most importantly, I am deeply indebted to my husband Ángel and my son Javier, as well as my family.

María A. Martín-Cabrejas

Contents

Part I: Introduction

Part II: Bioactive Compounds of Legumes

Food Chemistry, Function and Analysis No. 8
Legumes: Nutritional Quality, Processing and Potential Health Benefits
Edited by María Ángeles Martín-Cabrejas

Published by the Royal Society of Chemistry, www.rsc.org

Part I

Introduction

CHAPTER 1

Legumes: An Overview

M. A. MARTÍN-CABREJAS*

[a]Instituto de Investigación en Ciencias de la Alimentación, CIAL (CSIC-UAM), Universidad Autónoma de Madrid, Madrid, Spain; [b]Departamento de Química Agrícola y Bromatología, Facultad de Ciencias, Universidad Autónoma de Madrid, Madrid, Spain
*E-mail: maria.martin@uam.es

1.1 Introduction

Legumes belong to the family *Fabaceae* whose seeds represent a basic pillar of human nutrition since ancient times. Nowadays, legumes play an important role due to their nutritional and health associated benefits,[1–7] together with economic and environmental factors crucial in our sustainable future.[8] Legumes are safe for consumption, relatively inexpensive and also easily available, being present in the diet of millions of individuals around the world. From academic and governmental institutions, a great effort is being made to promote the value and use of legumes throughout the food system, and to make everyone aware of their benefits. To achieve this objective, it is necessary to increase their global production, integrating them into cropping systems, improving the research of new legume varieties, developing novel management strategies and overcoming the current market challenges.[9] In this sense, the Food and Agriculture Organization (FAO) of the United Nations (UN) announced the International

Food Chemistry, Function and Analysis No. 8
Legumes: Nutritional Quality, Processing and Potential Health Benefits
Edited by María Ángeles Martín-Cabrejas

Published by the Royal Society of Chemistry, www.rsc.org

Year of Pulses in 2016.[10] FAO-UN supported them on the following areas: (i) *pulses are highly nutritious*, (ii) *pulses are economically accessible and contribute to food security at all levels*, (iii) *pulses have important health benefits*, (iv) *pulses contribute to climate change mitigation and adaptation*, (v) *pulses promote biodiversity*.[10]

In general, legumes are characterized as providing high contents of proteins and complex carbohydrates such as dietary fiber and resistant starch, and low levels of lipids, as well as being an important source of minerals such as Fe, Zn and Ca.[10] They contain important levels of vitamins, such as folate, which reduces the risk of neural tube defects. In addition, they provide bioactive compounds that contribute to the prevention of chronic diseases.[11] Researchers have reported that these compounds are associated with a reduction in the risks of cardiovascular diseases (CVD), diabetes and some types of cancer, particularly colorectal cancer.[1,3–5,7]

Legumes are one of the most nutritious foods in the world and combined with other products are the basis of diet for a large part of the world's population, especially in poorer areas where meat, dairy products and fish are economically inaccessible.[10] Legumes are densely packed with proteins, twice the amount found in whole grain cereals. When combined with other foods, such as cereals, the protein quality of legumes is further increased.

The legume crops are essential for sustainable agriculture and they can adapt to climate change, being cultivated in arid climates with limited or erratic rainfall. Scientists are currently working on the development of legume varieties that can grow at temperatures that are 4 or 5 degrees higher than the usual temperature.[12] The goal is to adapt legumes to the effects of climate change such as rising temperatures. In addition, legumes can also fix biological nitrogen into the soil, increasing its fertility and allowing less use of agrochemicals and fertilizers. For example, 85 million hectares of pulses were grown worldwide that fixed from 3 to 6 tons of nitrogen in 2014. Consequently, these crops can help to achieve an adequate and efficient use of fertilizers, resulting in reduced greenhouse gas emissions.[10] One factor that is of concern today is the excessive accumulation of salts in the soil and the use of saline groundwater in legume production worldwide.[13] It is known that saline stress alters the photosynthetic process and hormonal regulation and contributes to the nutritional imbalance in legumes, decreasing yield and grain quality.[8] One solution would be to consider the development of transgenic crops and management strategies to achieve an increase in the yield of grain legumes in saline soils. In addition, legumes are also highly efficient in the use of water, especially in comparison with other sources of protein. For example, 1 kg of cooked beef requires 10 times more water than 1 kg of lentils.[10] Furthermore, intercropping with legumes increases biodiversity creating a more diverse landscape for animals and insects. Therefore, legumes must be preserved and enriched through the development of new varieties resistant to adverse crop conditions, improving their nutritional value and optimizing their health benefits.

Because legumes provide numerous benefits they may act as an affordable ally against malnutrition. Nowadays, they may be considered as a superfood for the future, enabling us to reach zero hunger at a time when one in five children under 5 years of age is chronically malnourished.[14] Legumes are more accessible to lower-income families and have a beneficial influence on physiological functions by improving the state of well-being and health, and reducing or preventing the risk of diseases. In fact, promising progress has been observed in the areas of South America and the Caribbean in the fight against hunger.[15] However, as the production of legumes is still lower than that of other basic products, such as cereals, starchy roots and vegetables, it is necessary to raise awareness about their benefits and encourage greater production of these crops. For this reason, it is convenient if the design of farming systems is local and based on regional expert knowledge.[8]

However, at a global level, the consumption of legumes has been reduced by an increase in the consumption of products of animal origin, especially in developed countries, which has led to a number of important health problems. The profile of the consumer has been changed and new preferences and needs in the food market have emerged. There are different reasons to develop new food products, such as the nutritional and health demands of the different sectors of the population and the reduction and/or elimination of certain nutrients, as well as the incorporation of others (dietary fiber, vitamins, minerals, antioxidants, polyunsaturated fatty acids, *etc.*), which can provide a health benefit and a better quality of life for consumers.[16]

In this sense, legumes are of special interest as a healthy and tasty alternative to the high-calorie diets prevalent in developed countries. In some population groups a low risk of suffering from chronic diseases, such as cardiovascular diseases and certain types of cancer, has been identified, and this has been attributed to a diversified eating style, either in food or in its components.[16]

1.2 Nutritional Quality of Legume Seeds

The chemical composition of the legumes varies between the different types of seeds. The average energy content of legumes is, in general, high (330–374 kcal/100 g dry matter).[10] The amount of proteins in legumes varies greatly depending on the species, and ranges between 20 and 35%. The protein content is considered high compared with cereals and is rich in the amino acid lysine and sulfur amino acids such as methionine, both of which are limited in most plant foods. Legumes in combination with other cereals such as corn and other products rich in starches and with low protein value significantly increase the nutritional value of the diet.[17] However, their nutritional value is poor compared to proteins of animal origin. Several factors are involved, such as the legume proteins have a storage function, their quaternary structure is more compact, which makes the action of digestive enzymes difficult, and the presence of protease inhibitors inhibits the activity of proteases, the digestive enzymes.[18,19] *A priori*, all these factors seem to affect their digestibility. However, during their traditional culinary preparation, the nutritional value can be improved because the thermal treatments inactivate the protease inhibitors as will be described in Chapter 10.

The processing induces structural changes in proteins, which facilitates the accessibility of the digestive enzymes, and favors protein metabolism.[18] In addition, the protein content is often overestimated, since legumes have a high proportion of non-protein nitrogen (peptides, free amino acids and other nitrogenous compounds) that is quantified as total protein and this should be taken into account for the nutritional studies.

On the other hand, carbohydrates in legumes are considered complex, which, unlike simple carbohydrates such as sugars, are slowly digested and absorbed by the intestine, which favors their consumption by diabetic individuals.[7] Starch is the predominant carbohydrate in legumes (75–80%), except in oilseeds; in peanuts the starch content accounts for one third of the total carbohydrates, while soybeans exhibit a very low starch content. Legumes are also rich in dietary fiber, and it is the soluble fiber fraction that is relevant (approximately 16–20%).[21–26] This type of fiber plays a leading role because it facilitates water adsorption in the intestine. Thus, the intestinal bolus increases, facilitating intestinal movement and preventing constipation.[20] Clinical trials and epidemiological studies have reported that dietary fiber contained in beans, helps to lower blood cholesterol levels. This fact is very important in the prevention and reduction of risk of cardiovascular diseases.[27] In this regard, legumes are also low in fat, with values from 1 to 4%, except for oilseeds that show mean values of 18% for soybeans and 50% for peanuts. The lipid fraction is characterized by a high content of triglycerides, with a high level of monounsaturated fatty acids (oleic acid, 18:1) and polyunsaturated fatty acids (PUFAs) (linoleic acids, 18:2 and a-linoleic acid 18:3). Recent studies have reported an interesting *n*-6/*n*-3 ratio of PUFAs, highlighting lentils and Azuki beans with values of 4.0 and 3.2 respectively, providing them a health-promoting benefit, such as reduced risk of CVD.[28]

In addition, legumes are low in sodium (3–41 mg per 100 g)[10] and high in potassium (616–2300 mg per 100 g),[10] very important factors for the maintenance of an adequate blood pressure. It should be noted that the legume-based meal varies according to the method of preparation for consumption. For example, cooked legumes may contain a higher sodium content due to the addition of salt and other condiments. These factors should be taken into account in preparing a healthy diet. Important minerals are present in legumes such as calcium (32–394 mg per 100 g), phosphorus (203–800 mg per 100 g), magnesium (58–472 mg per 100 g), iron (3.2–10 mg per 100 g), and zinc (1.6–6.3 mg per 100 g).[10] Calcium, phosphorus and magnesium are important for bone mineralization resulting in strong bones and helping to prevent osteoporosis.[29] Legumes also provide iron, a limiting mineral in diets and its lack produces anemia, a common health problem in children and women of reproductive age, mainly during pregnancy. Zinc is also very important for the proper growth of children and the maintenance of the immune system. However, the presence of phytic acid in legumes results in the formation of insoluble complexes with the divalent cations (Ca^{2+}, Fe^{2+} and Zn^{2+}) interfering in their absorption and thus, reducing their bioavailability. Although legumes exhibit a high content of Fe, the form of its presentation in plant foods (non-heme iron) shows a lower bioavailability than hemo iron from animal-derived foods. In addition, a high proportion of phosphorus is part of the chemical structure of phytic acid, so it is not available to the body.[30]

Legumes exhibit a great variability in their vitamin content and they are a source of B vitamins such as thiamine (0.3–1.6 mg per 100 g), the level being similar or even higher than that of cereals. They contain riboflavin (0.12–0.33 mg per 100 g) and niacin (traces-4.7 mg per 100 g), essential for energy metabolism.[10] In addition, the contribution of folic acid (0.4–2.1 mg per 100 g) is important as it is a critical nutrient during pregnancy and for the proper development of the neural tube of the fetus.[31] The level of vitamin C is rather low in legumes (0.4–27.7 mg per 100 g dry matter), its content is positively correlated to the antioxidant capacity of the water soluble components.[32] Numerous studies have shown the high increment of this vitamin during germination.[33] The content of liposoluble vitamins is low in legumes, except α-tocopherol (vitamin E) in soybeans and peanuts. The γ-tocopherol in legumes is the most abundant, the greatest levels being reported in peas, pigeon peas and lentils.[34]

In recent years, there has been a clear breakthrough in the discovery of new bioactive compounds and the development of functional food ingredients for the prevention of chronic diseases. Apart from an adequate nutrient content in carbohydrates, proteins, dietary fiber, vitamins and minerals, legumes also possess a variety of non-nutrient compounds considered bioactive with antioxidant, hypoglycemic, hypolipidemic and anticarcinogenic properties.[34] Among them, legumes display a great variety of phenolic compounds and, especially, a high proportion of condensed tannins.[35] A few decades ago tannins were considered as a non-nutritional factor, due to their ability to form complexes with proteins and divalent minerals, interfering in their absorption and digestive utilization; nowadays, they are part of the bioactive components of food because of their its antioxidant capacity. The phenolic profiles of legumes varied greatly as a consequence of the solvent used in their extraction.[36] The phenolic compounds identified in peas, beans, lentils and chickpeas have been hydroxybenzoic acids, aldehydes, hydroxycinnamic acids and derivatives, flavonol glycosides and isoflavones (daizdezein and genistein), which have been extensively described in Chapters 2 and 3. The concentrations of these compounds in the raw seeds are different depending on the type of legume.[36] Because of the processing, modifications in the content of these compounds have been detected. Different studies have shown that the amount of flavonol glycosides increases after cooking of chickpeas and after germination of peas, beans and lentils. In general, the composition among species of legumes is nutritionally very similar. The difference between phenolic compound levels leads to the legume's characteristic colour.[37] In the case of beans, a higher content of polyphenols and anthocyanins results in more colorful beans and therefore less colorful beans such as white beans have a lower content of these compounds.[38] These bean phenolic compounds present a great antioxidant capacity, corroborated in *in vitro* and *in vivo* assays, and they may reduce the capacity to develop chronic diseases such as cancer, diabetes and cardiovascular diseases.[38] In conclusion, it has been reported that black beans are superior to whites in the prevention of chronic diseases.

Some compounds of legume seeds have been considered as non-nutritional factors, ranging from oligosaccharides, tannins, and alkaloids to various groups of proteins with storage and no storage functions. Previously these components have been classified as anti-nutrients because of their low availability for metabolization and the fact that they may induce adverse effects if they are not inactivated by processing before consumption. On the other hand, the nutritional interest is clearly different for animal production, with fast levels of growth along with weight gain being important, while human nutrition targets a balanced diet. Recently, the anti-nutrient term has become settled as non-nutritional or bioactive compounds because they are involved in some health promoting effects in humans; they are described in Chapter 7.[39]

Among the legume non-nutritional factors, α-galactosides have received good deal of attention over many decades.[39] The presence of these indigestible oligosaccharides differs among legumes, reaching up to 10% dry matter. The family of raffinose oligosaccharides (α-galactosyl derivatives of sucrose) is fermented by bacteria in the colon, due to the lack of α-galactosidase, and produces gases (H_2, CO_2, and CH_4) that lead to an uncomfortable inflation of the abdomen known as flatulence and osmotic diarrhea.[40]

However, these compounds are water soluble, if legumes are subjected to soak for at least an hour before cooking, these compounds, along with other non-nutritional compounds contained in the raw grains, remain in the soaking water.[41,42] Soaking is more efficient when sodium bicarbonate is added to the soaking water due to the greater permeability obtained by the partial solubilization of the cell walls.[23] Germination or fermentation processes might also be applied to achieve the partial elimination of α-galactosides. Recently, these compounds have been attributed some of the beneficial properties of dietary fiber such as normalizing intestinal transit, increasing the number of colonies of lactobacilli and bifidobacteria by decreasing those enterobacteria present in the intestinal microflora, and reducing the N-nitroso compounds.

In this sense, these oligosaccharides are purified and marketed to improve the microbiota health. Recently, some studies have shown their characteristics as immune-stimulators in *in vivo* assays in addition to their prebiotic effects.[43]

Phytic acid (IP6) is the myosinophosphate hexaphosphate ester and represents the main form of phosphorus storage in legumes, reaching up to 3–5%. Phytic acid is considered a non-nutritive factor since the six phosphate groups of the IP6 molecule make it a strong chelating agent, reducing the bioavailability mainly of divalent cations, such as Zn^{2+}, Fe^{2+}, Mn^{2+}, Mg^{2+}, Ca^{2+}. Under the pH conditions of the gastrointestinal tract, insoluble metal–phytic complexes are formed, reducing their digestive utilization and subsequent assimilation in animals and humans. IP6 can also interact with proteins, leading to the formation of phytate–protein complexes that affect the solubility and digestibility of dietary proteins, as well as the inhibition of digestive enzymes such as pepsin, trypsin and α-amylase.[42] The cooking process reduces slightly the initial phytic acid content, although the decrease detected is dependent on the treatment conditions (temperature and time), the pH and the presence of proteins and the cations associated with the phytic acid molecule. The germination and fermentation of legume seeds allow the reduction of the phytic acid content, by increasing phytase activity, an enzyme that is found in a natural form in seeds and catalyzes the hydrolysis of IP6 to inositol plus orthophosphate.[22,25] The treatment with exogenous phytases applied during the manufacture of processed foods also reduces the IP6 content, which is of great interest to increase the mineral bioavailability in foods.[44]

Phytic acid exhibits antioxidant activity due to its chelating properties of iron, acting as a powerful inhibitor of the formation of hydroxyl iron radicals. Recent studies have attributed to phytic acid the potential to prevent colon cancer due to its roles in decreasing lipid peroxidation and hydroxyl radical formation, wherein it complexes with iron present in the colon. In addition, researchers have reported IP6 may increase blood NK cell activity in the colon tumor and may inhibit its growth and metastasis in rats.[45]

Enzyme inhibitors and lectins have been described as antinutritional factors over last decades.[1] Some studies have shown growth inhibition in different animal models and a reduction in starch and protein digestibility. α-Amylase inhibitors are found in most legumes, their presence displays an effect on starch digestibility, slowing the rate of its digestion and causing glucose suppression after legume intake. The protease inhibitors represent 0.2–2% by weight of soluble legume proteins. The best known are the inhibitors of trypsin and chymotrypsin.[46] Numerous studies carried out with rats, chickens and pigs showed that the presence of protease inhibitors exerts a negative effect on growth and can cause pancreatic hypertrophy, overstimulation of pancreatic enzyme secretion, as well as growth inhibition and thus, a potential reduction of metabolizable energy. These inhibitors stimulate the pancreas to synthesize larger amounts of chymotrypsin and trypsin; then a feedback mechanism (feedback) is involved. The traditional culinary treatments (soaking, cooking or under pressure cooking) lead to a significant reduction in activity due to their thermolabile character. However, recent studies associate the presence of protease inhibitors with the ability to prevent or inhibit certain tumor pathologies, with certain anti-inflammatory activities or with the reduction of the severity of degenerative and autoimmune diseases.[47] In addition, data have shown their potential in the treatment and/or the prevention of obesity and hypertension. In relation to lectins, they are a heterogeneous group of glycoproteins capable of reversibly binding to sugar residues of glycoproteins located on the cell surface of the entire digestive tract (stomach, small and large intestine). The consequences depend on the organ in question, thus, hyperplasia of the small intestine, hypertrophy of the colon can result. In addition, lectins can act as antitumor agents that induce programmed cell death by apoptosis and autophagic pathways.[48]

Phytosterols are present in legumes (up to 53 mg/100 g in lupins) and their profile depends on the variety of legumes. β-Sitosterol is the main phytosterol followed by campesterol and D5-avenasterol. These plant secondary metabolites exhibit potential effects on obesity, atherosclerosis and also reduce serum total cholesterol (TC) and low-density lipoprotein cholesterol (LDL-C) and cause a significant increase in high density lipoprotein cholesterol (HDL-C).[49] It has been observed in the intestinal lumen that the phytosterols compete with cholesterol in micelle formation, inhibiting its absorption. Phytosterols may regulate genes involved in the metabolism of cholesterol in mouse liver but not intestine.[50]

Saponins are widely distributed in all legumes, they are mainly found in lupins, lentils, chickpeas, soybeans, beans, and peas.[50] These secondary metabolites are a complex group of compounds, which contain a carbohydrate moiety attached to a triterpenoid or steroid, with several health-promoting effects. Saponins may influence the immune system to decrease the risks of cancers. In addition, literature has been found related to the lipidemic action of saponins.[50,51] Most studies suggest that they may reduce blood lipids, due to the formation of an insoluble complex with cholesterol, thus avoiding its absorption in the intestine. During processing, saponins may be dissolved in water and lost in the soaking, washing, and blanching supernatants. It is necessary to find in the food industry a technology of thermal processing that helps to improve the quality of legumes while maintaining the levels of saponins.[51]

With regard to alkaloids, they are mainly found in lupins, but also in peas, with very low contents in chickpeas and lentils.[50] Several biological activities may attribute to these compounds at very low levels to avoid any side effects based on the neurotoxicity of alkaloids. Recently, Wiedemann *et al.*, (2015) have found the enhancement of insulin secretion by the lupanine (lupin quinolizidine alkaloid), which potentiates glucose-stimulated insulin release by directly affecting the ATP-dependent K^+ channel.[52] These findings showed increases in insulin gene expression (*Ins-1* gene) and also in glucose tolerance in induced hyperglycemic rats. Therefore lupanine may be of relevance for supportive treatment of diabetes mellitus.

1.3 Legumes and Their Associated Health Benefits

The aim of this section is to highlight the associated health benefits of legumes, which have received rising interest from researchers. The consumption of legumes, alone or combined with other legumes, has been documented with the prevention and management of diabetes, the prevention of cardiovascular diseases and certain types of cancer.[53] It has been demonstrated that the physiological effects of legumes are different as a result of their chemical composition, in particular, the contents of polysaccharides (mainly, starch and dietary fiber fractions), proteins and bioactive compounds. Recently, Li *et al.*, (2017) conducted a meta-analysis of a prospective cohort study to assess the association between legume consumption and risk of CVD mortality and all-cause mortality.[54] The results obtained showed that a high consumption of legumes is associated with a lower risk of mortality from all causes. In addition, these results do not support an increase in the risk of mortality due to CVD in that population that habitually consumes legumes.[54]

1.3.1 Legumes, Diabetes and the Metabolic Syndrome

Numerous studies have associated the consumption of plant foods such as fruits, vegetables, whole grains, nuts and legumes with lower risks of insulin resistance and diabetes in either normal or diabetic individuals.[55] The chemical composition of legumes plays an essential role in energy regulation and weight control.[56] The metabolic syndrome may be ameliorated by the soybean intake, which leads to an improvement of lipid metabolism. There is evidence that the prevention and treatment of diabetes in humans might be attributed to legumes (lentils and chickpeas).[53–55] As stated before, legumes contain complex carbohydrates and starch resistant to digestive enzymes, which make them less digestible and their intestinal absorption as simple carbohydrates is slower.[53] Legumes might be recognized as a useful management tool for sugar levels in the blood, especially in diabetic individuals. It is well known that the consumption of legumes decreases the glycemic index. *In vitro* studies have showed that the peptides from the protein hydrolysis of 15 bean varieties exhibited biological activity which might help modulate hypertension and type II diabetes.[57,58]

There is still a scarcity of consistent studies indicating that legumes might protect against type 2 diabetes. The recent European study of the prevention with Mediterranean Diet (PREDIMED) evaluated the incidence of type 2 diabetes and the consumption of lentils, chickpeas, beans and peas in more than 3300 individuals without type 2 diabetes.[7] The conclusive results indicated that the consumption of a greater proportion of legumes, particularly lentils, had beneficial effects on the prevention of type 2 diabetes in those individuals with high cardiovascular risk.[59] Recently, the American Diabetes Association has recommended different nutrition strategies for the treatment of adults with type 2 diabetes, including many legume-eating patterns that are acceptable for diabetics.

1.3.2 Legumes and Cardiovascular Diseases

To date, studies have demonstrated that legume intake has a positive effect on cardiovascular risk, decreasing the blood levels of triacylglycerides, TC, LDL-C and increasing levels of HDL-C. It is necessary to point out that all the results obtained are highly dependent of the number of legume-servings consumed, the duration of the studies (from weeks to years) as well as the type of legume consumed. All these factors seem to be decisive in decreasing of cholesterol (TC, LDL-C) levels.

Epidemiological and clinical studies have shown positive effects of the consumption of beans in reducing the risk of coronary heart disease and cardiovascular disease.[53,60–63] This effect may be attributed to the high content of soluble fiber that reduces blood levels of triglycerides and cholesterol.[64] In this sense, a recent clinical assay suggests that black beans may also decrease cardiovascular risk by reducing inflammation.[65] The dietary fiber levels and properties of antioxidant capacity of the black beans might explain these findings. In addition, inflammatory biomarkers have been inversely correlated with legume intake, the low glycemic index of legumes being the possible mechanism involved in the modulation of the inflammation process. Clemente and Olías[1] have reported the potential role of resistant carbohydrates (dietary fiber, resistant starch and oligosaccharides) in intestinal microbiota modulation, motility, glucose homeostasis and cholesterolemia.

Because legumes are rich in potassium, magnesium, and dietary fiber they have a positive impact on blood pressure management. Researchers have found, in eight trials involving >500 people, half of whom were overweight or obese, reductions in blood pressure in subjects that had eaten legumes. Both systolic and mean arterial blood pressure were significantly reduced in individuals who consumed slightly less than 1 cup of legumes each day for 10 weeks.[66] Interestingly, Venn *et al.*[67] found decreases in blood pressure and triglyceride levels in addition to the reductions in waist circumference in 113 obese subjects when they consumed two servings of legumes and four servings of whole grains per day for 18 months in place of refined carbohydrate foods.

1.3.3 Legumes and Cancer

Legumes contain a wide range of bioactive compounds that, either themselves or their metabolites produced after fermentation by the intestinal microbiota, exhibit anticancer properties.[68] In epidemiological studies, legumes have been associated with a lower risk of colorectal, prostate, stomach and even pancreatic cancer.[53,69,70] There are several prospective cohort studies in relation to the effect of legumes on different type of cancers; however, the results are still inconsistent.

Regarding the incidence of prostate cancer, studies have indicated an inverse association between risk of prostate cancer and legume intake,[71,72] while the studies of Kirsh *et al.*[73] do not seem to corroborate it. However, recently a meta-analysis of prospective cohort studies published to date shows that the risk of prostate cancer was reduced by 3.7% if individuals ingested 20 g per day of legumes. Therefore, these results seem to suggest that low incidence of prostate cancer might be associated with a high level of legumes, although the bioactive compounds and metabolic mechanisms involved are still unknown.

One of the main bioactive compounds that exhibit higher anticancer activity are the flavonoids, especially isoflavones (see Chapters 3 and 15). The ability of flavonoids to form complexes with metal ions plays an important role in their antioxidant activity. There is a specific relationship between their chemical structures and their antioxidant activity.[74] Flavonoids in legumes not only inhibit tumor cell growth, but also induce cellular differentiation. The dietary fiber of legumes would not only promote a good intestinal functioning, thanks to the colonic fermentation products from the non-digestible polysaccharide fraction, but it would also provide a protective effect on the proliferation of cancer cells.[75] This health-promoting benefit results from the short chains fatty acid production, such as butyrate, which inhibits growth, induces apoptosis and also, promotes differentiation in colon rectal cancer. Dietary fiber, resistant starch and α-galactosides are modulators of intestinal microbiota and motility, glucose homeostasis and cholesterolemia. However, further studies with a larger sample size are needed to corroborate these results. In addition, researchers suggested that other legume components such as vitamin E, vitamin B6, and minerals such as selenium inhibit carcinogenesis of the colon at an early stage.[76] Recent studies suggest that the protective effect might also be related to the action of peptides in common bean fractions, which can inhibit, *in vitro*, the cellular proliferation of human colorectal cancer cells.[77] Additionally, legume lectins have exhibited anticancer activity because they have the ability to recognize and bind to specific glycoconjugates, which are present in tumor cells. Literature has described the mechanisms involved, which depend on the type of tumor cell and lectin content.[78] Therefore, there is an evident interest in legume lectins as a promising anticancer treatment.

Given the above, it is considered that the role of legumes in the prevention and management of chronic diseases such as metabolic syndrome, obesity, cardiovascular disease, gastrointestinal health and cancer have been documented. There is solid evidence suggest an outstanding association between diets rich in legumes and a lower risk of suffering numerous cancers. Thus, the incidence of cancer could be reduced by changing the diet pattern.

1.4 Conclusion

This overview has documented the potential health-promoting benefits associated with legume intake. These effects derive mainly from the concentrations and properties of starch, proteins, dietary fiber, vitamins, minerals and bioactive compounds in legumes. Resistant carbohydrates (dietary fiber, resistant starch and oligosaccharides) contribute to gastrointestinal function, they exert beneficial prebiotic effects in the large intestine. On the other hand, their hydrolyzed proteins may produce peptides with biological activities and their contents of minerals and vitamins may contribute to the prevention of deficiency-related diseases. The presence of bioactive compounds in legumes seems to play a relevant role in the cardiometabolic risk prevention. Legumes need to be processed before their consumption and as a consequence, their nutritional value is improved. Nowadays, there is increasing interest and research activity in legumes because they are an ideal element to include in the human diet. Therefore, legumes have gained attention for incorporation in the formulation of functional foods. This strategy will help towards increasing legume consumption and improving human diet throughout the world. The possible effects associated with legume intake might be considered as a management tool to resolve the global problem of obesity and chronic malnutrition. Certainly, legumes are essential to eradicate hunger in the world, consolidate healthy eating habits, and protect the environment as mandated by FAO-UN. The time for nutritious food is now.

References

1. A. Clemente and R. Olias, *Curr. Opin. Food Sci.*, 2017, **14**, 32.
2. D. Giugliano, A. Ceriello and K. Esposito, *J. Am. Coll. Cardiol.*, 2006, **48**, 677.
3. A. Tortosa, M. Bes-Rastrollo, A. Sanchez-Villegas, F. J. Basterra-Gortari, J. M. Nuñez-Cordoba and M. A. Martinez-Gonzalez, *Diabetes Care*, 2007, **30**, 2957.
4. C. Razquin, J. A. Martinez, M. A. Martinez-Gonzalez, M. T. Mitjavila, R. Estruch and A. Marti, *Eur. J. Clin. Nutr.*, 2009, **63**, 1387.
5. L. A. Bazzano, A. M. Thompson, C. H. Tees, M. T. Nguyen and D. M. Winham, *Nutr., Metab. Cardiovasc. Dis.*, 2011, **21**, 94.

6. Y. Aguilera, M. Dueñas, I. Estrella, T. Hernández, V. Benitez, R. M. Esteban and M. A. Martín-Cabrejas, *J. Agric. Food Chem.*, 2010, **58**, 10101.
7. N. Becerra-Tomás, A. Díaz-López, N. Rosique-Esteban, E. Ros, P. Buil-Cosiales, D. Corella, R. Estruch, M. Fitó, L. Serra-Majem, F. Arós, R. M. Lamuela-Raventós, M. Fiol, J. M. Santos-Lozano, J. Díez-Espino, O. Portoles and J. Salas-Salvadó, *Clin. Nutr.*, 2018, **37**, 906.
8. M. Reckling, G. Bergkvist, C. A. Watson, F. L. Stoddard, P. M. Zander, R. L. Walker, A. Pristeri, I. Toncea and J. Bachinger, *Front. Plant Sci.*, 2016, **7**, 669.
9. D. Murphy-Bokern, F. Stoddard and C. Watson, in *Legumes in Cropping Systems*, ed. D. Murphy-Bokern, F. Stoddard and C. Watson, CABI International, Wallingford, 2017, pp. 1–256.
10. http://www.fao.org/, last accessed June 2018.
11. B. Singh, J. P. Singh, A. Kaurb and N. Singh, *Food Res. Int.*, 2017, **101**, 1.
12. R. Schultze-Kraft, I. M. Rao, M. Peters, R. J. Clements, C. Bai and G. Liu, *Trop. Grasslands*, 2018, **16**, 1.
13. M. Farooq, N. Gogoi, M. Hussain, S. P. Barthakur, N. Bharadwaj, H. M. Migdadi, S. S. Alghamdi and K. H. M. Siddique, *Plant Physiol. Biochem.*, 2017, **118**, 199.
14. *UNICEF/WHO/World Bank in Levels and Trends in Child Malnutrition*, United Nations Children's Fund, the World Health Organization and World Bank Group, Washington DC, 2018, pp. 1–16.
15. FAO, IFAD, UNICEF, WFP and WHO, in *The State of Food Security and Nutrition in the World 2017. Building Resilience for Peace and Food Security*, Food and Agriculture Organization of the United Nations, Rome, 2017, pp. 1–132.
16. A. Vasileska and G. Rechkoska, *Procedia - Soc. Behav. Sci.*, 2012, **44**, 363.
17. R. Bressani, D. Navarrete and L. Elías, *Plant Foods Hum. Nutr.*, 1984, **34**, 109.
18. J. Boye, F. Zareb and A. Pletch, *Food Res. Int.*, 2010, **43**, 414.
19. H. F. Erbersdobler, C. A. Barth and G. Jahreis, *Sci. Res.*, 2017, **64**, 134.
20. S. Otles and S. Ozgoz, *Acta Sci. Pol., Technol. Aliment.*, 2014, **13**, 191.
21. M. A. Martín-Cabrejas, N. Ariza, R. M. Esteban, E. Mollá, K. Waldron and F. J. López-Andréu, *J. Agric. Food Chem.*, 2003, **51**, 1254.
22. M. A. Martín-Cabrejas, B. Sanfiz, A. Vidal, E. Mollá, R. M. Esteban and F. J. López-Andréu, *J. Agric. Food Chem.*, 2004, **52**, 261.
23. M. A. Martín-Cabrejas, A. Aguilera, V. Benitez, E. Mollá, F. J. López-Andréu and R. M. Esteban, *J. Agric. Food Chem.*, 2006, **54**, 7652.
24. M. A. Martín-Cabrejas, M. F. Díaz, A. Aguilera, V. Benitez, E. Mollá and R. M. Esteban, *Food Chem.*, 2008, **107**, 1045.
25. A. Aguilera, M. A. Martín-Cabrejas, V. Benitez, E. Mollá, F. J. López-Andréu and R. M. Esteban, *J. Food Compos. Anal.*, 2009, **22**, 678.
26. A. Aguilera, R. M. Esteban, V. Benitez, E. Mollá and M. A. Martín-Cabrejas, *J. Agric. Food Chem.*, 2009, **57**, 10682.
27. E. J. Reverri, J. M. Randolph, F. M. Steinberg, C. T. Kappagoda, I. Edirinsinghe and B. M. Burton-Freeman, *Nutrients*, 2015, **7**, 6139.

28. G. Caprioli, F. Giusti, R. Ballini, G. Sagratini, P. Vila-Donat, S. Vittori and D. Fiorini, *Food Chem.*, 2016, **192**, 965.
29. S. R. Drago, in *Nutraceutical and Functional Food Components*, ed. C. M. Galanakis, Elsevier, 2017, pp. 129–157.
30. S. Kumar, A. K. Verma, M. Das, S. K. Jain and P. D. Dwivedi, *Nutrition*, 2013, **29**, 821.
31. A. D. Gernand, K. J. Schulze, C. P. Stewart, K. P. West and P. Christian, *Nat. Rev. Endocrinol.*, 2016, **12**, 274.
32. M. Moriyama and K. Oba, *J. Nutr. Sci. Vitaminol.*, 2008, **54**, 1.
33. M. J. Shohag, Y. Wei and X. Yang, *J. Agric. Food Chem.*, 2012, **60**, 9137.
34. R. Amarowicz and R. B. Pegg, *Eur. J. Lipid Sci. Technol.*, 2009, **111**, 1053.
35. R. Amarowicz, *Eur. J. Lipid Sci. Technol.*, 2007, **109**, 549.
36. B. J. Xu and S. K. Chang, *J. Food Sci.*, 2007, **72**, 159.
37. A. S. M. Golam Masum Akond, L. Khandaker, J. Berthold, L. Gates, K. Peters, H. Delong and K. Hossain, *J. Food Technol.*, 2011, **6**, 385.
38. I. Hayat, A. Ahmad, T. Masud, A. Ahmed and S. Bashir, *Crit. Rev. Food Sci. Nutr.*, 2014, **54**, 580.
39. M. C. Vaz Patto, R. Amarowicz, A. N. A. Aryee, J. I. Boye, H.-J. Chung, M. A. Martín-Cabrejas and C. Domoney, *J. Crit. Rev. Plant Sci.*, 2015, **34**, 105.
40. F. L. Da Silva, V. M. Guimaraes, E. G. de Barros, M. A. Moreira, L. A. Dos Santos Dias, M. G. de Almeida Oliveira, I. C. Jose and S. T. Rezende, *Plant Foods Hum. Nutr.*, 2006, **61**, 87.
41. S. Queiroz-Kda, A. C. de Oliveira, E. Helbig, S. M. Reis and F. Carraro, *J. Nutr. Sci. Vitaminol.*, 2002, **48**, 283.
42. S. Kumar, A. K. Verma, M. Das, S. K. Jain and P. D. Dwivedi, *Nutrition*, 2013, **29**, 821.
43. W. Van den Ende, *Front. Plant Sci.*, 2013, **4**, 247.
44. A. Pusztai, S. Bardocz and M. A. Martín-Cabrejas, in *Recent Advances of Research in Antinutritional Factors in Legume Sedes and Oilseeds*, 2004, p. 87.
45. Z. Zhang, Y. Song and X.-L. Wang, *World J. Gastroenterol.*, 2005, **11**, 5044.
46. A. Clemente, G. Sonnante and C. Domoney, *Curr. Protein Pept. Sci.*, 2011, **12**, 358.
47. A. Rostami and A. R. Kennedy, *Jp Pat.* 6767564, 2004.
48. L. L. Fu, C. C. Zhou, S. Yao, J. Y. Yu, B. Liu and J. K. Bao, *Int. J. Biochem. Cell Biol.*, 2011, **43**, 1442.
49. R. Sultana, R. S. Singh, P. Ratnakumar, N. Verma, S. K. Chaturvedi, A. K. Chaudhary, C. V. Sameer and M. W. Siddiqui, in *Plant Secondary Metabolites*, ed. M. W. Siddiqui and K. Prasad, Apple Academic Press, Oakville, 2017, vol. 1 (7), p. 215.
50. E. D. Jesch, J. Y. Lee and T. P. Carr, *FASEB J.*, 2008, **22**, 700.
51. J. Shi, K. Arunasalam, D. Yeung, Y. Kakuda, G. Mittal and Y. Jiang, *J. Med. Food*, 2004, **7**, 67.
52. M. Wiedemann, C. M. Gurrola-Díaz, B. Vargas-Guerrero, M. Wink, P. M. García-López and M. Düfer, *Molecules*, 2015, **20**, 19085.
53. V. Messina, *Am. J. Clin. Nutr.*, 2014, **100**, 437s.

54. H. Li, J. Li, Y. Shen, J. Wang and D. Zhou, *BioMed Res. Int.*, 2017, **1**, 20.
55. R. Villegas, Y. T. Gao, G. Yang, H. L. Li, T. A. Elasy, W. Zheng and X. O. Shu, *Am. J. Clin. Nutr.*, 2008, **87**, 162.
56. K. J. Acheson, *Nutrition*, 2010, **26**, 141.
57. A. M. Hutchins, D. M. Winham and S. V. Thompson, *Br. J. Nutr.*, 2012, **108**, 52.
58. L. Mojica and E. de Mejía, *Plant Foods Hum. Nutr.*, 2015, **70**, 105.
59. J. Salas-Salvadó, M. Bulló, N. Babio, M. A. Martínez-González, N. Ibarrola-Jurado, J. Basora, R. Estruch, M. I. Covas, D. Corella, F. Arós, V. Ruiz-Gutiérrez and E. Ros, *Diabetes Care*, 2011, **34**, 14.
60. L. A. Bazzano, J. He, L. G. Ogden, C. Loria, S. Vupputuri, L. Myers and P. K. Whelton, *Arch. Intern. Med.*, 2001, **161**, 2573.
61. V. Ha, J. L. Sievenpiper, R. J. de Souza, V. H. Jayalath, A. Mirrahimi, A. Agarwal, L. Chiavaroli, S. B. Mejia, F. M. Sacks, M. Di Buono, A. M. Bernstein, L. A. Leiter, P. M. Kris-Etherton, V. Vuksan, R. P. Bazinet, R. G. Josse, J. Beyene, C. W. Kendall and D. J. Jenkins, *Can. Med. Assoc. J.*, 2014, **186**, 252.
62. D. M. Winham, A. M. Hutchins and C. S. Johnston, *J. Am. Coll. Nutr.*, 2007, **26**, 243.
63. L. A. Bazzano, A. M. Thompson, M. T. Tees, C. H. Nguyen and D. M. Winham, *Nutr., Metab. Cardiovasc. Dis.*, 2011, **21**, 94.
64. S. Hosseinpour-Niazi, P. Mirmiran, M. Hedayati and F. Azizi, *Eur. J. Clin. Nutr.*, 2015, **69**, 592.
65. E. J. Reverri, J. M. Randolph, F. M. Steinberg, C. T. Kappagoda, I. Edirinsinghe and B. M. Burton-Freeman, *Nutrients*, 2015, **7**, 6139.
66. V. H. Jayalath, R. J. de Souza, J. L. Sievenpiper, V. Ha, L. Chiavaroli, A. Mirrahimi, M. Di Buono, A. M. Bernstein, L. A. Leiter, P. M. Kris-Etherton, V. Vuksan, J. Beyene, C. W. Kendall and D. J. Jenkins, *Am. J. Hypertens.*, 2014, **27**, 56.
67. B. J. Venn, T. Perry, T. J. Green, C. M. Skeaff, W. Aitken, N. J. Moore, J. I. Mann, A. J. Wallace, J. Monro, A. Bradshaw, R. C. Brown, P. M. Skidmore, K. Doel, K. O'Brien, C. Frampton and S. Williams, *J. Am. Coll. Nutr.*, 2010, **29**, 365.
68. X. Sanchez-Chino, C. Jiménez-Martínez, G. Dávila-Ortiz, I. Alvarez-Gonzalez and E. Madrigal-Bujaidar, *Nutr. Cancer*, 2015, **67**, 401.
69. Y. Wang, Z. Wang, L. Fu, Y. Chen and J. Fang, *PLoS One*, 2013, **8**, e67335.
70. B. Zhu, Y. Sun, L. Qi, R. Zhong and X. Miao, *Sci. Rep.*, 2015, **5**, 8797.
71. A. Diallo, M. Deschasaux, P. Galan, S. Hercberg, L. Zelek, P. Latino-Martel and M. Touvier, *Br. J. Nutr.*, 2016, **15**, 1579.
72. A. G. Schuurman, R. A. Goldbohm, E. Dorant and P. A. van den Brandt, *Cancer Epidemiol., Biomarkers Prev.*, 1998, **7**, 673.
73. V. A. Kirsh, U. Peters, S. T. Mayne, A. F. Subar, N. Chatterjee, C. C. Johnson and R. B. Hayes, *J. Natl. Cancer Inst.*, 2007, **99**, 1200.
74. G. Cao, E. Sofic and R. L. Prior, *Free Radical Biol. Med.*, 1997, **22**, 749.
75. R. K. Cruz-Bravo, R. G. Guevara-González, M. Ramos-Gómez, B. D. Oomah, A. Wiersma, R. Campos-Vega and G. Loarca-Piña, *Genes Nutr.*, 2014, **9**, 359.

76. A. A. Feregrino-Perez, C. Pinol-Felis, X. Gomez-Arbones, R. G. Guevara-Gonzalez, R. Campos-Vega, J. Acosta-Gallegos and G. Loarca-Piña, *Plant Foods Hum. Nutr.*, 2014, **69**, 248.
77. G. Luna-Vital, D. A. Gonzalez de Mejía, E. Dia and G. Loarca-Piña, *Food Chem.*, 2014, **15**, 157.
78. T. Yau, X. Dan, C. C. W. Ng and T. B. Ng, *Molecules*, 2015, **20**, 3791.

Part II

Bioactive Compounds of Legumes

CHAPTER 2

Leguminous Seeds as a Source of Phenolic Acids, Condensed Tannins, and Lignans

R. AMAROWICZ*[a] AND R. B. PEGG[b]

[a]Institute of Animal Reproduction and Food Research, Department of Chemical and Physical Properties of Food, 10 Tuwima Street, Olsztyn, 10-748, Poland; [b]The University of Georgia, College of Agricultural and Environmental Sciences, 100 Cedar Street, Athens, GA, 30602-2610, USA
*E-mail: r.amarowicz@pan.olsztyn.pl

2.1 Introduction

The seeds of legumes are an important source of nutrients for human nutrition such as protein, starch, dietary fiber, and minerals.[1,2] They are widely consumed throughout the world, and are recognized as a staple foodstuff in many Latin American and African countries.[3] Legumes also contain a number of bioactive compounds including phenolics, which possess marked antioxidant properties, and are very important from nutritional and technological points of view.[4] Phenolic acids and condensed tannins (*i.e.* proanthocyanidins or PACs) next to flavonoids, form the main groups of phenolic compounds typical for legumes. Lignans are the minor phenolic compounds present in legumes. Because of their biological activity, they are also valuable phenolic compounds of legumes.

Food Chemistry, Function and Analysis No. 8
Legumes: Nutritional Quality, Processing and Potential Health Benefits
Edited by María Ángeles Martín-Cabrejas

Published by the Royal Society of Chemistry, www.rsc.org

This chapter reviews the chemical profiles and levels of phenolic acids, PACs and lignans of leguminous seeds. The influence of technological processing and germination on the content of phenolic acids and PACs is also discussed.

2.2 Phenolic Acids

2.2.1 Chemical Structure

Phenolic acids endogenous to legumes are divided into benzoic acid and *trans*-cinnamic acid derivatives. Figures 2.1 and 2.2 depict the chemical structures of the main phenolic acids typically found in legumes. Noteworthy is that the majority of phenolic acids in legumes are linked *via* ester, ether, or acetal bonds either to cellulose, protein, lignin, or smaller organic molecules such as glucose, quinic acid and maleic acid. The chemical structure of the ester of caffeic acid and (−)-quinic acid is known as chlorogenic acid; this is depicted in Figure 2.3.

2.2.2 Biological Activity

Phenolic acids can also possess antimicrobial activity in plants and foods.[5] The hydroxy groups attached to aromatic ring structures in phenolic compounds are thought to cause inhibitory action,[6] as these moieties can interact with the cell membrane of bacteria to disrupt membrane structures resulting in the leakage of cellular components.[7] For example, caffeic acid is more effective as an antimicrobial agent than *p*-coumaric acid due to the number of –OH groups on the phenolic ring.[8] Cueva *et al.*[9] grouped phenolic acids based on their chemical structure. According to these authors, the position of –OH substitution on the aromatic ring as well as the length of the saturated side chain were found to be influencing factors for antimicrobial activity.

Ferulic acid possesses a wide spectrum of antimicrobial activity. It showed inhibition towards Gram-positive bacteria, Gram-negative bacteria and yeasts.[10] It also demonstrated strong inhibitory effects on the growth of some human gastrointestinal microflora, including *Escherichia coli*, *Klebsiella pneumoniae*, *Enterobacter aerogenes*, *Citrobacter koseri*, *Pseudomonas aeruginosa*, *Helicobacter pylori* and *Shigella sonnei*.[11–13] The antimicrobial mechanism of ferulic acid was attributed to its inhibition of arylamine *N*-acetyltransferase in the bacteria.[12] Caffeic acid has been reported to possess bactericidal activity towards *Pseudomonas aeruginosa* and *Staphylococcus aureus*.[14]

Treatment of U87 and U25 glioma cells with gallic acid inhibited cell viability in a dose-dependent manner. Therefore, gallic acid may possibly be a valuable candidate for the treatment of brain tumors. It induces apoptosis and neurosis of HeLa cervical cancer cells.[15] From the experiments of Kratz *et al.*,[16] gallic acid was shown as an anti-herpes simplex viris-2 (anti-HSV-2) agent. Gallic acid is also a potent anti-inflammatory

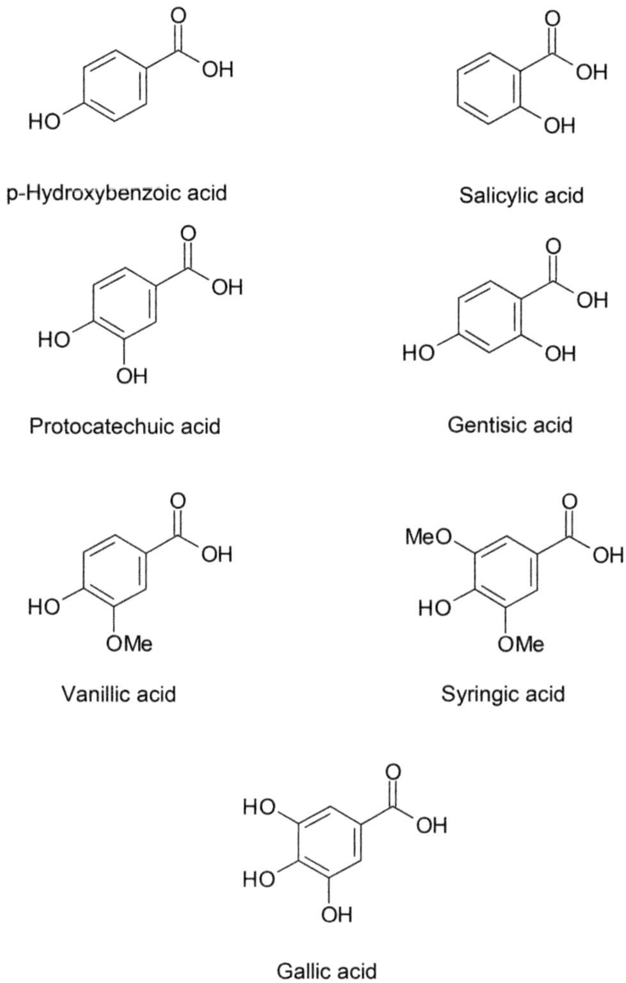

Figure 2.1 Chemical structure of hydrobenzoic acids.

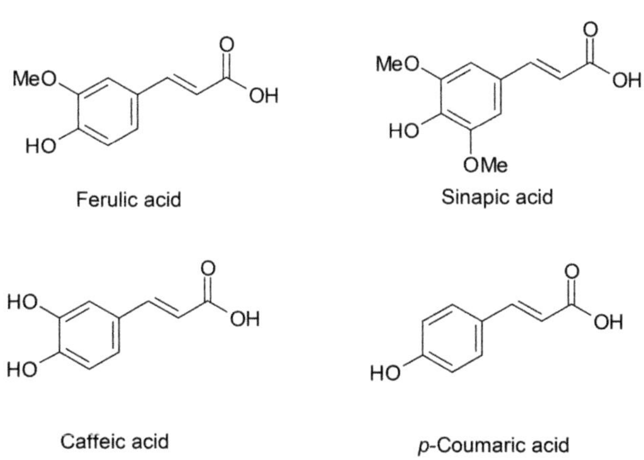

Figure 2.2 Chemical structure of hydroxycinnamic acids.

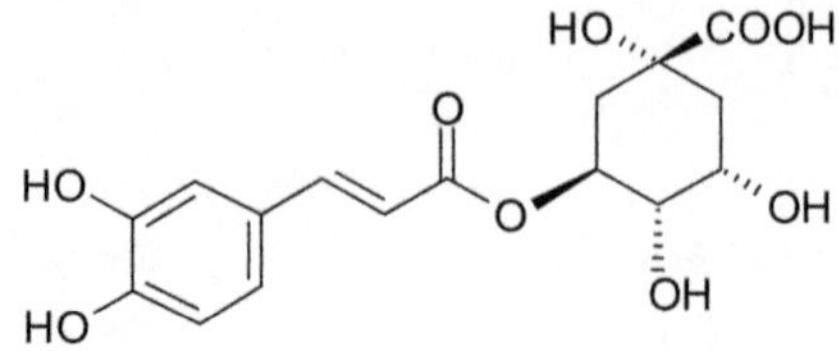

Figure 2.3 Chemical structure of chlorogenic acid.

compound.[17] Syringic and vanillic acids are hepatoprotective *via* their suppression of immune-mediated liver inflammation.[18] In experiments of Kawabata *et al.*[19] concerning azoxymethane (AOM)-induced colon cancer, rats which were fed with ferulic acid during the initiation phase exhibited lower incidences of colonic carcinomas than those which were administered AOM alone. The antihyperglycemic effect of sinapic acid in albino rats was reported by Kanchana *et al.*[20] Pari and Jalaludeen[21] described a protective effect from these phenolic acids against arsenic-induced toxicity in rats. *p*-Hydroxybenzoic acid has been reported to possess an estrogenic activity in human breast tissue.[22]

2.2.3 Antioxidant Activity

The antioxidant activity of phenolic acids has been reported in a plethora of studies. The authors of such works used several different methods to characterize antioxidation such as $ABTS^{\bullet+}$ (2,2′-azino-bis(3-ethylbenzothiazoline-6-sulfonic acid) radical cation), $DPPH^{\bullet}$ (2,2-diphenyl-1-picrylhydrazyl radical), β-carotene-linolenic acid, inhibitor of lipid peroxidation in rat brain homogenates, and inhibiting of lipid peroxidation induced by superoxide anion.[23–30]

The radical-scavenging activities of phenolic acids depend on the number of hydroxy moieties attached to the aromatic ring of the benzoic or *trans*-cinnamic acid molecule. In the study of Karamać *et al.*,[31] gallic acid, with three free hydroxy groups, was observed to be the most bioactive phenolic acid. Dihydroxylation of the aromatic ring afforded a high antiradical power (ARP) value for gentisic, caffeic, and protocatechuic acids. As expected, dihydroxy acids were more active than monohydroxy counterparts. Two methoxy moieties attached to the aromatic ring at positions 3 and 5 increased the radical-scavenging activity; that is, syringic acid was found to be more active than *p*-hydroxybenzoic acid.

2.2.4 Content in Legumes

The content of individual phenolic acids in various legumes is presented in Table 2.1. The main phenolic compounds of legumes comprise ferulic, *p*-coumaric, and sinapic acids. Ferulic acid is also present as glycoside, feruloyl aldaric, and quinic acid.[41–43] The presence of *p*-coumaroyl malic acid was repored in red beans, adzuki beans, and green lentils.[44–46] Small

Table 2.1 Content of phenolic acids in legume seeds or their extracts.

Material	Phenolic acid	Unit	Content	References
Pea	Gallic	mg/100 g seeds	0.53 ± 0.01	32
	p-Coumaric		6.12 ± 1.14	
	Ferulic		11.66 ± 1.93	
	Sinapic		3.92 ± 0.19	
Fractions of black turtle bean extract	Gallic	μg g^{-1} fraction	2644 ± 13	33
	Ferulic		3021 ± 4	
	Sinapic		641 ± 45	
	Syringic		788 ± 11	
Cranberry bean	*p*-Coumaric	mg g^{-1} extract	0.56 ± 0.01	34
	Ferulic		3.10 ± 0.09	
Dry common beans from Italy (12 samples)	Sum of phenolic acids determined by HPLC	mg g^{-1} seeds DM[a]	0.022–0.084	35
Four bean cultivars from Mexico	Sum of phenolic acids determined by HPLC	mg/100 g bean flour	12.95–22.66	36
10 bean cultivars from Mexico	*p*-Hydroxybenzoic	μg g^{-1} seeds DM	5.7–13.8	37
	Vanillic		5.2–14.11	
	p-Coumaric		4.8–7.1	
	Ferulic		20.4–36.0	
4 bean wild cultivars from Mexico	*p*-Hydroxybenzoic		8.3–11.3	
	Vanillic		7.4–16.6	
	p-Coumaric		3.2–5.6	
	Ferulic		24.6–28.4	
Bean Negro 8025 flour	Gallic	μg g^{-1} flour	146.9; 105.8; 184.6	38
	Chlorogenic		150.8; 94.2; 170.6	
	Caffeic		11.55; 8.41; 5.92	
	p-Coumaric		—; 3.79; 1.94	
	Ferulic		4.16; —; —	
	Gallic		113.3; 65.6; 115.1	

(continued)

Table 2.1 (*continued*)

Material	Phenolic acid	Unit	Content	References
Bean Madero Flour	Chlorogenic		137.3; 119.7; 16.57	
	Caffeic		7.71; 5.91; 16.66	
	p-Coumaric		—; —; —	
	Ferulic		1.37; —; —	
White bean	Vanillic	μg g^{-1} extract	377	39
	Gallic		2602	
	Chlorogenic		3782	
	Sinapic		146	
Common bean	Caffeic	mg/100 g seeds	0.18 ± 0.05	32
	p-Coumaric		11.10 ± 0.16	
	Ferulic		26.06 ± 2.19	
	Sinapic		9.55 ± 0.24	
Beans of three Polish cultivars	Chlorogenic	mg/100 g seeds	17.31; 19.20; 17.87	40
	Caffeic		0.49; 0.73; 0.37	
	Ferulic		4.41; 5.51; 6.44	
	p-Coumaric		0.89; 4.02; 3.95	
Kidney bean (water extract)	Ferulolyl hexoside acid	μg g^{-1} extract	1.26 ± 0.03	41
	Ferulolyl aldaric acid		0.79 ± 0.04	
	p-Hydroxybenzoic		3.48 ± 0.10	
	p-Coumaric		4.19 ± 0.32	
	Ferulic		10.56 ± 2.46	
	Feruolyl quinic acid		1.36 ± 0.42	
Kidney bean	Ferulolyl hexoside acid	μg g^{-1} seeds	4.03 ± 0.43	42
	p-Coumaroyl aldaric acid		6.43 ± 0.61	
	Sinaopyl aldaric acid		39.14 ± 5.54	
	Ferulic		2.71 ± 0.06	

Jack bean	*p*-Coumaric	mg/100 g seeds	1.84 ± 0.31	32
	Ferulic		11.72 ± 2.10	
	Sinapic		8.33 ± 0.61	
Cow bean	*p*-Coumaric	mg/100 g seeds	7.84 ± 0.25	32
	Ferulic		19.48 ± 2.30	
	Sinapic		6.22 ± 0.51	
Rice bean	Gallic	mg/100 g seeds	0.72 ± 0.02	32
	Caffeic		0.15 ± 0.02	
	p-Coumaric		3.79 ± 0.16	
	Ferulic		18.23 ± 1.49	
	Sinapic		3.49 ± 0.33	
Mung bean	*p*-Coumaric	mg/100 g seeds	5.76 ± 0.22	32
	Ferulic		19.86 ± 1.25	
	Sinapic		7.85 ± 0.46	
Broad bean	*p*-Coumaric	μg g^{-1} extract	406 ± 30	43
	Ferulic		224 ± 21	
Broad bean	Caffeic	mg/100 g seeds	0.78 ± 0.03	32
	p-Coumaric		1.68 ± 0.07	
	Ferulic		10.56 ± 1.58	
	Sinapic		2.58 ± 0.23	
Red bean	*p*-Coumaric acid	μg g^{-1} extract	6.19 ± 0.56	44
	p-Coumaric acid derivative		9.58 ± 0.92	
	p-Coumaroyl malic acid		8.82 ± 0.74	
	Ferulic		75.98 ± 7.01	
	Protocatechuic		45.73 ± 5.23	
	p-Hydroxybenzoic		84.75 ± 6.31	
	Vanillic		24.21 ± 1.64	
	Sinapic		0.25 ± 0.03	
Winged bean	*p*-Coumaric	mg/100 g seeds	1.08 ± 0.27	32
	Ferulic		10.39 ± 2.01	
	Sinapic		6.74 ± 0.38	

(*continued*)

Table 2.1 (*continued*)

Material	Phenolic acid	Unit	Content	References
Adzuki bean	Protocatechuic	μg g^{-1} extract	67.6 ± 4.01	45
	p-Coumaric		31.3 ± 1.96	
	p-Coumaroyl malic acid		4.57 ± 0.69	
Adzuki bean	Gallic	mg/100 g seeds	1.12 ± 0.15	32
	Caffeic		0.20 ± 0.07	
	p-Coumaric		2.68 ± 0.51	
	Ferulic		15.41 ± 1.36	
	Sinapic		6.80 ± 0.26	
Lima bean	*p*-Coumaric		3.25 ± 0.68	32
	Ferulic		15.12 ± 2.15	
	Sinapic		6.74 ± 0.15	
Hyacinth bean	Gallic	mg/100 g seeds	0.96 ± 0.10	32
	p-Coumaric		4.63 ± 0.88	
	Ferulic		19.05 ± 2.75	
	Sinapic		3.25 ± 0.16	
Green lentil	*p*-Coumaric	μg g^{-1} extract	37.3 ± 1.9	46
	p-Coumaroyl malic acid		6.4 ± 0.6	
	Ferulic		10.1 ± 0.9	
Red lentil	*p*-Coumaric	μg g^{-1} extract		47
	Ferulic			
	Sinapic			
	Gallic			
	p-Hydroxybenzoic			
Lentil cotyledon	Protocatechuic	μg g^{-1}	0.73–0.84	48
	p-Hydroxybenzoic		0.93–1.43	
	Ferulic		0.51–0.90	
	cis-*p*-coumaric		0–0.49	
	trans-*p*-coumaric		2.65–4.55	
	Coumarylmalic		0.36–9.22	
	Coumarylglycolic		1.03–2.84	

Lentil coat	Gallic	μg g^{-1}	1.22–4.94	48
	Protocatechuic		0–11.11	
	trans-p-coumaric		11.78–29.45	
Chickpea	Caffeic	mg/100 g seeds	0.13 ± 0.01	32
	p-Coumaric		4.59 ± 0.53	
	Ferulic		9.10 ± 1.29	
	Sinapic		4.47 ± 0.41	
Vetch	*p*-Coumaric	mg/100 g seeds	2.67 ± 0.19	32
	Ferulic		12.39 ± 1.63	
	Sinapic		2.27 ± 0.30	

[a]Abbreviation: DM, dry matter.

quantities of gallic acid were found in peas,[32] black turtle beans,[33] rice beans,[32] adzuki beans,[32] hyacinth beans,[32] red lentils,[47] and lentil coats.[48] Relatively high contents of gallic acid have been reported in the flours from Negro and Madro beans, as determined by Ramírez-Jiménez *et al.*[38] The presence of chlorogenic acid was determined for beans of three Polish cultivars[40] and white beans.[39] Flour prepared from navy beans contained phenolic acids in the form of soluble esters.[49]

Alkaline hydrolysis of the navy bean flour liberated ferulic, *p*-coumaric, and syringic acids. The content of these phenolic acids in navy beans was found to be greater than those in mung beans, field beans, lentils, faba beans, and pigeon peas.[49] Caffeic acid, *p*-coumaric acid, ferulic acid, and sinapic acid were also determined in the extracts of green- and yellow-podded bean varieties.[50] In a bean hull extract, Madhujith *et al.*[51] reported *p*-coumaric, ferulic, vanillic, caffeic, and sinapic acids as the main phenolic acids. Different amounts of gallic, protocatechuic, and *p*-coumaric acids in lentil coats were determined and reported by Dueñas *et al.*[48]

2.2.5 Changes in Content of Phenolic Acids in Leguminous Seeds During Technological Processing and Germination

The content of chlorogenic acid in bean extrudates increased compared to the raw bean starting material. The changes of caffeic, ferulic, and *p*-coumaric acids in the beans during extrusion were not clear and depended on the bean cultivar in question.[40] Solid-state fermentation (STF) of beans with *Bacillus subtilis* and liquid-state fermentation (LSF) with *Lactobacillus plantarum* increased the content of phenolic acids in the extracts. For STF the greater increase was observed for feruloyl hexoside acid, feruloyl aldaric acid, and *p*-hydroxybenzoic acid; whereas for LTF, it was sinapoyl aldaric acid.[41]

Dueñas *et al.*[42] investigated the effect of germination and elicitation (500 μM ascorbic acid, 50 μM folic acid, and 5 mM glutamic acid) on the phenolic composition and bioactivity of kidney beans. Elicitor treatment reduced the concentration of a large number of hydroxycinnamic compounds in the sprouts. Application of vitamin C enhanced the content of sinapic acid derivatives by 72%. Elicitation resulted in a significant decrease in the content of *p*-hydroxybenzoic acid, which was not detected in elicited sprouts.

2.3 Condensed Tannins (Proanthocyanidins)

2.3.1 Chemical Structure

Condensed tannins or PACs are flavan-3-ol-based biopolymers that, at high temperatures in alcohol solutions of strong mineral acids, release anthocyanidins and catechins as terminal end groups. The most common method

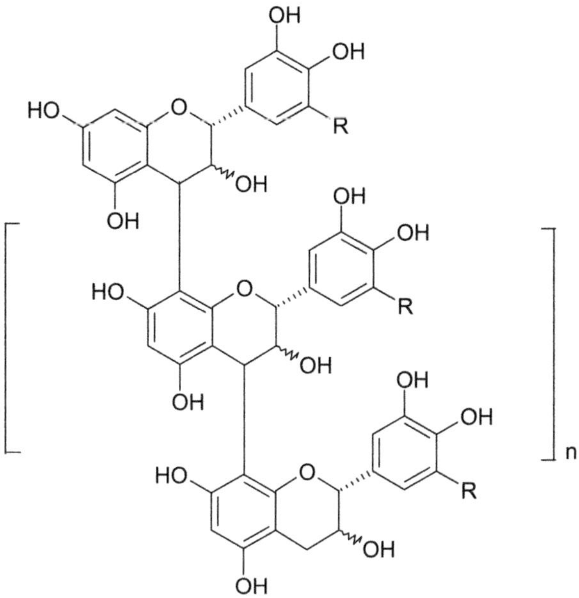

Figure 2.4 Chemical structure of condensed tannins (proanthocyanidins).

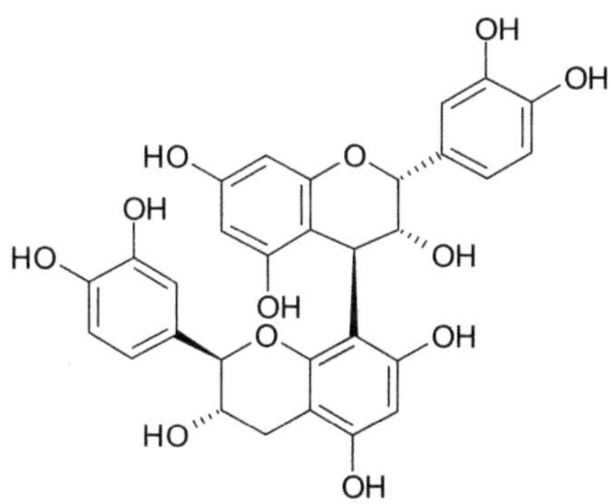

Figure 2.5 Chemical structure of procyanidin B_1.

employed for PACs analysis in beans is the vanillin/HCl method.[52] The basic chemical structure of PACs with a B-type linkage is depicted in Figure 2.4. The chemical structure of procyanidin B_1, which is a molecule with a 4→8 bond ((−)-epicatechin-(4β→8)-catechin), is presented in Figure 2.5.

PACs of black beans, blackeye peas, pinto beans, small red beans, and red kidney beans are of a B-type linkage, composed of epicatechin/catecin, with presence of sugars in the molecules. The average degrees of polymerisation of PACs of pinto beans, small red beans, and red kidney beans are 8.3, 12.0, and 6.7, respectively.[53] The PAC oligomers (dimers, trimers, tetramers, pentamers, hexamers) in seed coats of black jamapa beans (*Phaseolus vulgaris* L.)

are composed of catechin/epicatechin, gallocatechin/epigallocatechin, and afzelechin/epiafzelechin.[54]

2.3.2 Biological Activity

Several plant extracts containing high contents of PACs exhibited strong antimicrobial activities against bacteria such as *Proteus mirabilis*, *Micrococcus luteus*, *Bacillus licheniformis*, *Nocardia asteroids*, *Salmonella typhimurium*, *Bacillus subtilis*, and *Staphylococcus aureus*.[55–57] According to Howell *et al.*[58] PACs elicit *in vitro* anti-adhesion activity of *Escherichia coli* from uroepithelial cells.

Several studies highlighted the possible mechanisms for PAC mediation of blood pressure *via* the influence on angiotension I (Ang-I) converting enzyme (ACE). PACs aid in lowering plasma cholesterol levels, inhibit the oxidation of LDL, and activate endothelial nitric oxide synthase to prevent platelet adhesion and aggregation, which can contribute to blood clot formation.[59–62]

PAC-rich extracts effectively decreased matrix metalloproteinase activity in DU145 human prostate cancer cells *in vitro*[63] and acted as inhibitors of later stages of carcinogenesis.[64] According to Matchett *et al.*,[65] PACs may employ multiple mechanisms in down-regulating matrix metalloproteinase activity with possible involvement of protein kinase C and mitogen-activated kinase pathways.

2.3.3 Antioxidant Activity

PACs can function as primary antioxidants (*i.e.*, they can donate hydrogen atoms or electrons) and also as secondary antioxidants.[66] PACs have the capability to chelate Fe(II) and interfere with one of the reaction steps in the Fenton reaction and thereby retard lipid oxidation.[67] The inhibition of lipid peroxidation by tannin constituents can occur *via* the inhibition of cyclooxygenase activity.[68] The antioxidative properties of tannin-rich extracts have been observed in meat model system studies. The PAC fraction separated from a green tea crude extract using Sephadex LH-20 column chromatography showed inhibition of meat lipid oxidation when applied at levels of 50–500 ppm.[69]

The tannins fraction separated from acetonic extracts of peas, red beans, red lentils, green lentils, adzuki beans, and broad beans and peas demonstrated stronger antioxidant activity than the fractions of low-molecular-weight phenolics separated from the same respective plant material.[43–47] In the cited studies, the authors employed methods such as reducing power, antiradical activity against $DPPH^{\bullet}$ and $ABTS^{\bullet+}$, and antioxidant activity in a β-carotene-linoleate model system. For extracts prepared from leguminous seeds (*i.e.*, faba beans, broad beans, adzuki beans, red beans, red lentils, and green lentils), statistically significant correlations were reported between the total antioxidant activity and the PACs content.[70] Procyanidins B_1 and B_3 from adzuki beans acted as antioxidants in an aqueous system and exhibited antiradical activity against peroxyl radicals.[71,72]

2.3.4 Content in Legumes

Tables 2.2 and 2.3 provide a synopsis of tannin contents in legume seeds and seed coats, respectively. The Tables summarize research data of selected studies from 1995 to the present day, and indicate the method

Table 2.2 Content of condensed tannins in legume seeds or extracts.[a]

Legume	Method	Unit	Content	References
Faba bean (colour)	PVPP & F–C	g kg^{-1} seeds	10.3–21.0	73
Faba bean (white)			0.04–0.23	
Faba bean	PVPP & F–C	g kg^{-1} seeds	14.5	74
Faba bean (T+)	PVPP & F–C	g kg^{-1} DM	9; 10	75
Faba bean	PVPP & F–C	g GAE/100 g extract	0.9–1.9	76
Faba bean	Vanillin	mg CE/kg seeds	309–959	77
Broad bean extract	Vannillin	A_{500}/mg extract	5.96 ± 0.25	43
Broad bean tannin fraction			121 ± 13	
Lima bean	Vanillin	mg CAE/g seeds DM	0.0–9.0	78
Lentil	Vanillin	mg/100 g	269–378	79
Lentil	Vanillin	mg CE/g seeds	3.73–10.20	80
Red lentil	Vanillin	mg CE/g seeds	3.00–5.82	81
Green lentil			3.36–7.80	
Green lentil extract	Vanillin	A_{500}/mg extract or fraction	93 ± 2	46
Green lentil tannin fraction			252 ± 4	
Red lentil extract	Vanillin	A_{500}/mg extract or fraction	70 ± 2	47
Red lentil tannin fraction			129 ± 4	
Adzuki bean extract	Vanillin	A_{500}/mg extract or fraction	136	45
Adzuki bean tannin fraction			213	
Chickpea	PVPP & F–C	9.48 TAE/g extract	9.48	82
Chickpea flour	Vanillin	mg g^{-1} flour	5.12	83
Cowpea	PVPP & F–C	mg g^{-1} seeds	2.2–6.3	84
Cowpea flour	Vanillin	mg CE/g flour	2.13	83
Yellow pea	Vanillin	mg CE/g seeds	0.22–0.59	80
Green pea			0.23–0.61	
Pinto bean			3.23	
Navy bean			0.47	
Red kidney bean			2.87	
Small red bean			5.26	
Pink bean			3.55	
Pea	Vanillin	mg/100 g	294; 367	79
26 varieties of bean from Mexico	Vanillin	mg CE/g seeds DM	5.9–21.5	85
Black turtle bean extract and its fractions	Vanillin	mg CE/g extract mg CE/g fraction	40.69 ± 0.75 571.21–906.12	33

(continued)

Table 2.2 (*continued*)

Legume	Method	Unit	Content	References
Beans cultivated in Mexico	Vanillin	mg CE/g seeds	16.8–38.1	86
Beans from Mexico	Vanillin	mg CE/100 g seeds	2619	87
Black bean (cooked)	Vanillin	mg CE/g seeds DM	54.6 ± 0.3	88
10 common beans from Italy	Vanillin	µg CE/g seeds DM	0.021–2.36	89
19 traditional Italian legumes	Vanillin	mg CE/100 g seeds	9.4–105	90
Yellow field pea flour	Vanillin	mg CE/100 g DM	2.4 ± 0.1	91
Two bean cultivars	Vanillin	mg CE/g DM	2.15 ± 0.04 0.42 ± 0.00	92
Green gram (*P. aureus*)	Vanillin	mg CE/100 g	512.5 ± 31.44	93
Red gram (*C. cajan*)			256.6 ± 5.98	
Bengal gram (*C. arientinum*)			274.5 ± 24.79	
Lentil			329.7 ± 57.03	
Moth bean (*Vigna aconitifolia*)	PVPP & F–C	mg/100 g seeds	1.91 ± 0.12	94
Kidney bean	Vanillin	g CE/kg seeds DM	3.59 ± 0.05	95
Pea (3 cultivars)	Vanillin	g CE/kg seeds DM	13.5; 14.5; 13.8 1500	96
Phaseolus calcaratus	Vanillin	mg CE/100 g seeds	1370	97

[a]Abbreviations: PVPP, polyvinylpolypyrrolidone; F–C, Folin-Ciocalteu; DM, dry matter; GAE, gallic acid equivalent; A_{500}, absorbance at 500 nm wavelength; CE, (+)-catechin equivalents; and TAE, tannic acid equivalents.

Table 2.3 Content of condensed tannins in legume seed coats.[a]

Legume	Method	Unit	Content	References
Pigeon pea	Vanillin	mg CE/g	71.8	98
Red cowpea			41.7	
Adzuki bean			13.8	
Broad bean			18.0	
Red kidney bean			46.7	
Violet red kidney bean			57.9	
Big speckled kidney bean			79.5	
Small speckled kidney bean			79.9	
Panda bean			12.5	
Big runner bean			28.1	
Small runner bean			44.7	
Brown string bean			5.26	
Climbing bean			19.0	
Big rice bean			22.5	
Small rice bean			26.7	
Pea			7.56	
Bean	Vanillin	mg CE/g	60.6–369.3	87
Brazilian and Peruvian bean cultivars	Vanillin	mg CE/g	0–449	99

[a]Abbreviation: CE, (+)-catechin equivalents.

of tannin analysis performed; most of which were conducted by the vanillin-HCl assay. Furthermore, the findings are presented on a number of different bases including per g or kg seed or flour of the legume in question, as well as on the basis of prepared extracts or fractions. Some results report the mass of the seeds on a dry matter basis, while others are based on as delivered. To that end, care must be exercised when comparing the findings from the listed studies.

The results show that a wide range of tannin contents in legumes exist, the largest coming from studies more than 20 years ago on coloured faba beans with levels ranging between 10.3 and 21.0 g kg^{-1} seeds,[73] and by Alonso *et al.*,[96] who reported tannin contents in three pea cultivars ranging from 13.5 to 14.5 g catechin equivalents (CE)/kg seeds as dry matter. Makkar *et al.*[73] also analyzed white faba beans, but determined tannin levels of only 0.04 to 0.23 g kg^{-1} seeds. More recent findings by Aquino-Bolaños *et al.*,[85] who analysed 26 varieties of beans from Mexico, reported tannin contents of 5.9 to 21.5 g CE/kg seeds as dry matter, which is comparable to the older findings of Khalil and Mansour[74] and Makkar *et al.*[73] Yet, lentils and cowpeas have been found to possess lesser contents of tannins typically ~2–6 g kg^{-1} seeds.

As one would expect, the levels of tannins in the seed coats of legumes from peas, beans, and cowpeas is much higher than in the legumes themselves. Table 2.3 lists values for the seed coats determined by the vanillin-HCl assay. A recent investigation by Gan *et al.*[98] reported values from a low of 5.26 g CE/kg seed for brown string beans to a high of 79.9 g CE/kg seed for small speckled kidney beans. Many of the values reported for the selected seed coats are > 20 g CE/kg seed. In their study, Mojica *et al.*[87] reported tannin contents in bean coats ranging from 60.6 to 369 g gallic acid equivalents/kg seed coat in the Brazilian Pinto-Saltillo and BRS-Horizonte varieties, respectively. Of course, those legumes with colour to their seed coats tend to possess more anthocyanins, glycosidic flavanols, as well as tannins.

The types of PAC and their contents in selected legumes including beans, lentils, as well as seed coats and flours are presented in Table 2.4. The degree of polymerization determined for the various PACs will have important biological consequences during the digestion process with regard to their bioavailability. Generally, the legumes examined possessed tannins in the form of either dimers or trimers. For broad beans, the PAC in the greatest quantity was determined to be a digallate procyanidin dimer at 676 ± 52 followed by a procyanidin gallate at 457 ± 31 μg g^{-1} extract.[48] Lower levels of procyanidin dimers and trimers existed for the red beans, adzuki beans, red lentils, and green lentils analyzed by Amarowicz *et al.*[43–47] Research by Dueñas *et al.*[48] reported higher levels of procyanidins, but these were for lentil coats and not the lentils themselves. The greatest individual tannin content determined in the lentil coats was for procyanidin B_3, which ranged between 445 to 870 μg g^{-1} coat.[48]

Table 2.4 Content of individual proanthocyanidins in legume seeds or extracts.

Material	Compounds	Unit	Content	References
Broad bean	Prodelphinidin dimer	μg g^{-1} extract	88 ± 9	43
	Digallate procyanidin dimer		676 ± 52	
	Procyanidin gallate (1)		457 ± 31	
	Procyanidin gallate (2)		68 ± 6	
Red bean	Prodelphinidin dimer	μg g^{-1} extract	13.91 ± 1.92	44
	Prodelphinidin dimer		71.13 ± 7.14	
	Prodelphinidin dimer		5.98 ± 0.62	
	Procyanidin dimer (1)		34.19 ± 2.01	
	Procyanidin dimer		29.65 ± 2.03	
	Procyanidin dimer		26.72 ± 2.12	
	Procyanidin dimer		10.36 ± 0.85	
	Procyanidin trimer digallate		4.90 ± 0.05	
	Procyanidin tetramer		7.99 ± 0.61	
Red lentil	Procyanidin gallate	μg g^{-1} extract	83.29 ± 3.69	47
	Procyanidin dimer digallate		154.8 ± 4.14	
	Prodelphinidin dimer (1)		6.8 ± 0.19	
	Prodelphinidin dimer (2)		20.68 ± 1.23	
	Procyanidin dimer (1)		18.7 ± 1.14	
	Procyanidin dimer (2)		26.6 ± 1.28	
	Procyanidin dimer (3)		48.2 ± 2.11	
	Procyanidin trimer (1)		39.3 ± 1.19	
	Procyanidin trimer (2)		122.4 ± 3.66	
Green lentil	Procyanidin dimer gallate	μg g^{-1} extract	31.9 ± 2.4	46
	Digallate procyanidin dimer		7.3 ± 0.3	
	Digallate procyanidin dimer		14.1 ± 0.6	
	Prodelphinidin dimer		5.3 ± 0.1	
	Procyanidin dimer		14.1 ± 1	
	Procyanidin dimer		100.2 ± 5.7	
	Procyanidin trimer		18.6 ± 1.1	
Adzuki bean	Procyanidin gallate	μg g^{-1} extract	12.4 ± 1.06	45
	Procyanidin dimer (1)		213 ± 13.2	
	Procyanidin dimer (2)		15.9 ± 1.56	
	Procyanidin dimer (3)		25.3 ± 1.53	
	Procyanidin dimer (4)		16.0 ± 0.89	
	Procyanidin trimer (1)		41.8 ± 1.11	
	Procyanidin trimer (2)		42.4 ± 1.09	
Lentil coat	10 different procyanidins D	μg g^{-1} coat	Traces – 507.30	48
	3 different procyanidins T		5.18–59.07	
	Procyanidin B_1		41.73–324.65	
	Procyanidin B_2		131.67–209.68	
	Procyanidin B_3		445.31–869.88	

Table 2.4 (*continued*)

Material	Compounds	Unit	Content	References
Lentil flour	Prodelphinidin dimer	μg g^{-1} flour	8.03 ± 0.98	100
	Prodelphinidin dimer		6.02 ± 1.00	
	Prodelphinidin trimer		31.05 ± 1.56	
	Procyanidin digallate dimer		1.26 ± 0.15	
	Procyanidin B_2		13.80 ± 0.96	
	Procyanidin B_3		10.61 ± 1.07	
	Procyanidin trimer		13.16 ± 1.12	
	Procyanidin trimer		0.29 ± 0.05	
	Procyanidin trimer		1.61 ± 0.23	

2.3.5 Changes in the Content of Phenolic Acids in Leguminous Seeds During Technological Processing and Germination

The total tannin content in field peas, faba beans, and chickpeas were greatly reduced irrespective of the extrusion preconditioning and drying treatment employed.[101] Dehulling of leguminous seeds reduced the content of condensed tannins. In the case of faba beans and common beans, a decrease by 92.3 and 93.3%, respectively, was reported by Alonso *et al.*[101] Much lower results (11.2–28.65%) were obtained however for dehulled peas.[96]

The effect of soaking on the content of PACs has been was investigated by several studies. Soaking of faba beans and bean seeds resulted in a decrease in the content of PACs by 47.7 and 24.2%, respectively. Soaking of *Phaseolus angularis* and *Phaseolus calcaratus*, indigenous to China, decreased the content of PACs by 42.2 and 35.3%, respectively.[97] The migration of tannins from the cotyledons of legumes into the soaking or cooking water was observed by Mkanda *et al.*[102]

In several experiments, a reduction in the content of PACs in legumes during heating has been observed. For instance, autoclaving at 121 °C for 10 min decreased the content of condensed tannins in two common bean cultivars from 2.15 to 0.02 mg epicatechin equivalent/g (biofortified *carioca* bean) and from 0.42 mg epicatechin equivalent/g to 0.03 mg epicatechin equivalent/g (commercial bean).[92] The same process decreased the level of PACs in *Phaseolus calcaratus* and *Phaseolus calcaratus* by 68.8 and 59.9%, respectively.[97] According to Ramírez-Cárdenas *et al.*,[103] when cooking is performed without using the soaking water, the reduction in the tannins content of the beans is more extensive than when the seeds are cooked with the soaking water. The effective reduction of the tannins content during domestic processing (*i.e.*, soaking first followed by cooking) was reported by Mariotto-Cezar *et al.*[104] According to Fernandes *et al.*,[105] the most effective procedure to decrease the content of tannins in leguminous seeds is soaking first and then discarding the soaking water before cooking. From the research of Luo *et al.*,[76] faba bean seeds were processed by dry heating in a hot air oven at 150 °C for 30 min

(T_1); soaking for 12 h at 25 °C followed by autoclaving for 20 min at 120 °C (T_2); soaking with autoclaving with 0.1% of sodium bicarbonate (T_3); soaking with autoclaving with 1% ash solution (T_4); and soaking with autoclaving with 1% sugar (T_5). In the extract so obtained from the T_1 process, the PACs content was determined to be lower than that of the control extract obtained from untreated seeds. The extracts of T_4 and T_5 preparation exhibited the highest content of PACs. Counterpart results were reported by Siddhuraju[94] for moth bean (*Vigna aconitifolia*): the content of condensed tannins in dry heated seeds (open hot plate at 125 °C for 25 min) was 1.31 g/100 g whereas in untreated seeds they were only 1.91 g/100 g. The total tannin levels in field peas, faba beans, and chickpeas were markedly reduced irrespective of the extrusion preconditioning and drying treatment employed.[101] Alonso *et al.*[95] reported that extrusion of faba beans and kidney beans decreased the content of PACs by 54.4 and 83.3%, respectively. Steaming decreases the PAC content in immature faba beans compared to the corresponding immature raw faba bean parts (42.3%). Boiling resulted in decreased PAC contents in immature and mature faba beans as well (*i.e.*, 41.3 and 22%, respectively).[106] It has been suggested that the reduction in the PACs content might be due to dissolution of these compounds in the cooking water.[107]

Germination of beans for 24–72 h decreased the tannins content by 43.5–71.6%. For faba beans, a reduction of 55.9–60.0% was reported.[95] In experiments of germination of *Phaseolus angularis* for 24 and 48 h, a decrease of the PACs by 50.7 and 63.7% resulted. For *Phaseolus calcaratus*, decreases of 53.4 and 63.7% were obtained, respectively.[97] Germination of peas for 72 h afforded a decrease in the PACs content of condensed tannins by 13.3–26.1%.[96]

2.4 Lignans

2.4.1 Chemical Structure

Lignans are diphenol compounds formed from phenylalanine *via* dimerization of substituted cinnamic alcohols. The biochemical reactions for the formation of these secondary plant metabolites is catalyzed by oxidative enzymes and is often regulated by dirigent proteins. The chemical structures of four main lignans present in legumes are depicted in Figure 2.6.

2.4.2 Biological Activity

Many studies have confirmed the biological activity of lignans.[108] Enterolignans, enterodiol and enterolactone are formed by the action of intestinal bacteria on lignans found in plants after consumption. Because some of the effects of the estrogen are reduced by enteradiol and enterolactone, lignans are commonly referrred to as phytoestrogens. Due to their structural similarity to 17-β-estradiol, at normal estradiol levels in organisms, lignans are an antagonist of this hormone (Hutchins and Slavin, 2003).[109] Yet, in

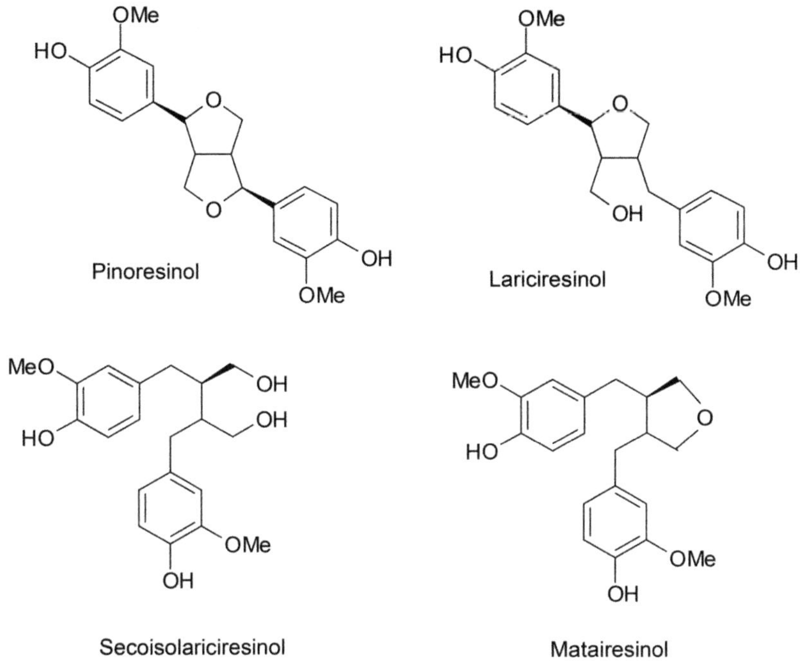

Figure 2.6 Chemical structure of lignans.

postmenopausal woman, when the level of estradiol is low, they can act as weak estrogens. Lignans are able to inhibit the binding of estrogen and testosterone cell receptors.[110] A number of studies have shown that lignan can be protective against breast and prostate cancer.[111–115] It is worth emphasizing, that in the majority of the studies mentioned, flaxseed lignans were the ones investigated.

2.4.3 Antioxidant Activity

Numerous experiments have conformed the antioxidant activity of lignans. For example, secoisolariciresinol diglucoside (SDG) was an active antioxidant in both hydrophobic and hydrophilic *in vitro* systems.[116] Secoisolariciresinol (SECO) and SDG exhibited antioxidant activities in triacylglycerols of sunflower oil and in emulsion systems; activity of SECO was greater than that of SDG in both model systems.[117] SECO and SDG retarded the degradation of canola oil, as measured by induction time, in a concentration-dependent manner.[118] In experiments of Prasad,[119] SDG showed its capability to scavenge hydroxyl radicals. Flaxseed lignans were also scavengers of the DPPH radical.[120] SECO and matairesinol exhibited antioxidant activities, as determined using the ferric-reducing antioxidant power (FRAP) method and the results were comparable to that of nordihydroquaiaretic acid (NDGA) and greater than that of ascorbic acid.[121] Hu *et al.*[122] reported that physiological concentrations of SDG and SECO were protective against 2,2′-azo-bis(2-amidinopropane) dihydrochloride (AAPH)-initiated peroxyl radical plasmid DNA damage and phosphatidylcholine liposome lipid peroxidation.

2.4.4 Content in Leguminous Seeds

Table 2.5 reports the lignan contents in selected legumes including lentils, chickpeas, beans, and peas. Furthermore, the data for fresh, dried, and cooked legumes in the Table are presented in most cases on a μg/100 g dry matter basis. The lignans determined included isolariciresinol, lariciresinol,

Table 2.5 Content of lignans in legume seeds.[a]

Legume	Compound	Unit	Content	References
Green lentil	Isolariciresinol	μg/100 g DM	129.60 ± 8.50	123
	Lariciresinol		163.89 ± 10.79	
	Secoisolariciresinol		174.15 ± 6.15	
	Pinoresinol		175.59 ± 10.45	
	Matairesinol		49.47 ± 7.13	
Red lentil	Isolariciresinol		117.33 ± 19.00	
	Lariciresinol		176.87 ± 16.83	
	Secoisolariciresinol		81.81 ± 10.48	
	Pinoresinol		119.87 ± 0.58	
	Matairesinol		47.31 ± 0.00	
Chickpea	Isolariciresinol		6.54 ± 1.70	
	Lariciresinol		22.22 ± 2.31	
	Secoisolariciresinol		—	
	Pinoresinol		8.54 ± 0.45	
	Matairesinol		—	
Bean	Secoisolariciresinol	μg/100 g DM	22	124
	Matairesinol		<1	
Bean (heated)	Secoisolariciresinol		19	
	Matairesinol		<1	
Broad bean (fresh)	Secoisolariciresinol		20	
	Matairesinol		<1	
Broad bean (cooked)	Secoisolariciresinol		21	
	Matairesinol		<1	
Butter bean (dried)	Secoisolariciresinol		141	
	Matairesinol		2	
Butter bean (cooked)	Secoisolariciresinol		22	
	Matairesinol		<1	
French bean	Secoisolariciresinol		94	
	Matairesinol		<1	
French bean (cooked)	Secoisolariciresinol		108	
	Matairesinol		<1	
Haricot bean (dried)	Secoisolariciresinol		106	
	Matairesinol		—	
Haricot bean (Cooked)	Secoisolariciresinol		22	
	Matairesinol		<1	
Kidney bean	Secoisolariciresinol		88	
	Matairesinol		<1	

Table 2.5 (*continued*)

Legume	Compound	Unit	Content	References
Kidney bean (cooked)	Secoisolariciresinol		26	
	Matairesinol		<1	
Runner bean (trimmed & strung)	Secoisolariciresinol		26	
	Matairesinol		<1	
Runner bean (trimmed, strung & cooked)	Secoisolariciresinol		17	
	Matairesinol		—	
Chickpea	Secoisolariciresinol		<1	
Chickpea (cooked)	Matairesinol		2	
Green bean (frozen)	Secoisolariciresinol		<1	
	Matairesinol		3	
Green bean (frozen & cooked)	Secoisolariciresinol		38	
	Matairesinol		—	
Red lentil	Secoisolariciresinol		2	
	Matairesinol		1	
Red lentil (cooked)	Secoisolariciresinol		<1	
	Matairesinol		<1	
Mung bean	Secoisolariciresinol		289	
	Matairesinol		—	
Mung bean (cooked)	Secoisolariciresinol		42	
	Matairesinol		<1	
Pea	Secoisolariciresinol		1	
	Matairesinol		<1	
Pea (cooked)	Secoisolariciresinol		—	
	Matairesinol		<1	
Black bean	Lariciresinol	μg/100 g	1.4	125
	Secoisolariciresinol		4.6	
	Pinoresinol		0.2	
	Matairesinol		0.1	
Broad bean	Lariciresinol		6.2	
	Secoisolariciresinol		1.2	
	Pinoresinol		4.2	
	Matairesinol		0.1	
Chickpea	Lariciresinol		1.8	
	Secoisolariciresinol		0.6	
	Pinoresinol		0.5	
	Matairesinol		0.1	
Kidney bean	Lariciresinol		1.8	
	Secoisolariciresinol		3.2	
	Pinoresinol		0.4	
	Matairesinol		1.1	
Lentil	Lariciresinol		16.3	
	Secoisolariciresinol		1.2	
	Pinoresinol		9.0	
	Matairesinol		0.2	
Lima bean	Lariciresinol		3.9	
	Secoisolariciresinol		10.1	
	Pinoresinol		0.5	
	Matairesinol		0.1	

(*continued*)

Table 2.5 (*continued*)

Legume	Compound	Unit	Content	References
Mung bean	Lariciresinol		12.5	
	Secoisolariciresinol		29.9	
	Pinoresinol		0.9	
	Matairesinol		0.4	
Red bean	Lariciresinol		2.2	
	Secoisolariciresinol		2.4	
	Pinoresinol		0.3	
	Matairesinol		0.3	
White bean	Lariciresinol		2.8	
	Secoisolariciresinol		29.9	
	Pinoresinol		0.6	
	Matairesinol		0.2	
Kidney bean	Secoisolariciresinol	μg/100 g	56–153	126
	Matairesinol		Traces	
Pigeon pea	Secoisolariciresinol		21–58	
	Matairesinol		0	
Chickpea	Secoisolariciresinol		7–8	
	Matairesinol		0	
Pea	Secoisolariciresinol		3–13	
	Matairesinol		0	
Black gram	Secoisolariciresinol		46–240	
	Matairesinol		71–262	
Cowpea	Secoisolariciresinol		195–196	
	Matairesinol		0	
Lentil	Secoisolariciresinol		0–7	
	Matairesinol		Traces	

[a]Abbreviation: DM, dry matter.

secoisolariciresinal, pinoresinal, and matairesinol. The study of Durazzo *et al.*[123] reported the highest content of individual lignans with green lentils possessing secoisolariciresinol and pinoresinol levels of ~175 μg/100 g dry matter basis. Contents of 82 and 120 μg/100 g dry matter basis for these respective lignans were found in the red lentils counterpart, but the greatest lignan present was lariciresinol at 177 μg/100 g dry matter basis. Chickpeas, which were also analyzed in this study, possessed extremely low levels of lignans ranging from 6.5 to 22 μg/100 g dry matter basis. For some of the specific lignan types, their content was below the detectable limit of the assay employed.

Beans, for the most part, contain low levels of lignans like that of chickpeas. Levels of selected lignans, as reported by Kuhle *et al.*[124] ranged from <1 to ~100 μg/100 g dry matter basis. The lignan findings of Thompson *et al.*[125] were also quite low for the bean, chickpea, and lentils analyzed in this study, the greatest content of 29.9 μg secoisolariciresinol/100 g being found in white beans. The results from this study are on a fresh weight basis, not a dry matter basis, and hence, it would be expected to be much lower.

References

1. M. C. Vaz Patto, R. Amarowicz, A. N. A. Aryee, J. I. Boye, H. -L. Chung, M. A. Martín-Cabrejas and C. Domoney, *Crit. Rev. Plant Sci.*, 2015, **34**, 105.
2. I. Hayat, A. Ahmad, T. Masud, A. Ahmed and S. Bashir, *Crit. Rev. Food Sci. Nutr.*, 2014, **54**, 580.
3. S. K. Dutta, D. Chatterjee, D. Sarkar, S. B. Singh, T. Boopathi, R. Kuotsu, K. Vikramjeet, R. S. Akoijam, S. Saha, S. Vanlalhmangaihaa, A. Malsawmzualia, S. Chowdhury and S. Lungmuana, *Indian J. Tradit. Knowl.*, 2016, **15**, 313.
4. R. Amarowicz and R. B. Pegg, *Eur. J. Lipid Sci. Technol.*, 2008, **110**, 865.
5. R. Gyawali and I. A. Salam, *Food Control*, 2014, **46**, 412.
6. P. K. Lai and J. Roy, *Curr. Med. Chem.*, 2004, **11**, 6.
7. J. Xue, P. M. Davidson and Q. Zhong, *J. Agric. Food Chem.*, 2013, **61**, 12720.
8. R. Gyawali and S. A. Ibrahim, *Food Control*, 2014, **46**, 412.
9. C. Cueva, M. Moreno-Arribas, P. J. Martín-Álvarez, G. Bills, M. F. Vicente, A. Basilio, C. L. Rivas, T. Requena, J. M. Rodríguez and B. Bartolomé, *J. Food Sci.*, 2013, **78**, M270.
10. Y. C. Jeong, H. M. Jae and H. P. Keun, *Korean J. Food Sci. Technol.*, 2000, **32**, 1403.
11. M. F. Tsou, C. F. Hung, H. F. Lu, L. T. Wu, S. H. Chang, H. L. Chang, G. W. Chen and J. G. Chung, *Microbios*, 2000, **101**, 37.
12. H. H. Lo and J. G. Chung, *Anticancer Res.*, 1999, **19**, 133.
13. C. I. Nilsson, *Rapid Commun. Mass Spectrom.*, 1999, **19**, 1061.
14. B. Fernandez de Simon, J. Perez-Ilzarbe, T. Hernandez and C. E. I. Gomez-Cordoves, *J. Agric. Food Chem.*, 1992, **40**, 1531.
15. Y. Lu, F. Jiang, H. Jiang, K. Wu, X. Zheng, Y. Cai, M. Kotakowski, M. Chopp and S. -S. To, *Eur. J. Pharmacol.*, 2010, **641**, 102.
16. J. M. Kratz, C. R. Andrighetti-Fröhner, D. J. Kolling, P. C. Leal, C. C. Cirne-Santos, R. A. Yunes, R. J. Nunes, E. Trybala, T. Bergström, I. C. Frugulhetti, C. R. Barardi and C. M. Simões, *Mem. Inst. Oswaldo Cruz*, 2008, **103**, 437.
17. M. -F. Maggi-Capeyron, P. Ceballos, J. -P. Cristol, S. Delbose, C. Le Doucen, M. Pons, C. L. Léger and B. Descomps, *J. Agric. Food Chem.*, 2001, **49**, 5646.
18. A. Itoh, K. Isoda, M. Kondoh, M. Kawase, M. Kobayashi, M. Tamesada and K. Yagi, *Biol. Pharm. Bull.*, 2009, **32**, 1215.
19. K. Kawabata, T. Yamamoto, A. Hara, M. Shimizu, Y. Yamada, K. Matsunaga, T. Tanaka and H. Mori, *Cancer Lett.*, 2000, **157**, 15.
20. G. Kanchana, W. J. Shyni, M. Rajadurai and R. Periasamy, *J. Pharm.*, 2011, **5**, 3338.
21. L. Pari and M. A. Jalaludeen, *Chem. Biol. Interact.*, 2011, **194**, 40.
22. D. Pugazhendhi, G. S. Pope and P. D. Darbre, *J. Appl. Toxicol.*, 2005, **25**, 301.
23. E. Graff, *Free Rad. Biol. Med.*, 1992, **13**, 435.

24. W. Brand-Williams, M. E. Cuvelier and C. Berset, *Lebensm.-Wiss. Technol.*, 1995, **28**, 25.
25. N. Nenandis, H. -Y. Zhang and M. Z. Tsimidou, *J. Agric. Food Chem.*, 2003, **51**, 1874.
26. S. Toda, M. Kumura and M. Ohnishi, *Planta Med.*, 1991, **57**, 8.
27. M. Karamać, L. Koleva, V. D. Kancheva and R. Amarowicz, *Molecules*, 2017, **22**, 527.
28. M. Karamać, A. Buciński, R. B. Pegg and R. Amarowicz, *Czech J. Food Sci.*, 2005, **32**, 64.
29. O. Koroleva, A. Torkova, I. Nikolaev, E. Khrameeva, T. Fedorova, M. Tsentalovich and R. Amarowicz, *Int. J. Mol. Sci.*, 2014, **15**, 16351.
30. R. Re, N. Pellegrini, A. Proteggente, A. Panalla, M. Yang and C. Rice-Evans, *Free Rad. Boil. Med.*, 1999, **26**, 1231.
31. M. Karamać, A. Kosińska and R. B. Pegg, *Pol. J. Food Nutr. Sci.*, 2005, **55**, 165.
32. Y. Yao, X. Cheng, L. Wang, S. Wang and G. Ren, *Int. J. Mol. Sci.*, 2011, **12**, 7048.
33. Y. Tan, S. H. C. Chang and Y. Zhang, *Food Chem.*, 2017, **214**, 259.
34. P. X. Chen, H. Zhang, M. F. Marcone, K. P. Pauls, R. Liu, Y. Tang, B. Zhang, J. B. Renaud and R. Tsao, *J. Funct. Foods*, 2017, **38**, 675.
35. D. Heimler, P. Vignolini, M. G. Dini and A. Romani, *J. Agric. Food Chem.*, 2015, **53**, 3053.
36. M. R. Moreno-Jiménez, V. Cervantes-Cardoza, J. A. Gallegos-Infante, R. F. González-Laredo, I. Estrella, T. J. García-Gasca, E. Herrera-Carrera, J. O. Díaz-Rivas and N. E. Rocha-Guzmán, *Food Res. Int.*, 2015, **76**, 79.
37. L. Díaz-Batalla, J. M. Widholm, G. C. Fahey Jr., E. Castaño-Tostado and O. Paredes-López, *J. Agric. Food Chem.*, 2006, **54**, 2045.
38. A. K. Ramírez-Jiménez, R. Renoso-Camacho, S. Mendoza-Díaz and G. Loarca-Piña, *Food Chem.*, 2014, **161**, 254.
39. K. Huber, P. Brigide, E. B. Bretas and S. G. Canniatti-Brazaca, *PLoS One*, 2014, **9**, e99325.
40. J. Korus, D. Gumul and K. Czechowska, *Food Technol. Biotech.*, 2007, **45**, 139.
41. R. I. Limón, E. Peñas, M. I. Torino, C. Martínez-Villaluenga, M. Dueñas and J. Frias, *Food Chem.*, 2015, **172**, 343.
42. M. Dueñas, C. Martínez-Villaluenga, R. I. Limón, E. Peñas and J. Frias, *Food Res. Int.*, 2015, **70**, 55.
43. R. Amarowicz and F. Shahidi, *J. Funct. Foods*, 2017, **38**, 656.
44. R. Amarowicz, M. Karamać, M. Dueñas and R. B. Pegg, *Nat. Prod. Comm.*, 2017, **12**, 541.
45. R. Amarowicz, I. Estrella, T. Hernández and A. Troszyńska, *J. Food Lipids*, 2008, **15**, 119.
46. R. Amarowicz, I. Estrella, T. Hernández, A. Troszyńska, A. Kosińska and R. B. Pegg, *Food Chem.*, 2010, **121**, 705.
47. R. Amarowicz, I. Estrella, T. Hernández, M. Dueñas, A. Troszyńska, A. Kosińska and R. B. Pegg, *Int. J. Mol. Sci.*, 2009, **10**, 5513.

48. M. Dueñas, T. Hernández and I. Estrella, *Eur. Food Res. Technol.*, 2002, **49**, 478.
49. F. W. Sosulski and K. J. Dąbrowski, *J. Agric. Food Chem.*, 1984, **32**, 131.
50. S. Weidner, A. Król, M. Karamać and R. Amarowicz, *CyTA – J. Food*, 2018, **16**, 373.
51. T. Madhujith, R. Amarowicz and F. Shahidi, *J. Am. Oil Chem. Soc.*, 2004, **81**, 691.
52. M. L. Price, S. van Scoyoc and L. G. Butler, *J. Agric. Food Chem.*, 1978, **26**, 1214.
53. L. Gu, M. A. Kelm, J. F. Hammerstone, G. Beecher, J. Holder, D. Haytowitz and R. L. Prior, *J. Agric. Food Chem.*, 2003, **51**, 7513.
54. X. Aparicio-Fernandez, G. G. Yousef, G. Loarca-Pina, E. De Mejia and M. A. Lila, *J. Agric. Food Chem.*, 2005, **53**, 4615.
55. M.-A. Jabri, K. Rtibi, A. Ben-Said, C. Aouadhi, K. Hosni, M. Sakly and H. Sebai, *J. Pharm. Pharmacol.*, 2016, **68**, 264.
56. D. Shahwar, M. A. Raza, M. A. S. Mughal, M. A. Abbasi and V. U. Ahmad, *J. Chem. Soc. Pak.*, 2010, **32**, 357.
57. T. Hatano, M. Kusuda, K. Inada, T. O. Ogawa, S. Shiota, T. Tsuchiya and T. Yoshida, *Phytochemistry*, 2005, **66**, 2047.
58. A. B. Howell, J. D. Reed, S. G. Krueger, R. Winterbottom, D. G. Cunningham and M. Leahy, *Phytochemistry*, 2005, **66**, 2047.
59. K. J. Murphy, A. K. Chronopoulos, I. Singh, M. A. Francis, H. Moriarty, M. J. Pike, A. H. Turner, N. J. Mann and A. J. Sinclair, *Clin. Nutr.*, 2003, **77**, 1466.
60. W. S. Jeong and A. N. Y. Kong, *Pharmaceut. Biol.*, 2004, **42**, 84.
61. S. E. Rasmussen, H. Frederiksen, K. S. Krogholm and L. Poulsen, *Mol. Nutr. Food Res.*, 2005, **49**, 159.
62. P. M. Aron and J. A. Kennedy, *Mol. Nutr. Food Res.*, 2008, **52**, 79.
63. M. D. Matchett, S. L. MacKinnon, M. I. Sweeney, K. T. Gottschall-Pass and R. A. R. Hurta, *Biochem. Cell Biol.*, 2005, **83**, 637.
64. T. E. B. Kraft, B. M. Schmidt, Y. Knight, M. Cuendet, Y.-H. Kang, J. M. Pezzut, D. S. Seigler and M. A. Lila, *J. Food Sci.*, 2005, **70**, S159.
65. M. D. Matchett, S. L. MacKinnon, M. I. Sweeney, K. T. Gottschall-Pass and R. A. R. Hurta, *J. Nutr. Biochem.*, 2006, **17**, 117.
66. R. Amarowicz, *Eur. J. Food Lipids Technol.*, 2007, **109**, 549.
67. M. Karamać, A. Kosinska and R. Amarowicz, *Bromatol. Chem Toksykol.*, 2006, **39**, 257.
68. Y. J. Zhang, D. L. DeWitt, S. Murugesan and M. G. Nair, *Chem. Biodiver.*, 2004, **1**, 426.
69. R. Amarowicz, R. B. Pegg, G. A. Dykes, A. Troszynska and F. Shahidi, *ACS Symp. Series*, 2005, **909**, 94.
70. R. Amarowicz, A. Toszynska, N. Barylko-Pikielna and F. Shahidi, *J. Food Lipids*, 2004, **11**, 278.
71. T. Ariga and M. Hamano, *Agric. Biol. Chem.*, 1990, **54**, 2499.
72. T. Ariga, I. Koshiyama and D. Fukushima, *Agric. Biol. Chem.*, 1988, **52**, 2717.

73. H. P. S. Makkar, K. Becker, H. Abel and E. Pawelzik, *J. Sci. Food Agric.*, 1997, **75**, 511.
74. A. H. Khalil and E. H. Mansour, *Food Chem.*, 1995, **54**, 177.
75. G. Duc, P. Marget, R. Esnault, J. Le Guen and D. Bastianelli, *J. Agric. Sci.*, 1999, **133**, 185.
76. Y. -W. Luo, Q. Wang, J. Li, Y. Wang, J. Xiao-Xiao and H. Zheng-Ping, *Adv. J. Food Sci. Technol.*, 2015, **7**, 361.
77. C. Baginsky, Á. Peña-Neira, A. Cáceres, T. Hernández, I. Estrella, M. Morales and R. Pertuzé, *J. Food Comp. Anal.*, 2013, **31**, 1.
78. T. S. Agostini-Costa, A. F. P. Teodoro, R. B. N. Alves, L. R. Braga, I. F. Ribeiro, J. P. Silva, L. G. Quintana and M. L. Burle, *Ciencia Rural*, 2015, **2015**(45), 335.
79. A. L. Jin, J. A. Ozga, D. Lopes-Lutz, A. Schieber and D. M. Reinecke, *Food Res. Int.*, 2012, **46**, 528.
80. B. J. Xu, S. H. Yuan and S. K. C. Chang, *J. Food Sci.*, 2007, **72**, S167.
81. B. Zhang, Z. Deng, D. D. Ramdath, Y. Tang, P. X. Chen, R. Liu and R. Tsao, *Food Chem.*, 2015, **172**, 862.
82. S. Nithiyanantham, S. Selvakumar and P. Siddhuraju, *J. Food Comp. Anal.*, 2012, **27**, 52.
83. Y. N. Sreerama, V. B. Sashikala and V. M. Pratape, *Food Chem.*, 2012, **133**, 156.
84. L. O. Ojwang, L. Yang, L. Dykes and J. Awika, *Food Chem.*, 2013, **139**, 35.
85. E. N. Aquino-Bolaños, Y. D. García-Díaz, J. L. Chavez-Servia, J. C. Carrillo-Rodríguez, C. Carrillo-Rodríguez, A. M. Vera-Guzmán and E. Heredia-García, *Emir. J. Food Agric.*, 2016, **28**, 581.
86. E. G. de Mejía, S. H. Guzmán-Maldonado, J. A. Acosta-Gallegos, R. Reynoso-Camacho, E. Ramírez-Rodríguez, J. L. Pons-Hernández, M. M. González-Chavira, J. Z. Castellanos and J. D. Kelly, *J. Agric. Food Chem.*, 2003, **51**, 5962.
87. L. Mojica, A. Meyer, M. A. Berhow and E. G. de Mejía, *Food Res. Int.*, 2015, **69**, 38.
88. M. Hernández-Salazar, P. Osario-Diaz, G. Loarca-Piña, R. Reynoso-Camacho, J. Tovar and L. Bello-Pérez, *J. Food Sci. Agric.*, 2010, **90**, 1417.
89. E. Doria, B. Campio, F. Sparvoli, A. Tava and E. Nielsen, *J. Food Comp. Anal.*, 2012, **26**, 72.
90. J. A. Curiel, R. Coda, I. Centomani, B. De Las Rivas, R. Muñoz, J. M. Guisán, J. M. Mancheño, A. Cardelle-Cobas, M. Gobbetti and C. G. Rizzello, *Int. J. Food Microb.*, 2015, **20**, 7874.
91. Z. Ma, J. I. Boye and X. Hu, *LWT -- Food Sci. Technol.*, 2018, **92**, 147.
92. M. O. Silva, P. Brigide, N. M. Viva de Toledo and S. G. Canniatti-Braazaca, *Braz. J. Food Technol.*, 2018, **21**, e2016072.
93. S. Khandelwal, *Food Res. Int.*, 2010, **43**, 526.
94. P. Siddhuraju, *Food Chem.*, 2006, **99**, 149.
95. R. Alonso, A. Aguirre and F. Marzo, *Food Chem.*, 2000, **68**, 159.
96. R. Alonso, E. Orúe and F. Marzo, *Food Chem.*, 1998, **63**, 505.

97. C. -F. Chau and P. C. -K. Cheung, *J. Agric. Food Chem.*, 1997, **45**, 4773.
98. R. Gan, Z. Q. Deng, A. X. Yan, N. P. Shah, W. Y. Lui, C. L. Chan and H. Corke, *LWT -- Food Sci. Technol.*, 2016, **73**, 168.
99. L. G. Ranilla, M. I. Genovese and F. M. Lajolo, *J. Agric. Food Chem.*, 2007, **55**, 90.
100. M. Dueñas, T. Hernández and I. Estrella, *Food Chem.*, 2007, **101**, 90.
101. S. Adamidou, I. Nengas, K. Grigorakis, D. Nikolopoulou and K. Jauncey, *Cereal Chem.*, 2011, **88**, 80.
102. A. V. Mkanda, A. Minnaar and H. L. Kock, *J. Sci. Food Agric.*, 2007, **87**, 2869.
103. L. Ramírez-Cárdenas, A. J. Leonel and N. M. B. Costa, *Cienc. Tecnol. Aliment.*, 2008, **28**, 200.
104. T. C. Mariotto-Cezar, S. R. M. Coelho, D. Christ, V. Schoenninger and A. J. B. Almieda, *J. Food Agric. Environ.*, 2013, **11**, 268.
105. A. C. Fernandes, W. Nishida and R. P. C. Proença, *Int. J. Food Sci. Technol.*, 2010, **45**, 1365.
106. S. Boukhanouf, H. Louaileche and D. Perrin, *Int. Food Res. J.*, 2016, **23**, 954.
107. H. Han and B. -K. Baik, *Int. J. Food Sci. Technol.*, 2008, **43**, 1971.
108. A. Touré and X. Xueming, *Compr. Rev. Food Sci. Food Safety*, 2010, **9**, 261.
109. A. M. Hutchin and J. L. Slavin in *Flaxseed in Human Nutrition*, eds. L. U. Thompson and S. C. Cunnane, AOCS Press, Champaign, 2003, p. 126.
110. M. E. Martin, M. Haourigui and C. Pelissero, *Life Sci.*, 1996, **58**, 429.
111. H. Adlercreutz, Y. Mousavi, J. Clark, K. Hockerstedt, E. Hämälainen, K. Wähälä, T. Mäkelaä and T. Hase, *J. Steroid Biochem. Mol. Biol.*, 1992, **41**, 331.
112. N. W. Saarinen, A. Warri and S. I. Makela, *Nutr. Cancer*, 2000, **36**, 207.
113. N. M. Saarinen, P. E. Penttinen, A. I. Smeds, T. T. Hurmerinta and S. I. Makela, *J. Steroid Biochem. Mol. Biol.*, 2005, **93**, 209.
114. W. Demark-Wahnefried, D. T. Price, T. J. Polascik, C. N. Robertson, E. E. Anderson, D. F. Paulson, P. J. Walther, M. Gannon and R. T. Vollmer, *Urology*, 2004, **58**, 47.
115. W. Demark-Wahnefried, C. N. Robertson, P. J. Walther, T. J. Polascik, D. F. Paulson and R. T. Vollmer, *Urology*, 2004, **63**, 900.
116. D. D. Kitts, Y. V. Yuan, A. N. Wijewickreme and L. U. Thompson, *Mol. Cell. Biochem.*, 1999, **202**, 91.
117. A. Slavova-Kazakova, M. Karamać, V. Kancheva and R. Amarowicz, *Molecules*, 2016, **21**(1), 21010017.
118. F. S. Hosseinian, A. D. Muir, N. D. Westcott and E. S. Krol, *J. Am. Oil Chem. Soc.*, 2006, **83**, 835.
119. K. Prasad, *Mol. Cell. Biochem.*, 1997, **168**, 117.
120. P. C. Eklund, O. K. Langvik, J. P. Warna, T. O. Salmi, S. M. Willfor and R. E. Sjoholm, *Org. Biomolec. Chem.*, 2005, **3**, 3336.
121. H. B. Niemeyer and M. Metzler, *J. Food Eng.*, 2003, **56**, 255.

122. C. Hu, Y. V. Yuan and D. D. Kitts, *Food Chem. Toxicol.*, 2007, **45**, 2219.
123. A. Durazzo, V. Turfani, E. Azzini, G. Maiani and M. Carcea, *Food Chem.*, 2014, **149**, 666.
124. G. G. C. Kuhnle, C. Dell'Aquila, S. M. Aspinall, S. R. Runnswick, A. M. C. P. Joosen, A. A. Mulligan and S. A. Bingham, *Food Chem.*, 2009, **116**, 542.
125. L. U. Thompsom, B. A. Boucher, Z. Liu, M. Cotterchio and N. Kreiger, *Nutr. Cancer*, 2006, **54**, 184.
126. W. Mazur and H. Adlercreutz, *Pure Appl. Chem.*, 1998, **70**, 1759.

CHAPTER 3

Phenolic Compounds: Flavonoids in Legumes

M. I. DIAS, M. CAROCHO, L. BARROS AND I. C. F. R. FERREIRA*

Centro de Investigação de Montanha (CIMO), Instituto Politécnico de Bragança, Campus de Santa Apolónia, 5300-253 Bragança, Portugal
*E-mail: iferreira@ipb.pt

3.1 Importance of the Fabaceae Family

3.1.1 Historic and Biological Aspects

The Fabaceae family, also known as Papilionoideae, or by their common name, legume, pea or bean, is the third largest family of flowering plants (angiosperms), behind Asteraceae and Orchidaceae, which evolved about 60 million years ago in the late Cretaceous. There are reports of domestication of lentils (*Lens culinaris* L.) in the geographical location of Iran, dating back 8500 to 9000 years, although it was already considered a food source in prehistoric Amazonian regions. About 3000 years ago, bean (*Phaseolus vulgaris* L.) and soybean (*Glycine max* L.) were also domesticated, becoming staple foods in the Americas and Asia. The Romans where the first to use Fabaceae plants to improve soil and as pasture.[1] The Fabaceae family is composed of about 700 genera and more than 19 000 species, they span the whole globe with a wide range of habitats, from the tropical rain forests to underwater environments, but also dry forests, temperate shrublands with xeric

Food Chemistry, Function and Analysis No. 8
Legumes: Nutritional Quality, Processing and Potential Health Benefits
Edited by María Ángeles Martín-Cabrejas

Published by the Royal Society of Chemistry, www.rsc.org

climates and tempered areas of the planet. Contrarily, they do not thrive in mesic temperate shrublands, namely arctic and alpine regions. These preferences for growth are related to their high demanding nitrogen metabolism, in which they fixate this element through various mechanisms, from root-nodulating with symbiotic bacteria, arbuscular mycorrhizae, ectomycorrhizae, to taking up nitrogen in its inorganic forms. Annually, 21 Mt of nitrogen are fixed by legume-rhizobia symbioses, which returns 5 to 7 Mt of this element to soils, thus saving 8–12 billion US dollars in annual fertilization. They are divided into three families, the caesalpinioids, mimosoids, and papilionoids, although some debate about this division is fostered by new phylogenetic studies.[2–4] The flowers of the legume family are pentamerous with five sepals, five petals, two whorls of five stamens each, and a single carpel, although some species can display a reduction or multiplication of the floral organs. Furthermore, flowers in the legume family are radially symmetrical.[3] The other two most important families of plants for mankind are the Gramineae with around 10 000 species and the Brassicaceae with about 3500 species, which combined would not reach the 19 000 species found in the legume family.[5]

3.1.2 Economic and Nutritional Interest

It is widely known that the consumption of legumes is important due to nutritional richness, but also for overall health, thanks to a high range of bioactive molecules. Still, in Europe and most developed countries, legumes are used for animal feed, and more than half of the world's production of grain legume is processed. The high amounts of protein and micronutrients make these vegetables desirable, especially in developing countries, and, regarding their healthiness, a study carried out with the Mediterranean diet, which consumes an average of 20 g of legumes daily, found that there is a 10% reduction in all-cause mortality in these diet consumers.[2,6] Although having thousands of species, only a dozen hold outstanding commercial interest as food crops, making beans and peas (*Pisum sativum* L.) 25% of the whole production of the family of Fabaceae. In the European Union (EU), three countries accumulate 60% of the legume consumption, namely Spain, France and the United Kingdom.[7] In terms of global trade, the legume family is only second to the Poaceae, and saw a doubling of worldwide exports between 1990 and 2012, according to the Food and Health Organization (FAO), which translated into an increase of 6.6 to 13.4 million tonnes reaching a value of 9.5 billion US dollars.[2] Although there has been an increase in legume plantation worldwide, the amount of cereal plantation in the same period is four times higher, mostly due to changes in varieties and agronomic improvements, while the increase in legumes has been due to an increase in effective plantation. Soybeans (*Glycine max* (L.) Merr.), chickpea (*Cicer arietinum* L.), and lentil have increased their production in line with the increase of land plantation, while faba (*Vicia faba* L.)

bean and peas yields are increasing with a decrease in their harvested area.[6] In 2004, more than 300 million metric tonnes of grain legumes were produced on 190 million planted acres, which is only 13% of the total land under cultivation worldwide.[5]

The health benefits of many plants, Fabaceae included, are attributed to both nutritional, but also non-nutritional compounds. Seeds of grain legumes contain at least 20 to 30% protein, being the ideal complement of many diets rich in cereals and root crops in terms of carbohydrates.[5] Nutritionally, the richness of Fabaceae comes mainly from unsaturated fatty acids, enzymes, proteins and dietary fiber. With regard to amino-acid concentration, legumes tend to be quite rich in lysine while being quite poor in sulfur-containing amino-acids and tryptophan, while cereals seem to be poor in lysine, hence their co-production in many countries. Essential minerals can also be obtained from consumption of Fabaceae, along with many other plant secondary metabolism compounds that have antitumor activity, and blood cholesterol reducing activity, which is why they are advisable for insulin-dependent diabetics. Other evidence has linked the consumption of legumes with a reduction not only in the incidence of cardiovascular diseases, obesity, inflammatory diseases and cancer, but also gut disorders.[5,8] Apart from having highly nutritious seeds, some compounds that accumulate in legumes have very interesting antimicrobial potential.[7] Other industrial applications of legumes are their use as flours, and they are used in liquid form to produce milks, yogurts, infant formula, and oils. The most important oil producing Fabaceae plants are soybean, followed by peanut, palm, canola, sunflower, cotton, olive, coconut and maize.[1] Glycyrrhizin, a food additive is obtained from the legume liquorice (*Glycyrrhiza glabra* L.) and is used as a natural sweetener. Plants from this family have also been used to manufacture biodegradable plastics, oils, gums, dyes and inks. Furthermore, legumes are extremely important in folk medicine, and many of their known empiric effects have been corroborated through modern science.[1,9]

3.2 Flavonoids Importance in the Plant Kingdom

Of the non-nutrients present in the Fabaceae, one of the most important are phenolic compounds which comprise about 8000 different molecules, widely and diversely spread throughout the plant kingdom. Phenolic compounds are, like many other groups of molecules, produced by the secondary metabolism of plants. The secondary metabolism produces compounds and promotes plant behaviours that are not vital to the plants survival, while it is the primary metabolism that promotes plant growth, development and reproduction. The secondary metabolism handles plant fitness, protection against viral, microbial (bacteria and fungi) and herbivore (slugs, snails, insects and vertebrates) attacks, ultra-violet radiation, attraction of pollinators, allelopathy and interplant signalling. The compounds produced by the secondary metabolism can be divided in three groups, the terpenoids/isoprenoids, the nitrogen or sulphur containing compounds, and finally the phenolic compounds.[10]

The phenolic compounds are synthesized through the shikimate/phenylpropanoid pathways, and produce a wide range of different molecules. The occurrence of these compounds can be in free forms, glycosylated (combined with a sugar), acylated (linked to an acyl group), but also as oligomeric and polymerized structures, like hydrolysable and condensed tannins, phlorotannins and lignins. In terms of their location within the vegetable cells, they tend to be found in walls or linked to carbohydrates, and proteins. Although many authors use the term "*polyphenols*" and "*phenolic compound*" indiscriminately, the former should only be used when describing a compound with at least two phenolic rings.[11]

Phenolic compounds are characterized by their structure and divided into various classes, namely simple phenols (catechol, resorcinol, vanillin) with a C6 carbon structure, hydroxybenzoic acids (C6–C1 - *p*-hydroxybenzoic acid, syringic acid, gallic acid), hydroxycinnamic acids (C6–C3 - caffeic acid, sinapic acid, chlorogenic acid), coumarins (C6–C3 - scopoletin, umbelliferone, aesculetin), naphtoquinones (C6–C4 - juglone, pumbalgin), xanthones (C6–C1–C6 - mangostin, mangiferin), stilbenes (C6–C2–C6 - resveratrol, piceid, e-viniferins), anthraquinones (C6–C2–C6 - emodin, physcion), chalcones (phloretin, arbutin, butein), condensed tannins ($(C6–C3–C6)_n$ - procyanidins, prodelphinidins)), hydrolysable tannins (gallotannins, ellagitannins) and flavonoids (C6–C3–C6) which are further divided in subclasses with slight structural differences. Within the flavonoids there are 6 subgroups: the flavan-3-ols (or flavanols) which are one of the most complex subclasses, being characterized by not being planar due to a saturated C element in the heterocyclic C ring. They are responsible for astringency, sourness and bitterness in fruits. Flavonols (quercetin, kaempferol, myricetin) are one of the most widespread subclasses throughout the kingdom, having a high amount of variation throughout the year, due to seasons and other environmental factors. They typically have a conjugation in the third carbon of the C-ring. Flavones (apigenin, luteolin, chrysin) are also a major subclass, that derive from the subclass flavanonols (direct biosynthetical precursor), which displays a subtraction of two hydrogen atoms, mainly catalyzed by flavone synthase. The flavanones (hesperidin, naringenin, eriodictyol) are a rather minor subclass, occurring mainly in fruits, especially in citrus, and are characterized by the absence of a double bond at the chiral center of the C ring. Isoflavones (genistein, daidzein, glycitein), also known as phytoestrogens, due to their phytoestrogenic effects differ from the other subclasses by having the B-ring attached to the C3 carbon, instead of the C2. Finally, the last subclass of flavonoids are the anthocyanins (cyanidin, delphinidin, malvidin), which can have conjugation in positions 3, 5, 7, 3′ and 5′, and tend to accumulate in flowers and fruits, instead of leaves and roots. They are responsible for red, blue, purple and other colours in plants.[11–14]

3.3 Species of Cultural and Economic Interest in the Fabaceae Family

As stated above, despite having about 19 000 species, only a fraction of the Fabaceae family species are of commercial, economic and cultural importance. Henceforth, eleven species of this family will be briefly described and their content of individual flavonoids will be tabled.

Cicer arietinum (L.) is the Latin name of chickpeas, which are a valuable legume all around the world, given their high content of protein, minerals and vitamins, being also important as livestock feed. More than 70% of their worldwide production is based in India, although they are produced and consumed all over the world. There are two types of chickpeas, Kabuli cultivars, which are white to cream coloured, while the Desi cultivars are darker and show a wrinkled shell.[15]

Phaseolus vulgaris L., the common bean is probably the best-known representative of the legume family, given its high consumption worldwide. It offers high quantities of protein, starch and dietary fiber. One of the main benefits of beans is that they are a cheaper source of proteins when compared to animal counterparts. China, India and Brazil are the biggest producers of this species, both in dry and green forms.[16]

Cowpea (*Vigna unguiculata* L. Walp.) is another very important legume, just below beans and chickpeas in terms of importance. Its main advantage in relation to them is its tolerance to drought and heat, making it a very environmentally friendly crop and well adapted to climate change. Furthermore, it is of extreme importance due to its contribution of dietary protein for populations living in arid regions of the globe, and thus, West Africa is the biggest producer worldwide, with 87% of its total production.[17]

Vicia faba L., also known as the faba bean is grown world-wide in cropping systems both as a grain and a green-manure legume. It is a very important part of cropping due to is ability to fix nitrogen in the soil, diversification of systems, thus reducing pests and weed build up, reduction of the consumption of fossil fuel (decrease in external fertilization), production of food and feed rich in protein. Still, faba bean production has increased to only 2.6 M acres in 2008.[18]

Lens culinaris L., lentils, also an edible pulse, known as red dhal, masur and split peas, are especially important in developing countries for their protein quantities, dietary fiber, complex carbohydrates, vitamins and minerals. They rank fifth in terms of global production, with a yield of 850 to 1100 kg per acre. This legume is also quite resistant to drought, and interestingly, unlike most legumes, the biggest producer is not a developing country, instead, Canada has been leading in its plantation.[19]

Lupin, or *Lupinus albus* L. is quite different to the previous species listed, it has been a consolidated food in the Mediterranean basin for centuries, but only recently has it been used as a food source in human nutrition. Apart from the bioactive compounds, it is a good source of protein, constituting around 35 to 40% of its weight. Dietary fiber and carbohydrates are also quite abundant, although it is mainly composed of oligosaccharides, and negligible quantities of starch.[20]

Vigna radiata L. R. Wilczek, known commonly as mung bean, moong bean, mongo, green gram or just mung, is a legume with high impact in Asia, although it has been consumed around the world for about 3500 years. Mung plants, a short duration crop, can grow in various types of soils, having a high adaptability, even in arid or semi-arid conditions. The fruits, beanlike, ellipsoidal, are round and can vary from green to yellow, brown or black.[21]

Arachis hypogaea L. or peanut, known around the world as goober, groundnut, earthnut, and pinder, is a very important source of vegetable protein, accounting for 11% of the total world protein supply, thanks to 70% of its constitution being protein. There are various commercial applications of peanuts, namely peanut butter and peanut oil, among others. Still, peanuts are among the top ten most common food allergens in the world.[22,23]

Peas (*Pisum sativum* L.), are, together with lentils, one of the oldest crops to be domesticated, being grown in the Middle East 9000 years ago. Today, they are grown in 84 countries worldwide, given their high resilience to soil changes, allowing successful production in both irrigated and semi-arid soils. There are five main types of peas, the Austrian winter pea, green pea, maple pea, marrowfat pea, and yellow pea. Nutritionally, dietary fiber and protein are the main constituents of these legumes, which are quite low in sodium and fat and rich in folates, minerals, vitamins and potassium. Reduction in the incidence of colon cancer, diabetes, cholesterol and other diseases have been linked to the consumption of peas. Canada is the leading producer of peas worldwide, being also a leading producer of transformed foods with an origin in peas, namely flour, starch or vegetable protein concentrates.[24]

Soybean (*Glycine max* L.), takes up 7% of the arable land worldwide, and is a crop used for human food and animal feed, but also to produce vegetable oil and flour. Since 1970, the increase in plantation of soybean has surpassed all other major crops. The first historical evidence of soybean domestication dates back to 1700 B.C., in North-eastern China, although today, the biggest producers are the USA, Brazil and Argentina. Soybean is the biggest source of vegetable protein in the world, although only 38% of its nutritional composition is actually protein, the other major components being the precursors of soy oil.[25]

The winged bean (*Psophocarpus tetragonolobus* L.) is a widely known but scarcely cultivated species of the Fabaceae family, also known as asparagus pea and goa bean. The whole plant is edible, and it can be consumed in salads or cooked like spinach. The seeds are very similar to soybeans and are consumed in the same form. The biggest producers and consumers of winged beans are Asian countries, especially the ones near the Equator, where conditions are optimal for the plant to thrive. Although underexploited, the plant could be a future staple legume.[26]

Although legumes are known widely for their nutritional contribution to human and animal feed, a large body of evidence shows that they are rich in bioactive compounds that display medicinal properties, namely phenolic compounds, vitamins, tocopherols and other molecules.

3.4 Variation of Flavonoids in Fabaceae Species

3.4.1 Beneficial Effects and Physiological Roles of the Most Abundant Flavonoids in Legumes

The role of phenolic compounds in the grains and seeds of most Fabaceae species is still poorly understood. However, what experts know, is that the accumulation of flavonoids in the seed coats of legumes have important physiological roles in defence and symbiotic relationships, and also during imbibition and germination processes.[27] Many of the flavonoids present in legume seeds are involved in seed nodulation, releasing signals to increase plant–microbe interactions, and can also function as phytoalexins, phytoestrogens and anti-bloat agents (mainly proanthocyanidins, important for legume forage).[28]

Lepiniec *et al.*[29] reported that there are five classes of flavonoids very abundant in legumes, flavan-3-ols (or flavanols), flavanones, flavonols, isoflavones and anthocyanidins. The same pattern was observed by the authors in the eleven legume species described above. Some major compounds found in these species are exemplified in Table 3.1, divided into the respective class of flavonoids to which they belong.

All of these flavonoid classes have recognized beneficial health effects, as well as important physiological roles in legumes, wherein type, amount,

Table 3.1 Examples of the most commonly found flavonoids in legumes.

Flavonoid Class	Examples	Name
Flavanols	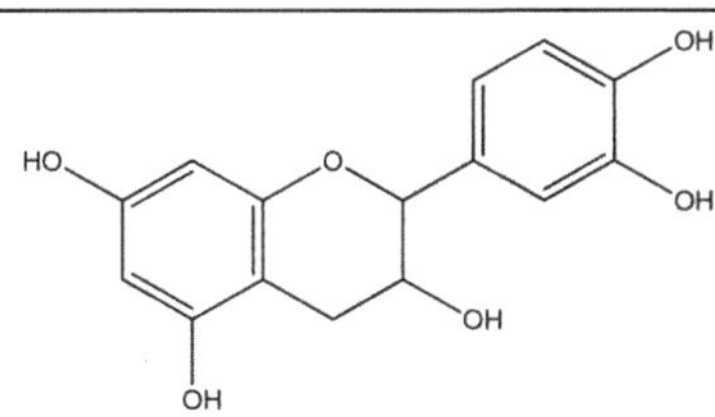	(+)-catechin
Flavanones	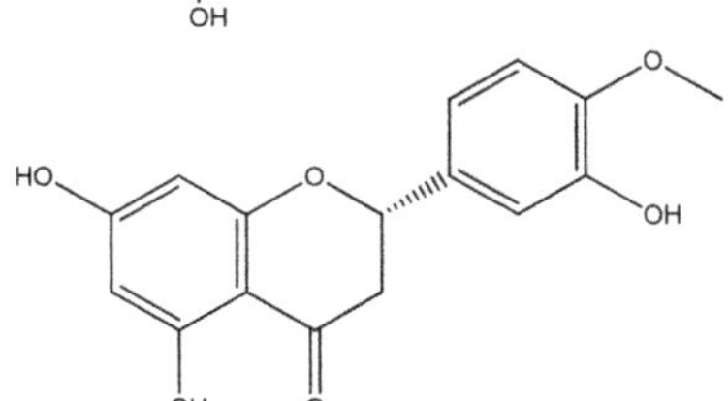	Hesperetin
	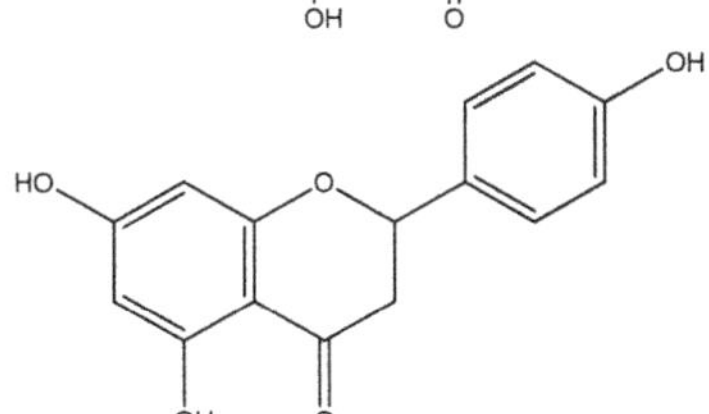	Naringenin

(continued)

Table 3.1 (*continued*)

Flavonoid Class	Examples	Name
Flavonols	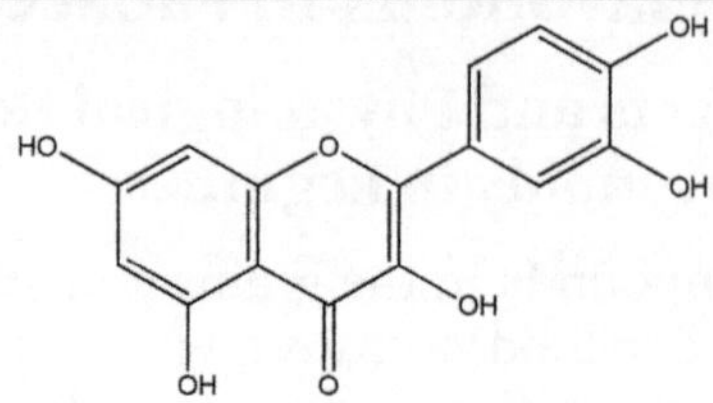	Quercetin
	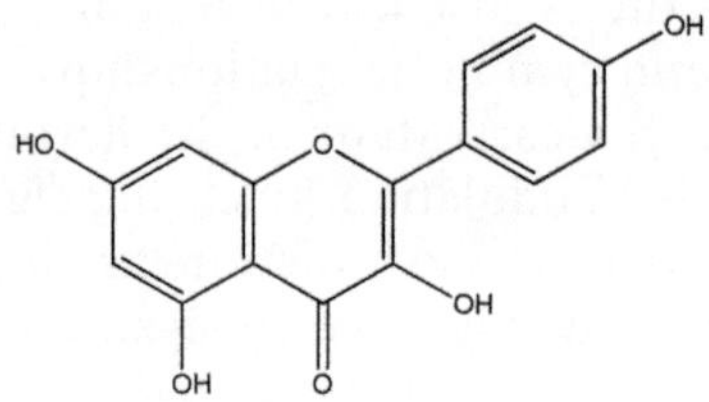	Kaempferol
Isoflavones	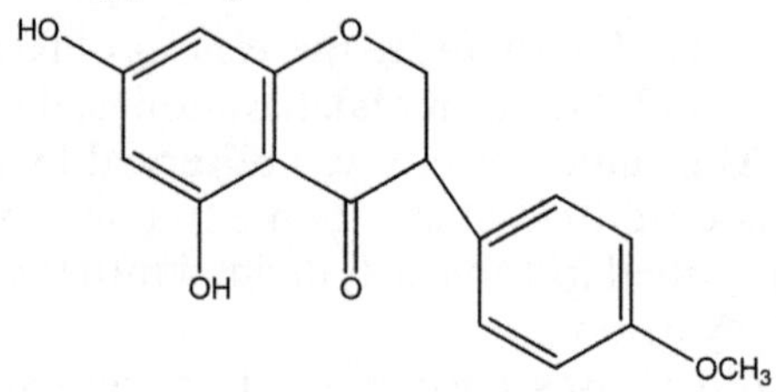	Biochanin A
	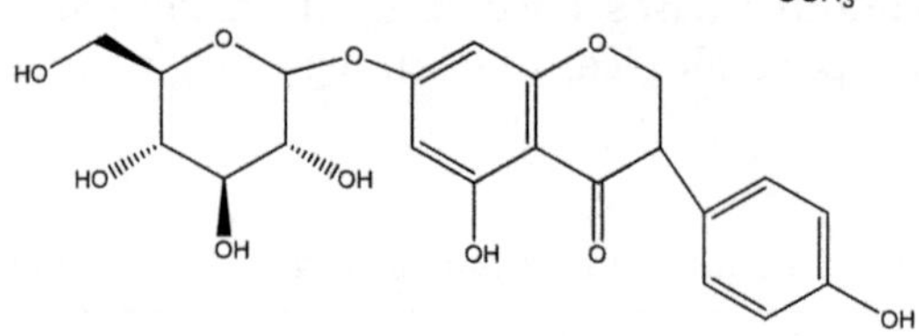	Genistin
	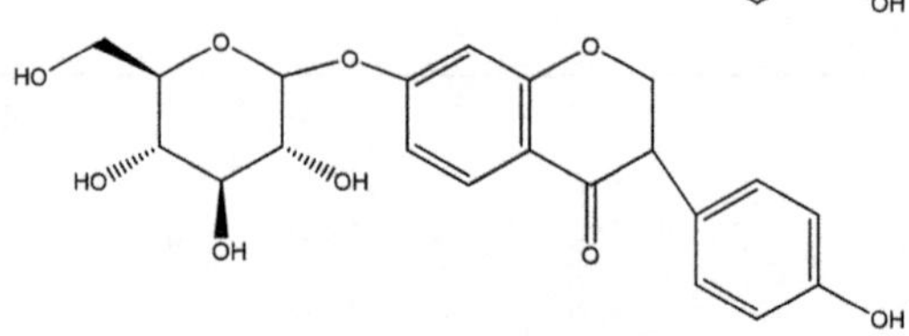	Daidzin
Anthocyanidins	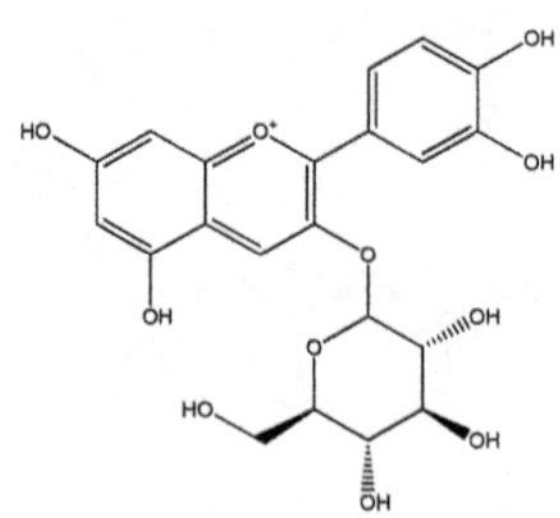	Cyanidin 3-*O*-glucoside
	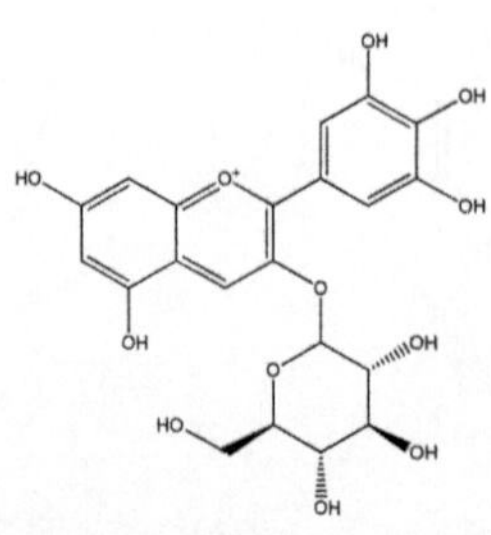	Delphinidin 3-*O*-glucoside

and localization of many of them will be influenced by the plant species and developmental stage of the tissues.[29]

Flavan-3-ols (or flavanols) have a characteristic 2-phenyl-3,4-dihydro-2*H*-chromen-3-ol skeleton that, depending on the OH's position, can present very different bioactive properties. The cardiovascular effects of these type of compounds has been extensively studied as has their bioavailability in foods, and because of their chemical structure it is difficult for the researchers to reach a satisfactory conclusion. Their monomeric forms, (+)-catechin and (−)-epicatechin, are among the most widespread and known flavan-3-ols. However, they can also be present in their polymerized forms as oligomers (dimers to tetramers) or polymers (five or more units, known as condensed tannins – proanthocyanidins) in foodstuff.[30] These type of flavonoids are mostly present in fruits and fruit preparations such as juices and jams, but also in tea and some cereals. Their presence in legumes is not widely reported, however they have already been described in *C. arietinum*,[31] *P. vulgaris*,[32–36] *V. unguiculata*,[37–40] *V. faba*,[41–43] *L. culinaris*,[43–50] *V. radiata*,[51,52] *A. hypogaea*,[53] *P. sativum*,[54,55] and *P. tetragonolobus*.[56] The major drawback of legumes as sources of flavan-3-ols is that the vast majority of them are found in the skin of the seeds, which are prone to be discarded or lost during food processing.[30]

Flavanones are mainly made up of three main abundant aglycones (hesperetin, naringenin, and eriodictyol and their glycosylated molecules), being present in a very limited range of foodstuffs, such as citrus fruits,[57] some aromatic herbs and legumes.[58] They are widely distributed in the Compositae, Leguminosae and Rutaceae families,[57] but their presence in legumes is scarce, derivatives of naringenin only being reported in common bean and faba bean,[36,42,59] hesperitin derivatives in common bean and soybean[36,59] and eriodyctiol derivatives in lupin bean and soybean.[60,61] Among the bioactive potential of flavanones, they stand out as having anti-inflammatory, anti-cancer, antimutagenic, cardiovascular, anti-proliferative, vasorelaxant and vasoprotective effects.[57]

Flavonols is the class with the highest number of compounds in the 11 legume species studied. This class is abundantly present in all types of foodstuff, such as fruits, legumes, vegetables, aromatic plants, tea, among others.[62] The major representatives of this class are quercetin and kampferol derivatives and their glycosylated and methylated (isorhamnetin is the methylated form of quercetin) molecules, the majority being found in Nature with an hydroxylation in the C-3′ and C-4′ positions of the flavonoid C-ring structure.[63] These types of compounds have been reported in common beans,[32,64–68] cowpeas,[37–40,69] faba beans,[42] lentils,[44,48,50,68,70] lupin beans,[60] mung beans,[52,71–73] peanut,[74] and peas,[42,54,75–77] in its most varied forms. Of the eleven legume species discussed in this chapter, it was not possible to find any reference referring to the presence of flavonols in chickpea, soybean and winged bean. The beneficial health effects of flavonols are widely known, especially on the mechanisms that regulate cancer and also against various cardiovascular diseases.[78] Quercetin, for example, is a powerful scavenger of reactive species of oxygen, also presenting anti-inflammatory and genotoxic effects.[79] Moreover, myricetin another widely found flavonol, has been used on several fronts, either as preservative to extend lipophilic foodstuffs shell-life, or as a bioactive compound with strong antioxidant capacity, and in that respect its effects against Parkinson's and Alzheimer's diseases have been studied.[99] It has been reported in common bean[32] and in cowpeas.[40]

Isoflavones are, without a doubt, the class of flavonoids most abundant in legumes. Numerous authors reported the presence of these compounds in chickpeas,[80–83] common beans,[65] mung beans,[52,84] peanuts,[85] and soybeans.[86–90] In the case of soybean there are many articles that report the presence of isoflavones, but because of space limitations we report only on the work of a limited number of authors. Their chemical structure is very similar to naturally occurring and synthetic oestrogens, and anti-oestrogens, which allows isoflavones to bind to estrogenic molecular receptors.[91,92] However, its action has an antagonistic effect when compared to oestrogens, due to the absence of a specific lipophilic region, and therefore isoflavones can present many biological functions, being used to relieve pre- and pro-menstrual symptoms, against breast and prostate cancer, and also in some cardiovascular diseases and osteoporosis.[91,92] Its primary function in plant tissues is to protect against abiotic factors, but also to regulate some mechanisms of action. The vast majority of isoflavones are present in their β-glycosylated form and may also have acetyl or malonyl bonds.[93] Daidzin, genistin and glycitein are aglycones of these type of compounds, and their presence is largely influenced by the cultivation and climatic conditions in which the plants are cultivated or planted.[93] Genistin (genistein 7-*O*-β-D-glucoside) found in high quantities in soybean, is one of the most important isoflavones due to its vast biological activity against numerous types of carcinomas, prostate, colon, liver, lung, ovarian, bladder, neuroblastoma, brain and gastric. Its action is focused on decreasing proliferation and metastasis, induction of apoptosis and chemosensitivity.[94] Formononetin, a 4-methylated form of daidzein, was only identified in chickpea.[80,83] This compound is known to be involved in antiproliferative actions of cancer cells and its effective inhibition in the proliferation of human colorectal cancer has been proven.[95] Biochanin A, a methylated isoflavone, only found in chickpea[80–82] and peanut[85] has shown to be a more powerful inhibitor of fatty acid amide hydrolase (FAAH), the enzyme responsible for pain in animals, when compared to genistein.[96]

The presence of anthocyanidins has also been reported in legume species, cyanidin and delphinidin glycosides being the most abundant. The presence of cyanidin and delphinidin has been described in soybean[97] and common bean,[32,34] respectively. Although the major biological function of these types of compounds is to attract pollinators, these pigments can act as powerful antioxidants.[98] However, their study in legumes species has been underexplored.

Other classes of flavonoids that have been reported in legumes are flavones and stillbenoids. Apigenin derivatives (flavones) have been reported in cowpeas[38] and peas,[54,76,77] mostly being identified as *C*-glycosylated. These types of flavonoids are receiving increasing attention from researchers, due to their wide range of biological activities, such as anticancer, cytotoxic, hepatoprotective, antidiabetic, anti-inflammatory, antiviral and anti-ageing activities.[99] The only stilbenoid compound found in the eleven legume species was resveratrol in peas.[77] Stilbenoids function mainly as a defence mechanism of the plant to stress, mainly caused by fungal infections.[100] They may present other types of bioactivities, which in the case of resveratrol is characterized by the capacity to induce cell differentiation, serve as a mediator of anti-inflammatory mechanism and also exhibit anti-aging properties.[101]

3.4.2 Identification and Characterization of Flavonoids in Different Fabaceae Species

The analytical methods used to separate and identify phenolic compounds, such as flavonoids (Table 3.2), are increasing their sensitivity, reducing their analysis time and quantifying a wider range of compounds.[57] Although all the analytical methodologies reported for flavonoid analysis in legumes were based on high performance liquid chromatography, there are still many authors who based the identification of compounds on the UV spectrum and retention time of the molecules in comparison with commercial standard compounds, a method which does not meet the requirements of high sensitivity, selectivity and efficiency required for the study of bioactive compounds in natural matrices.[102]

This type of identification was used to identify flavonoids in chickpeas,[82,103] common beans,[32,65,67,104] cowpeas[37,39] faba beans,[42,105] lentils,[33,45,47] mung beans,[51,52,72,106] peanuts,[53,74,85] peas,[54,76,77] soybeans,[87–89,97,106,107] and winged beans.[56] Moreover, the identification using an HPLC-DAD (diode array detector) system, in some cases, does not allow the exact compound to be identified, leading only to the flavonoid class by using the UV spectra. This frequently happens with the identification of isoflavones performed by Lee *et al.*[107] in soybeans, and anthocyanins identified by Ranilla *et al.*[67] in common beans. It should be noted, however, that the great majority of the authors use chromatographic techniques coupled to mass spectrometry methodologies, allowing a tentative identification of the compound to be made using its molecular ion and fragmentation pattern. Nevertheless, there are many authors who do not report this data in their manuscripts. Few authors have adopted more advanced resolution techniques with high acquisition rates by coupling time-of-flight (TOF) technologies to the mass spectrometer,[102] as performed by Guardado-Félix *et al.*[83] in chickpeas for the identification of isoformononetin glycoside, malonylated formononetin glycosidea and formononetin, and

Table 3.2 Most commonly used HPLC methodologies and extraction solvents, and part of the plant used for the study of flavonoid composition in 11 different species of legumes.

Common name	Latin name	Origin	Plant part	Methodology	Extraction solvent	Ref.
Chickpea	*Cicer arietinum* L.	Italy	Grain	UPLC-DAD	Water : acetone (1 : 4 v/v)	103
		Mexico	Sprouts	HPLC-TOF/MS	Methanol : water (80 : 20, v/v)	83
		Spain	Grain	HPLC-DAD-MS	Methanol–HCl (1%) : water (80 : 20 v/v)	81
				HPLC-UV	Methanol–HCl (1%) : water (80 : 20 v/v)	82
		Turkey	Grain	HPLC-DAD-ESI/MS	Acetone : ethanol : water : acetic acid 0.1% (2 : 2 : 1; v/v)	80
Common bean	*Phaseolus vulgaris* L.	USA	Grain	HPLC-DAD	Hexane : ethyl acetate and methanol	104
		Brazil		HPLC-DAD	Methanol : water (70 : 30, v/v)	67
		Canada		HPLC-DAD/MS	Methanol : water (70 : 30 v/v)	33
		Italy		HPLC-DAD/MS	Acetonitrile : HCl (1 : 0.25 v/v)	66
		Mexico		HPLC-DAD	Methanol : water (80 : 20 v/v)	65
				HPLC-MS	Methanol (100%)	34
		Spain		HPLC-DAD-ESI/MS	Methanol : water (80 : 20 v/v)	36
				HPLC-DAD/MS	Methanol–HCl (1%) : water (80 : 20 v/v)	68
				HPLC-DAD/MS	Methanol–HCl (1%) : water (80 : 20 v/v)	35
				HPLC-TOF/MS	Methanol : water (80 : 20 v/v)	64
		USA		HPLC-DAD	Acetone : water : acetic acid (70 : 29.5 : 0.5 v/v/v)	32
Cowpeas	*Vigna unguiculata* (L.) Walp.	Tanzania	Seeds	HPLC-DAD-ESI/MS	Acetic acid, acetonitrile and water	108
		Indonesia	Leaves	HPLC-UV	Methanol, acetonitrile and chloroform	39
		South Africa	Seeds	HPLC-UV	—	37
				UPLC-DAD-ESI/MS	Formic acid, acetonitrile and water	40
		USA		PLC-TQD/MS	Methanol, formic acid and water	38
Faba bean	*Vicia faba* L.	—	Immature seeds	HPLC-DAD-ESI/MS	Acetic acid and water	41
		Greece	Aerial parts	HPLC-UV	Acetic acid, acetonitrile and water	105
		Serbia	Stems, leaves and hull	HPLC-MS	Formic acid, methanol and water	42

Lentils	*Lens culinaris* L.	Canada	Seeds	HPLC-DAD	Methanol–HCl (1%) : water (70 : 30 v/v)	33
		Poland	Seeds	HPLC-ESI/MS	Acetone : water (80 : 20 v/v)	43
				HPLC-MS	Ethanol : water (75 : 25 v/v)	48
			Aerial parts	UHPLC-MS	Methanol : water (80 : 20 v/v)	70
		Spain	Cotyledon	HPLC-DAD	Methanol–HCl (1%) : water (80 : 20 v/v)	45
			Grain	HPLC-DAD-MS	Methanol–HCl (1%) : water (80 : 20 v/v)	68
			Seeds	HPLC-DAD-ESI/MS	Methanol : water (80 : 20 v/v)	49
			Seed coat	HPLC-DAD	Methanol–HCl (1%) : water (80 : 20 v/v)	45
				HPLC-DAD-MS	Fractionated dealcoholized extract	46
			Aerial parts	HPLC-DAD	Acetone : water (70 : 30 v/v).	47
		USA	Seeds	HPLC-DAD-MS	Acetone : water : acetic acid (70 : 29.5 : 0.5 v/v/v)	50
Lupin bean	*Lupinus albus* L.	Germany	Roots	LC-ESI/MS	Methanol : water (80 : 20 v/v)	110
		Poland	Leaves, roots	LC-MS	Methanol : water (80 : 20 v/v)	60
Mung beans	*Vigna radiata* (L.) R. Wilczek	China	Seeds	UPLC-Q-TOF-MS	Formic acid, methanol and water	71
		Czech Republic		UPLC-ESI/MS	Methanol, formic acid and water	73
		India		HPLC-DAD	Trifluoroacetic acid, methanol and water	51
		Portugal		HPLC-DAD	Acetic acid	106
		South Asia		HPLC-DAD	Acetic acid, acetonitrile and water	52
		USA		HPLC-UV	Acetonitrile, formic acid and water	72
		Portugal	Sprouted seeds	HPLC-DAD	Acetonitrile and water	106
Peanut	*Arachis hypogaea* L.	USA	Seeds	HPLC-UV	Ethanol : water (80 : 20 v/v)	85
		Brazil	Skins	HPLC-DAD	Acetone : water (60 : 40 v/v)	53
		Myanmar	Seeds	HPLC-DAD	Methanol–HCl (1%) : water (80 : 20 v/v)	74
Peas	*Pisum sativum* L.	China	Sprouts	HPLC-DAD	Acetic acid, acetonitrile and water	77
		Portugal	Seeds	HPLC-DAD	Formic acid, methanol and water	54
		Canada	Seeds	HPLC-DAD-ESI/MS	Acetic acid, acetonitrile and water	55
		Germany	Seeds	HPLC-DAD-ESI/MS	Acetic acid, acetonitrile and water	75
			Leaves	HPLC-DAD-ESI/MS	Acetonitrile and water	69
		Malaysia	Seeds	HPLC-UV	Methanol, trifluoroacetic acid and water	76
		Serbia	Stems, leaves and hull	HPLC-MS	Formic acid, methanol and water	42

(*continued*)

Table 3.2 (*continued*)

Common name	Latin name	Origin	Plant part	Methodology	Extraction solvent	Ref.
Soybean	*Glycine max* L.	China	Seeds	HPLC-DAD	Acidified methanol : water (60 : 40 v/v)	97
		Korea	Seeds	HPLC-DAD	Acetonitrile:0.1 M HCl (5 : 1)	88 and 107
				HPLC-DAD	Methanol : water (50 : 50 v/v)	87 and 89
				HPLC-MS	Methanol : water (80 : 20 v/v)	90
			Leaves	UPLC-Q-TOF/MS	Methanol : water (70 : 30 v/v)	86
		Portugal	Sprouts	HPLC-DAD	Boiled water	106
		Spain	Seeds	HPLC-DAD-ESI/MS	Methanol–HCl (1%) : water (80 : 20 v/v)	61
Winged bean	*Psophocarpus tetragonolobus* (L.) D.C.	Malaysia	Mixed plant parts	HPLC-UV	Methanol, trifluoroacetic acid and water	56

Abu-Reidah *et al.*[64] for the identification of flavanones, flavonol derivatives and isoflavones in common beans, Tang *et al.*[71] tentatively identified kaempherol derivatives in mung beans, and Yuk *et al.*[86] identified malonyldaidzin and malonylgenistin compounds in soyabeans.

Regarding the solvents used to extract flavonoids, there is still some use of non-green solvents, such as for example acetone, hexane, ethyl acetate and chloroform.[32,39,47,50,80,103,104,108] The use of these extraction solvents tends to be reduced, respecting the principles of green chemistry, by using safer and more effective solvents and emerging extraction technologies, such as high hydrostatic pressure, ultrasonication, pulsed electric fields, pressurized and supercritical fluids.[109] Nevertheless, hydroalcoholic mixtures with methanol and ethanol (with or without acidification with hydrogen chloride, trifluoroacetic acid or acetic acid) continue to be the most commonly used solvent, since it allows a better extractability of these compounds, taking into account their polarity. Allowing the extraction of more polar compounds, such as the glycosylated form of flavonoids, which would not be fully extracted with the use of only organic solvents, and for that matter water mixtures, are necessary to increase extraction efficiency.[11] The acidification of the medium allows the phenol–phenolate equilibrium to shift toward the less polar phenyl form, facilitating the extraction. In the case of anthocyanins the acidification of the medium needs to be taken into consideration, since they are pH dependent, modifying their solubility characteristics and stability.[11] Silva *et al.*[106] was the only author reporting the use of boiled water for the extraction of flavonoides in *G. max*, *V. radiata* and *M. sativa*, demonstrating a greater extractability of glycosylated flavonoids.[11]

Table 3.2 also shows the origin of the species studied, which is a determining factor for the different phenolic profiles described later. The parts of the plants studied are mostly grains, which is the most consumed form of legumes. However, some authors report the phenolic profile in other parts of the plant, such as sprouts of chickpeas,[83] leaves of cowpeas,[39] stems, leaves and hulls of faba beans,[42,105] aerial parts of lentils,[47,70] leaves and roots of lupin beans,[60,110] sprouts and leaves of peas,[42,69,77] leaves of soybeans[86] and mixed plant parts of winged beans.[56]

Tables 3.3 and 3.4 describe the most abundant flavonoids in the eleven species of legumes prospected for this chapter. Of each author reporting the composition in flavonoids we describe only the three major flavonoids. The division of the plants into two tables was only for aesthetic purposes. The amount of work related to the phenolic composition of some species is enormous, the most studied ones being those that also have greater economic value, such as common beans, cowpeas, faba beans, lentils, mung beans and peas.

The first major drawback after analysis of the tables, is that the variety of the vegetable greatly influences its phenolic composition, since for the same flavonoid there are very distinct concentrations depending on the variety of the legume.

Table 3.3 Tentative identification, chromatographic characteristics (molecular ion and fragmentation pattern) and quantification (µg per g dw) of the most abundant flavonoids present in *C. arietinum*, *P. vulgaris*, *V. unguiculata*, and *V. faba*.[a]

Latin name	Tentative identification	Molecular ion (MW)	Fragmentation pattern	Variety (observations)	Quantification (µg per g dw)	Ref.
Cicer arietinum L.	Biochanin A	285	283	—	19.0	80
	Biochanin A 7-*O*-glucoside	445	283	Sinaloa and Castellano	10.3	81
	Biochanin A hexoside	445	283, 268	Kabuli blanco lechoso	1.6	82
	Biochanin A derivative	283	268	Kabuli blanco lechoso	3.3	82
		283	268, 239	Kabuli blanco lechoso	6.2	82
		—	267	Sinaloa and Castellano	18.6	81
	Biochanin B derivative	—	267	Sinaloa and Castellano	22.1	81
	(+)-Catechin	—	—	Blanco Sinaloa	178	31
				Sassano	147	31
	Formononetin	269	267	—	1	80
		269	—	Blanco Sinaloa	1145	83
	Isoformononetin glycoside	431	—	Blanco Sinaloa	458	83
	Malonylated formononetin glycoside	517	—	Blanco Sinaloa	962	83
Phaseolus vulgaris L.	Anthocyanins	—	—	FT Nobre	37	67
	Kaempferol 3-*O*-β-D-glucoside (astragalin)	—	—	—	2398*	104
	Kaempferol 3-*O*-glucoside	447	285	Sarconi	0.07	66
		447	285	Zolfino del Pratomagno	0.2	66
	(+)-Catechin	—	—	Turtle Eclipse	247	32
		289	—	Red Rider	44 400	33
	(+)-Catechin glucoside	451	289	Curruquilla	32.8	35
	Daidzen	—	—	Guerrero (G-12878)	17.7*	65
				Michoacán (G-12896-B)	19.3*	65
				Chiapas	11.6*	65
				Chihuahua	14.5*	65
				Durango (G-11025-B)	17.4*	65
				Jalisco (G-12915-A)	22.1*	65
				Morelos (G-10016)	20.9*	65
				Nayarit	15.5*	65
				Oaxaca (OaxNPort)	15.9*	65
				Sinaloa	20.6*	65
				Zacatecas (Jamapa)	24.3*	65

Delphinidin 3-*O*-glycoside	—	—	Black Jamapa	n.a.	34
	—	—	Turtle Eclipse	2196	32
(−)-Epicatechin	289	—	Red Rider	17 600	33
Flavanones	—	—	Helda	n.a.	64
			Perona	n.a.	64
			Strike	n.a.	64
Flavonol derivatives	—	—	Helda	n.a.	64
			Perona	n.a.	64
			Strike	n.a.	64
Hesperetin 7-*O*-rutinoside	609	301	Tolosana	12	36
Isoflavones	—	—	Helda	n.a.	64
			Perona	n.a.	64
			Strike	n.a.	64
Kaempferol	—	—	Guerrero (G-12878)	128*	65
			Michoacán (G-12896-B)	961*	65
			Chiapas	39.2*	65
			Chihuahua	31.5*	65
			Durango (G-11025-B)	552*	65
			Jalisco (G-12915-A)	77.4*	65
			Morelos (G-10016)	76.8*	65
			Nayarit	51.3*	65
			Oaxaca (OaxNPort)	101*	65
			Sinaloa	111*	65
			Zacatecas (Jamapa)	170*	65
Kaempferol 3-*O*-glucoside	447	285	Sarconi	0.4	66
			Zolfino del Pratomagno	0.5	66
Kaempferol 3-*O*-xylosylglucoside	579	285	Sarconi	0.2	66
			Zolfino del Pratomagno	0.1	66
Kaempferol derivatives	—	—	Jalo Precoce	30 700	67
Kaempferol dihexoside-rhamnoside	755	593, 285	Tolosana	1.1	82
Kaempferol 3-*O*-acetylglucoside	—	—	Pinto	29.9	32

(*continued*)

Table 3.3 (*continued*)

Latin name	Tentative identification	Molecular ion (MW)	Fragmentation pattern	Variety (observations)	Quantification (μg per g dw)	Ref.
	Kaempferol 3-*O*-glucoside	—	—	Pinto	147	32
	Myricetin	—	—	Turtle Eclipse	98.9	32
	Naringenin 7-*O*-neohesperidoside	579	271	Tolosana	11.1	60
	Proanthocyanidin dimers	—	—	Black Jamapa	n.a.	34
				Black Jamapa	n.a.	34
	Procyanidin dimer	577	289	Curruquilla	1.4	35
		577	425, 289	Tolosana	49.9	106
		865	289	Curruquilla	2.1	35
	Quercetin	—	—	Guerrero (G-12878)	90.2*	65
				Michoacán (G-12896-B)	57.8*	65
				Chiapas	40.6*	65
				Chihuahua	33.8*	65
				Durango (G-11025-B)	207*	65
				Jalisco (G-12915-A)	121*	65
				Morelos (G-10016)	54.2*	65
				Nayarit	31.2*	65
				Oaxaca (OaxNPort)	60.9*	65
				Sinaloa	107*	65
				Zacatecas (Jamapa)	235*	65
	Quercetin derivatives	—	—	FT Nobre	3.0	67
				Jalo Precoce	0.8	67
Vigna unguiculata (L.) Walp.	(+)-Catechin glucoside	451	289, 451	Agrinawa	174[a]	37
				Bechuana white	138[a]	37
				Glenda	151[a]	37
		—	—	IAR-48 (light brown)	169[a]	37

(+)-Catechin	289	289	Agrinawa	123[b]	37
			Bechuana white	209[b]	37
			Glenda	272[b]	37
	289	245, 205, 187, 151, 109	09CFV-CC27M (Light brown)	249[b]	37
			IAR-48 (light brown)	185[b]	37
			IT82D-889 (red)	243	38
			IT845-2246 (gold brown)	12.9	38
			IT97KK-1042-3 (red)	244	38
			IT98k-1092-1 (black)	16	38
(+)-Catechin glucoside	451	289 451	Agrinawa	10.7	38
			Bechuana white	12.9	39
			Glenda	104	40
	451	289 245	Agrinawa	770	38
			Blackeye	297	38
(+)-Catechin 7-*O*-glucoside	457	289, 179, 165, 151	09CFV-CC27M (Light brown)	62	38
			IAR-48 (light brown)	71	38
			IT82D-889 (red)	16.6	38
			IT845-2246 (gold brown)	75	38
			IT97KK-1042-3 (red)	44	38
			IT98k-1092-1 (black)	2.8	38
(Epi)afzelechin glycosides	435	273, 299, 137, 135, 125	09CFV-CC27M (light brown)	15	38
			IT82D-889 (red)	2553	38
			IT845-2246 (gold brown)	1022	38
			IT97KK-1042-3 (red)	1000	38
			IT98k-1092-1 (black)	1230	38
Apigenin	—	—	—	938	38
Kaempferol	—	—	—	324*	39
Kaempferol 3-*O*-glucoside	447	359, 285	Blackeye	1.7[c]	37
Myricetin 3-*O*-glucoside	479	316, 271, 179	Blackeye	3.85	40

(*continued*)

Table 3.3 (*continued*)

Latin name	Tentative identification	Molecular ion (MW)	Fragmentation pattern	Variety (observations)	Quantification (µg per g dw)	Ref.
	Quercetin	—	—		2666*	39
		301	301	Blackeye	31.2[d]	37
	Quercetin 3-*O*-arabinosyl-diglucoside	757	300 625	Unknown variety 1	1536	108
				Unknown variety 2	1563	108
	Quercetin 3-*O*-diglucoside	479	316, 271, 179 (6)	Agrinawa	32.0	40
	Quercetin 3-*O*-glucoside	463	301, 300, 271	Agrinawa	15.3	40
				Blackeye	2.9	40
	Quercetin 3-*O*-sophoroside-7-*O*-glucoside	787	625, 301	Unknown variety 1	1490	108
				Unknown variety 2	2687	108
	Quercetin 7-*O*-diglucoside	625	625, 300	Unknown variety 2	5338	108
				Unknown variety 1		108
Vicia faba L.	(−)-Epicatechin	—	289	Alargá	6581*	41
				HBP/SO A/2005	5586*	41
				HBP/SO B/2005	7546*	41
				Luz de Otoño	1443*	41
				Portuguesa INIA	5148*	41
				Retaca	3022*	41
				Súper Aguadulce Agrical	6308*	41
				Súper Aguadulce Anasac	6376*	41
				Verde Bonita	5450*	41
		289	—	—	6864*	41
		—	—	—	9535*	41
		—	289	Reina Mora	1472*	41
	(+)-Catechin	—	289	Alargá	1930*	41
				Luz de Otoño	8755*	41
				Portuguesa INIA	3227*	41
				Reina Mora	5976*	41
				Retaca	6279*	41
				Súper Aguadulce Agrical	9096*	41
				Súper Aguadulce Anasac	35	42
				Verde Bonita	864	43
		—	—	—	740	43

(+)-Catechin gallate	442	—	—	2060	43
Isorhamnetin	—	—	—	121	42
Naringenin	—	—	—	7.4	42
Procyanidin dimer	577	289	HBP/SO B/2005	13 182*	41
			Luz de Otoño	1745*	41
			Portuguesa INIA	5138*	41
			Súper Aguadulce Agrical	2525*	41
Procyanidin trimer	577	289	Alargá	7624*	41
	865	577, 289	Verde Bonita	5099*	41
Prodelphinidin dimer	577	289	HBP/SO A/2005	1121*	41
			HBP/SO B/2005	5352*	41
			Reina Mora	4894*	41
			Retaca	2135*	41
			Súper Aguadulce Anasac	1940*	41
	593	305, 289	HBP/SO A/2005	2856*	41
Digallate procyanidin dimer	881	577	—	78.8	42
Quercetin	—	—	—	278	42
Quecetin 3-*O*-rutinoside	—	—	—	190	42

[a]n.a. – not available; Results expressed in (−)-epicatechin equivalents (a), (+)-catechin equivalents (b), kaempferol equivalents (c) and quercetin equivalents (d); * – results converted to dry weight (dw) units using a medium value for moisture percentage in Fabaceae family grains (9.75%). Standard deviation of the quantified compound is not present since it was lower than 5% of the medium value.

Table 3.4 Tentative identification, chromatographic characteristics (molecular ion and fragmentation pattern) and quantification (µg per g dw) of the most abundant flavonoids present in *L. culinaris*, *L. albus*, *V. radiata*, *A. hypogaea*, *P. sativum*, *G. max*, and *P. tetragonolobus*.

Latin name	Tentative identification	Molecular ion (MW)	Fragmentation pattern	Variety (observations)	Quantification (µg per g dw)	Ref.
Lens culinaris L.	(−)-Epicatechin glucoside	451	289	Aldona	60	43
		—	—	Greenland	73.2	44
	(+)-Catechin	—	—	Castellana	0.9	45
				Castellana	1281	45
				Pardina	2.9	45
				Pardina	1475	45
				Vulgaris	7.3	47
	(+)-Catechin gallate	—	—	Greenland	67.2	44
	(+)-Catechin glucoside	—	—	Pardina	790	46
		451	289	Aldona	289	43
				Aldona	289	48
		451	289	Rubia de la Armuña	53.1	49
		451	—	Morton	n.a.	50
	B2 [epicatechin-(4 → 8)-epicatechin]	—	—	Vulgaris	5.7	47
	B3 [catechin-(4 → 8)-catechin]	—	—	Vulgaris	8.6	47
	Acylated kaempferol derivatives	—	—	Tina	n.a.	70
	Kaempferol	285	—	Rubia de la Armuña	1.6	68
	Kaempferol dihexoside	609	285	Rubia de la Armuña	1.4	68

	Kaempferol rhamnose hexose hexose rhamnose	901	755, 285	Rubia de la Armuña	743	49
	Kaempferol tetraglycoside	—	—	Greenland	279	44
	Kaempferol triglycoside	755	—	Morton	n.a.	50
	Procyanidin B3	—	—	Castellana	754	45
				Pardina	745	45
	Procyanidin D	—	—	Castellana	223	45
				Pardina	507	45
	Procyanidin dimer	577	289	Aldona	100	48
				Rubia de la Armuña	25.0	49
		593	—	Castellana	3800	46
				Pardina	4100	46
	Procyanidin trimer	897	—	Castellana	3700	46
				Pardina	3390	46
	Quercetin derivatives	—	—	Tina	n.a.	70
	Quercetin diglycoside	609	301	Aldona	114	48
	Quercetin hexoside acylated	505	301	Aldona	37.2	43
	Quercetin 3-*O*-glucoside	461	301	Rubia de la Armuña	1	68
Lupinus albus L.	Eriodictyol derivatives	—	—	—	n.a.	60
	Isorhamnetin derivatives	—	—	—	n.a.	60
	Genistein 6′-*O*-malonyl-*O*-glucoside	517	—	Amiga	n.a.	110
		679	—	Amiga	n.a.	110
	Genistein 7-*O*-diglucoside	593	—	Amiga	n.a.	110

(*continued*)

Table 3.4 (*continued*)

Latin name	Tentative identification	Molecular ion (MW)	Fragmentation pattern	Variety (observations)	Quantification (µg per g dw)	Ref.
Vigna radiata (L.) R. Wilczek	(+)-Catechin	—	—	Bari Mung 2	69.8*	51
				BU Mug 1	35.3*	51
				CN9-5	4.5	52
				Harsha	6.7	52
				KPS1	6.3	52
				KPS2	9.6	52
				NM92	5.8	52
				NM94	4.9	52
				Pant Mung 5	3.7	52
				Tainan Selection No.5	4.1	52
	Daidzein	—	—	—	5.6	52
				—	4.6	52
				—	970	52
	Genistein	—	—	—	253*	106
				—	1833*	106
	Genistin	—	—	—	531	52
				—	8.7	72
	Glycitein	—	—	—	477*	106
	Isorhamnetin	—	—	Zhonglv No.1	10	73
	Apigenin 6-*C*-glucoside	—	—	Zhonglv No.1	114	52
	Kaempferol	—	—	-	35.4	71
				—	155	71
	Kaempferol 3-*O*-rutinoside	—	—	Zhonglv No.	8.9	72
	Myricetin	—	—	—	8.9	72
	Pomiferin	—	—	—	33.3	71

	Quercetin	—	—	—	68.7	72
				—	6.2	73
				—	31.7	72
				Bari Mung 2	17.2	72
				BU Mug 1	4780	52
				CN9-5	3890	52
				Harsha	4800	52
				KPS1	4670	52
				KPS2	4310	52
				NM92	3930	52
				NM94	4260	52
				Pant Mung 5	4070	52
				Tainan Selection No.5	4330	52
	Quercetin 3-*O*-rutinoside	—	—	—	4770	52
				—	10	73
	Apigenin 8-*C*-glucoside	—	—	Zhonglv No.1	22.4	71
				Zhonglv No.1	67.1	71
Arachis hypogaea L.	Biochanin	—	—	Runner var	9.3	85
	Daidzein	—	—	Runner var	9.4	85
	Genistein	—	—	Runner var	0.7	85
	Kaempferol	—	—	Virginia spreading	2.3	74
	Proanthocyanidin A1	—	—	Runner 886	n.a.	53
	Proanthocyanidin A2	—	—	Runner 886	n.a.	53
	Quercetin	—	—	Virginia spreading	121	74
Pisum sativum L.	Apigenin	—	—	—	176	76
	Apigenin 6-*C*-glucoside	—	—	Esmeralda	0.7[a]	77
				Marqueta	25	54
				Montrebei	5.9	54
				Montsant	2.7	54

(*continued*)

Table 3.4 (*continued*)

Latin name	Tentative identification	Molecular ion (MW)	Fragmentation pattern	Variety (observations)	Quantification (µg per g dw)	Ref.
	Apigenin 8-*C*-glucoside	—	—	Esmeralda	7.7	54
				Marqueta	21	54
				Montsant	9	54
	Epigallocatechin-(4β → 2)-phloroglucinol	429	177, 261, 303	CDC Acer	8.5	54
				CDC Rocket	4875*	55
	Gallocatechin	305	137, 179, 219	CDC Acer	4875*	55
				CDC Rocket	994.5*	55
	Gallocatechin-(4α → 2)-phloroglucinol	429	177, 261, 303	CDC Acer	1023*	55
				CDC Rocket	3012*	55
	Quercetin 3-*O*-galactoside	—	—	Dunav	2929*	55
	Isorhamnetin	—	—	Dunav	48	42
				Tamiš	332	42
	Kaempferol	—	—	—	200	42
	Kaempferol 3-*O*-coumaroyl-sophorotriose	917	285, 447, 609	Navarro	820	75
	Luteolin 3,7-di-*O*-glucoside	—	—	Grisel	15.8	54
				Pixel	46	54
	Luteolin 6-*C*-glucoside	—	—	Esmeralda	11.3	54
				Marqueta	1.4	54
				Montrebei	3.7	54
				Montsant	2.4	54
	Myricetin	—	—	-	48.5	76

Quercetin	—	—	-	136	76
			Tamiš	279	42
Quercetin triglycoside	—	—	Salamanca	0.2	75
Quercetin 3-*O*-coumaroyl-triglycoside	—	—	Salamanca	0.7	75
Quercetin 3-*O*-feroluyl-triglycoside	—	—	Salamanca	0.19	75
Quercetin 3-*O*-coumaroyl-sophorotriose	933	301, 463, 625	James	600	75
			Salamanca	3710	75
Quercetin 3-*O*-coumaroyl-sophorotriose (isomer)	933	301, 463, 625	Salamanca	370	75
Quercetin 3-*O*-feruloyl-sophorotrioside	963	301, 463, 625	Gregor	900	75
			Salamanca	1530	75
			Starter	740	75
			James	1430	75
Quercetin 3-*O*-glucoside	—	—	Dunav	55	42
	—	—	Tamiš	133	42
Quercetin 3-*O*-sinapoly-sophorotrioside	993	301, 463, 625	Gregor	3000	75
			Navarro	1370	75
			Starter	1320	75
			James	980	75
Quercetin 3-*O*-sophorotriose	787	301, 463, 625	Gregor	1550	75
			Navarro	2660	75
			Starter	2360	75
Resveratrol	—	—	—	0.3[a]	77
Quercetin 3-*O*-rutinoside	—	—	—	0.8[a]	77

(*continued*)

Table 3.4 *(continued)*

Latin name	Tentative identification	Molecular ion (MW)	Fragmentation pattern	Variety (observations)	Quantification (μg per g dw)	Ref.
Glycine max L.	Cyanidin 3-*O*-glucoside	—	—	Merrill	5565	97
	Daidzin	417	255, 227, 137	Merrill	4.9	90
		—	—	Merrill	9.8	106
				Merrill	n.a.	88
				Merrill	556	106
	Eriodictyol 7-*O*-rutinoside	595	287	Merit	7.8	61
	Genistin	—	—	Merrill	n.a.	88
				Merrill	52.9	106
	Hesperetin 7-*O*-rutinoside	609	301	Merit	5.3	61
	Isoflavonoids	—	—	Iksan	4.3[b]	107
	Malonyldaidzin	503	255	Merrill	18.0	90
		503	255, 227	Merrill	<100	86
		—	—	Yellow	1757	87
				Saedanbaek	417	89
	Malonylgenistin	519	271, 215, 197,153,149	Merrill	8.6	90
		519	433, 271	Merrill	<100	86
		—	—	Merrill	n.a.	88
				Saedanbaek	572	89
				Yellow	1308	87
	Malonylglycitin	—	—	Saedanbaek	95	89
				Yellow	172	87
	Naringenin	271	—	Merit	2.2	61
Psophocarpus tetragonolobus (L.) D.C.	(+)-Catechin	—	—	—	12 900	56
	(−)-Epicatechin				8840	56
	Quercetin				1961	56

[a]n.a. – not available; results expressed without indication of the units used to quantify (a); results expressed in express in μmol g^{-1}, (b);* - results converted to dry weight (dw) units using a medium value for moisture percentage in Fabaceae family grains (9.75%). Standard deviation of the quantified compound is not present since it was lower than 5% of the medium value.

There are several examples of this observation in the tables, such as for the biochanin A derivative identified in chickpea varieties Kabuli Blanco Lechoso and Sinalao and Castellano that presented concentrations of 6.5 and 18.6 µg per g dw, respectively.[81,82] Formononetin identified by Wu *et al.*[80] and Guardado-Félix *et al.*[83] in chickpea showed concentrations of 1 and 1145 µg per g dw, respectively. The main reason for this difference could be due to the fact that one study is in sprouts[83] and another in the grains.[80] (+)-Catechin concentration was also very different in the varieties turtle eclipse (264.9 µg per g dw)[33] and red rider (44 400 µg per g dw)[32] of common beans. The extraction of flavonoids in the red rider variety was made by using a steaming process, which could increase the extractability of (+)-catechin. The same was observed for eleven different varieties of cowpeas regarding (+)-catechin glucoside and (+)-catechin 7-*O*-glucoside concentration, ranging from 16.6 to 770 and 15 to 2553 µg per g dw, respectively.[38,40] In lentils, (−)-epicatechin glucoside presented very different values for the varieties Aldona, Greenland and Castellana, ranging from 0.98 to 1281.3 µg per g dw.[43–45] The same was observed for (+)-catechin in lentils for the varieties Pardina and Vulgaris, with concentrations ranging concentrations from 2.89 to 1475 µg per g dw.[45,47] For procyanidins there was a considerable discrepancy in the concentrations according to the varieties of the species in question. Procyanidin dimer found in faba beans showed a concentration range of 2525.25–7624.50 µg per g dw for the varieties Luz de Otoño, Portuguesa INIA, Súper Aguadulce agrical and alargá.[41] The same was observed for procyanidin trimer also found in faba beans in the varieties Verde bonita, HBP/SO A/2005 and HBP/SO B/2005, ranging from 1121.25–5352.75 µg per g dw.[41] Procyanidin D found in Pardina and Aldona varieties of faba beans also showed different concentrations, 507.3 and 100 µg per g dw, respectively.[45] Luteolin-3,7-di-*O*-glucoside found in peas also showed very different concentrations for the Grisel (15.8 µg per g dw), Pixel (46 µg per g dw) and Esmeralda (11.3 µg per g dw) varieties.[54] The great variability of flavonoids between varieties can be explained by the complexity of the biosynthetic pathways of these type of compounds and their distinct location within plant tissues.[111,112]

3.5 Conclusions

Legumes are a very large family of plants that have made an immeasurable contribution to mankind, since the dawn of agriculture, and are still today staple foods throughout the world. The most commonly eaten vegetable foodstuffs in many countries are legumes, and although their production is basically to feed humans and farm animals, they also have a great number of bioactive molecules with medicinal potential for human health.

Phenolic compounds, and more specifically, flavonoids, detailed in this chapter, are one of those groups of molecules that science has used to benefit of human health and wellbeing. Their applications have been in medicine, cosmetics, food and other sciences that harness the hidden potential in the plant kingdom. Five classes of flavonoids are very abundant in legumes, flavonols, isoflavones, flavanones, flavones, and anthocyanidins in which the type, amount, and localization of many of them will be influenced by the plant species and developmental stage of the tissues. Also, a great part of the analytical methodologies reported for flavonoid analysis and identification is still based on the UV spectrum and retention time of the molecules, which does not meet the requirements of high sensitivity, selectivity and efficiency, necessary to meet today's academic and scientific requirements. Only a few authors have adopted more advanced resolution techniques with high acquisition rates by coupling mass spectrometers such as electrospray ionization (ESI) and TOF. As for the solvents applied to extract flavonoids, the use of acetone, hexane, ethyl acetate and chloroform is still being used by many authors. However, by respecting the principles of green chemistry, this tendency is ending, and more researchers are using safer and more effective solvents such as hydroalcoholic mixtures with methanol and ethanol.

The consumption of legumes has always been associated with improvement of health, and in the past decades, this empiric belief has been corroborated by science, using high performance methods to pinpoint the molecules and compounds that are responsible. So, today, the consumption of legumes is encouraged not only for nutrition reasons, which in itself already demonstrates the importance of these plants, but also for health reasons, namely to prevent many oncologic, circulatory, neurodegenerative diseases, and to maintain a healthy lifestyle and varied diet.

Still, a long road lies before mankind in terms of the Fabaceae family, in that only a handful of species are currently used for economic, nutritional, bioactive, cultural and historic reasons, when compared to the thousands of edible species that exist within the family. To try and predict the benefits of the consumption of other understudied plants would be a gamble, given the number of unused species. One of the most important factors that should encourage further studies in the Fabaceae, and, aiding their consumption, is that the probability of finding new compounds with incredible medicinal advantages is quite high, and so the study of flavonoids and other groups of molecules is an endless source of research.

Concomitantly, the cutting edge of detection methodologies increases abruptly, and peering into the depth of the chemical build of plants becomes easier and cheaper. Thus, it is expected to eventually find, somewhere in the 19 000-odd species of Fabaceae, either compounds that can greatly advance medicine, or a nutritious plant that can increase yields of production to feed humanity, and adapt to this climate changing world.

Acknowledgements

The authors are grateful to the Foundation for Science and Technology (FCT, Portugal) and FEDER under Programme PT2020 for financial support to CIMO (UID/AGR/00690/2013), M. Carocho (SFRH/BPD/114650/2016) grant and L. Barros contract. The authors are also grateful to FEDER-Interreg

España-Portugal programme for financial support through the project 0377_Iberphenol_6_E.

References

1. P. H. Graham and C. P. Vance, *Plant Physiol.*, 2014, **131**, 872.
2. N. Azani, M. Babineau, C. D. Bailey, H. Banks and A. R. Barbosa, *et al.*, *Taxon*, 2017, **66**, 44.
3. J. M. I. Hofer and T. H. Noel Ellis, *Curr. Opin. Plant Biol.*, 2014, **17**, 153.
4. M. F. Wojciechowski, M. Lavin and M. J. Sanderson, *Am. J. Bot.*, 2004, **91**, 1846.
5. P. Gepts, W. D. Beavis, E. C. Brummer, R. C. Shoemaker, H. T. Stalker, N. F. Weeden and N. D. Young, *Plant Physiol.*, 2005, **137**, 1228.
6. C. H. Foyer, H.-M. Lam, H. T. Nguyen, K. H. M. Siddique, R. K. Varshney, T. D. Colmer, W. Cowling, H. Bramley, T. A. Mori, J. M. Hodgson, J. W. Cooper, A. J. Miller, K. Kunert, J. Vorster, C. Cullis, J. A. Ozga, M. L. Wahlqvist, Y. Liang, H. Shou, K. Shi, J. Yu, N. Fodor, B. N. Kaiser, F.-L. Wong, B. Valliyodan and M. J. Considine, *Nat. Plants*, 2016, **2**, 16112.
7. M. C. Pina-Pérez and M. A. Ferrús Pérez, *Trends Food Sci. Technol.*, 2018, **72**, 114.
8. A. Clemente and R. Olias, *Curr. Opin. Food Sci.*, 2017, **14**, 32.
9. M. Carocho, P. Morales and I. C. F. R. Ferreira, *Food Chem. Toxicol.*, 2017, **107**, 302.
10. A. Aharoni and G. Galili, *Curr. Opin. Biotechnol.*, 2011, **22**, 239.
11. A. M. G.-P. C. Santos-Buelga, S. Gonzalez-Manzano and M. Dueñas, *Natural Products Isolation*, Humana Press, 2012, vol. 864.
12. M. Carocho and I. C. F. R. Ferreira, *Anticancer Agents Med. Chem.*, 2013, **13**, 1236.
13. D. Raffa, B. Maggio, M. V. Raimondi, F. Plescia and G. Daidone, *Eur. J. Med. Chem.*, 2017, **142**, 213.
14. V. Cheynier, *Phytochem. Rev.*, 2012, **11**, 153.
15. V. A. Bampidis and V. Christodoulou, *Anim. Feed Sci. Technol.*, 2011, **168**, 1.
16. F. G. B. Los, A. A. F. Zielinski, J. P. Wojeicchowski, A. Nogueira and I. M. Demiate, *Curr. Opin. Food Sci.*, 2018, **19**, 63.
17. J. M. Awika and K. G. Duodu, *J. Funct. Foods*, 2017, **38**, 686.
18. E. S. Jensen, M. B. Peoples and H. Hauggaard-Nielsen, *Field Crop. Res.*, 2010, **115**, 203.
19. M. Joshi, Y. Timilsena and B. Adhikari, *J. Integr. Agric.*, 2017, **16**, 2898.
20. A. Arnoldi, G. Boschin, C. Zanoni and C. Lammi, *J. Funct. Foods*, 2015, **18**, 550.

21. K. Ganesan and B. Xu, *Food Sci. Hum. Wellness*, 2018, **7**, 33.
22. S. K. Ghatak and K. Sen, *J. Ind. Eng. Chem.*, 2013, **19**, 369.
23. A. S. Chang, A. Sreedharan and K. R. Schneider, *Food Control*, 2013, **32**, 296.
24. F. Roy, J. I. Boye and B. K. Simpson, *Food Res. Int.*, 2010, **43**, 432.
25. G. L. Hartman, E. D. West and T. K. Herman, *Food Secur.*, 2011, **3**, 5.
26. P. Lepcha, A. N. Egan, J. J. Doyle and N. Sathyanarayana, *Plant Foods Hum. Nutr.*, 2017, **72**, 225.
27. P. A. Ndakidemi and F. D. Dakora, *Funct. Plant Biol.*, 2003, **30**, 729.
28. T. Aoki, T. Akashi and S.-I. Ayabe, *J. Plant Res.*, 2000, **113**, 475.
29. L. Lepiniec, I. Debeaujon, J.-M. Routaboul, A. Baudry, L. Pourcel, N. Nesi and M. Caboche, *Annu. Rev. Plant Biol.*, 2006, **57**, 405.
30. S. de Pascual-Teresa, D. A. Moreno and C. García-Viguera, *Int. J. Mol. Sci.*, 2010, **11**, 1679.
31. F. Fratianni, F. Cardinale, A. Cozzolino, T. Granese, D. Albanese, M. Di Matteo, M. Zaccardelli, R. Coppola and F. Nazzaro, *J. Funct. Foods*, 2014, **7**, 551.
32. B. Xu and S. K. C. Chang, *J. Agric. Food Chem.*, 2009, **57**, 4754.
33. P. X. Chen, H. Zhang, M. F. Marcone, K. P. Pauls, R. Liu, Y. Tang, B. Zhang, J. B. Renaud and R. Tsao, *J. Funct. Foods*, 2017, **38**, 675.
34. X. Aparicio-Fernandez, G. G. Yousef, G. Loarca-Pina, E. De Mejia and M. A. Lila, *J. Agric. Food Chem.*, 2005, **53**, 4615.
35. M. M. Pedrosa, C. Cuadrado, C. Burbano, M. Muzquiz, B. Cabellos, B. Olmedilla-Alonso and C. Asensio-Vegas, *Food Chem.*, 2015, **166**, 68.
36. A. López, T. El-Naggar, M. Dueñas, T. Ortega, I. Estrella, T. Hernández, M. P. Gómez-Serranillos, O. M. Palomino and M. E. Carretero, *Food Chem.*, 2013, **138**, 547.
37. F. B. Apea-Bah, J. C. Serem, M. J. Bester and K. G. Duodu, *Food Chem.*, 2017, **237**, 247.
38. L. O. Ojwang, L. Yang, L. Dykes and J. Awika, *Food Chem.*, 2013, **139**, 35.
39. N. Andarwulan, D. Kurniasih, R. A. Apriady, H. Rahmat, A. V. Roto and B. W. Bolling, *J. Funct. Foods*, 2012, **4**, 339.
40. A. M. Nderitu, L. Dykes, J. M. Awika, A. Minnaar and K. G. Duodu, *Food Chem.*, 2013, **141**, 1763.
41. C. Baginsky, álvaro Peña-Neira, A. Cáceres, T. Hernández, I. Estrella, H. Morales and R. Pertuzé, *J. Food Compos. Anal.*, 2013, **31**, 1.
42. F. Šibul, D. Orčić, M. Vasić, G. Anačkov, J. Nadpal, A. Savić and N. Mimica-Dukić, *Ind. Crops Prod.*, 2016, **83**, 641.
43. R. Amarowicz, I. Estrella, T. Hernández, S. Robredo, A. Troszyńska, A. Kosińska and R. B. Pegg, *Food Chem.*, 2010, **121**, 705.
44. B. Zhang, Z. Deng, Y. Tang, P. X. Chen, R. Liu, D. Dan Ramdath, Q. Liu, M. Hernandez and R. Tsao, *J. Funct. Foods*, 2017, **32**, 248.
45. M. Dueñas, T. Hernández and I. Estrella, *Eur. Food Res. Technol.*, 2002, **215**, 478.

46. M. Dueñas, B. Sun, T. Hernández, I. Estrella and M. I. Spranger, *J. Agric. Food Chem.*, 2003, **51**, 7999.
47. B. Bartolome, I. Estrella and T. HernaÂndez, *Z. Lebensm.-Unters. Forsch*, 1997, **205**, 290.
48. A. Troszyńska, I. Estrella, G. Lamparski, T. Hernández, R. Amarowicz and R. B. Pegg, *Food Res. Int.*, 2011, **44**, 3195.
49. A. López, T. El-Naggar, M. Dueñas, T. Ortega, I. Estrella, T. Hernández, M. P. Gómez-Serranillos, O. M. Palomino and M. E. Carretero, *J. Food Process. Preserv.*, 2017, **41**, 1.
50. Y. Zou, S. K. C. Chang, Y. Gu and S. Y. Qian, *J. Agric. Food Chem.*, 2011, **59**, 2268.
51. G. Hithamani and K. Srinivasan, *J. Agric. Food Chem.*, 2014, **62**, 11170.
52. R. M. Nair, D. Thavarajah, P. Thavarajah, R. R. Giri, D. Ledesma, R. Y. Yang, P. Hanson, W. Easdown, J. d. A. Hughes and J. D. H. D. Keatinge, *J. Food Compos. Anal.*, 2015, **39**, 23.
53. T. L. C. Oldoni, P. S. Melo, A. P. Massarioli, I. A. M. Moreno, R. M. N. Bezerra, P. L. Rosalen, G. V. J. Da Silva, A. M. Nascimento and S. M. Alencar, *Food Chem.*, 2016, **192**, 306.
54. S. C. Q. Magalhães, M. Taveira, A. R. J. Cabrita, A. J. M. Fonseca, P. Valentão and P. B. Andrade, *Food Chem.*, 2017, **215**, 177.
55. A. Jin, J. A. Ozga, D. Lopes-Lutz, A. Schieber and D. M. Reinecke, *Food Res. Int.*, 2012, **46**, 528.
56. R. A. Mustafa, A. A. Hamid, S. Mohamed and F. A. Bakar, *J. Food Sci.*, 2010, **75**, 35.
57. M. K. Khan, Zill-E-Huma and O. Dangles, *J. Food Compos. Anal.*, 2014, **33**, 85.
58. C. Manach, C. Morand, A. Gil-Izquierdo, C. Bouteloup-Demange and C. Rémésy, *Eur. J. Clin. Nutr.*, 2003, **57**, 235.
59. D. Thavarajah and P. Thavarajah, *Food Res. Int.*, 2012, **49**, 99.
60. A. Wojakowska, A. Piasecka, P. M. García-López, F. Zamora-Natera, P. Krajewski, Ł. Marczak, P. Kachlicki and M. Stobiecki, *Phytochemistry*, 2013, **92**, 71.
61. M. Dueñas, T. Hernández, S. Robredo, G. Lamparski, I. Estrella and R. Muñoz, *Pol. J. Food Nutr. Sci.*, 2012, **62**, 241.
62. P. C. Hollman and I. C. Arts, *J. Sci. Food Agric.*, 2000, **80**, 1081.
63. S. A. Aherne and N. M. O'Brien, *Nutrition*, 2002, **18**, 75.
64. I. M. Abu-Reidah, D. Arráez-Román, J. Lozano-Sánchez, A. Segura-Carretero and A. Fernández-Gutiérrez, *Phytochem. Anal.*, 2013, **24**, 105.
65. L. G. Espinosa-Alonso, A. Lygin, J. M. Widholm, M. E. Valverde and O. Paredes-Lopez, *J. Agric. Food Chem.*, 2006, **54**, 4436.
66. G. Dinelli, A. Bonetti, M. Minelli, I. Marotti, P. Catizone and A. Mazzanti, *Food Chem.*, 2006, **99**, 105.
67. L. G. Ranilla, M. I. Genovese and F. M. Lajolo, *J. Agric. Food Chem.*, 2009, **57**, 5734.
68. M. Dueñas, T. Sarmento, Y. Aguilera, V. Benitez, E. Mollá, R. M. Esteban and M. A. Martín-Cabrejas, *LWT--Food Sci. Technol.*, 2016, **66**, 72.

69. S. Neugart, S. Rohn and M. Schreiner, *Food Res. Int.*, 2015, **76**, 114.
70. J. Zuchowski, Ł. Pecio and A. Stochmal, *Molecules*, 2014, **19**, 18152.
71. D. Tang, Y. Dong, N. Guo, L. Li and H. Ren, *J. Sci. Food Agric.*, 2014, **94**, 1639.
72. M. L. Wang, A. G. Gillaspie, J. B. Morris, R. N. Pittman, J. Davis and G. A. Pederson, *Plant Genet. Resour. Characterisation Util.*, 2008, **6**, 62.
73. E. A. Prokudina, L. Havlíček, N. Al-Maharik, O. Lapčík, M. Strnad and J. Gruz, *J. Food Compos. Anal.*, 2012, **26**, 36.
74. M. M. Win, A. Abdul-Hamid, B. S. Baharin, F. Anwar and N. Saari, *Eur. Food Res. Technol.*, 2011, **233**, 599.
75. S. Bußler, W. B. Herppich, S. Neugart, M. Schreiner, J. Ehlbeck, S. Rohn and O. Schlüter, *Food Res. Int.*, 2015, **76**, 132.
76. K. H. Miean and S. Mohamed, *J. Agric. Food Chem.*, 2001, **49**, 3106.
77. H. K. Liu, Y. Y. Chen, T. T. Hu, S. J. Zhang, Y. H. Zhang, T. Y. Zhao, H. E. Yu and Y. F. Kang, *J. Funct. Foods*, 2016, **25**, 459.
78. P. C. H. Hollman and M. B. Katan, *Free Radical Res.*, 1999, **31**, 75.
79. A. W. Boots, *Health effects of quercetin : from mechanism to nutraceutical*, 2008.
80. Z. Wu, L. Song, S. Feng, Y. Liu, G. He, Y. Yioe, S. Q. Liu and D. Huang, *J. Agric. Food Chem.*, 2012, **60**, 8606.
81. Y. Aguilera, M. Dueñas, I. Estrella, T. Hernández, V. Benitez, R. M. Esteban and M. A. Martín-Cabrejas, *Plant Foods Hum. Nutr.*, 2011, **66**, 187.
82. C. Megías, I. Cortés-Giraldo, M. Alaiz, J. Vioque and J. Girón-Calle, *J. Funct. Foods*, 2016, **21**, 186.
83. D. Guardado-Félix, S. O. Serna-Saldivar, E. O. Cuevas-Rodríguez, D. A. Jacobo-Velázquez and J. A. Gutiérrez-Uribe, *Food Chem.*, 2017, **226**, 69.
84. D. Tang, Y. Dong, H. Ren, L. Li and C. He, *Chem. Cent. J.*, 2014, **8**, 1.
85. Y. C. Chukwumah, L. T. Walker, M. Verghese and S. Ogutu, *Ultrason. Sonochem.*, 2009, **16**, 293.
86. H. J. Yuk, Y. H. Song, M. J. Curtis-Long, D. W. Kim, S. G. Woo, Y. B. Lee, Z. Uddin, C. Y. Kim and K. H. Park, *J. Agric. Food Chem.*, 2016, **64**, 7315.
87. K. M. Cho, T. J. Ha, Y. B. Lee, W. D. Seo, J. Y. Kim, H. W. Ryu, S. H. Jeong, Y. M. Kang and J. H. Lee, *J. Funct. Foods*, 2013, **5**, 1065.
88. E. H. Kim, H. M. Ro, S. L. Kim, H. S. Kim and I. M. Chung, *J. Agric. Food Chem.*, 2012, **60**, 6045.
89. J. H. Lee, B. W. Lee, B. Kim, H. T. Kim, J. M. Ko, I. Y. Baek, W. T. Seo, Y. M. Kang and K. M. Cho, *J. Korean Soc. Appl. Biol. Chem.*, 2013, **56**, 605.
90. J. H. Lee, S. W. Jeong, Y. A. Cho, S. Park, Y. H. Kim, D. W. Bae, J. Il Chung, Y. S. Kwak, M. J. Jeong, S. C. Park, J. H. Shim, J. S. Jin and S. C. Shin, *J. Sci. Food Agric.*, 2013, **93**, 3081.
91. M. Duncan Alison, R. Phipps William and S. Kurzer Mindy, *Best Pract. Res., Clin. Endocrinol. Metab.*, 2003, **17**, 253.
92. K. D. R. Setchell and A. Cassidy, *J. Nutr.*, 1999, **129**, 758.
93. J. Vacek, B. Klejdus, L. Lojková and V. Kubán, *J. Sep. Sci.*, 2008, **31**, 2054.
94. M. Russo, G. L. Russo, M. Daglia, P. D. Kasi, S. Ravi, S. F. Nabavi and S. M. Nabavi, *Food Chem.*, 2016, **196**, 589.

95. J. Huang, M. Xie, P. Gao, Y. Ye, Y. Liu, Y. Zhao, W. Luo, Z. Ling, Y. Cao, S. Zhang, F. Gao and W. Tang, *Process Biochem.*, 2015, **50**, 912.
96. L. Thors, J. J. Burston, B. J. Alter, M. K. McKinney, B. F. Cravatt, R. A. Ross, R. G. Pertwee, R. W. Gereau IV, J. L. Wiley and C. J. Fowler, *Br. J. Pharmacol.*, 2010, **160**, 549.
97. R. F. Zhang, F. X. Zhang, M. W. Zhang, Z. C. Wei, C. Y. Yang, Y. Zhang, X. J. Tang, Y. Y. Deng and J. W. Chi, *J. Agric. Food Chem.*, 2011, **59**, 5935.
98. M. Wink, *S. Afr. J. Bot.*, 2013, **89**, 164.
99. J. Xiao, E. Capanoglu, A. R. Jassbi and A. Miron, *Crit. Rev. Food Sci. Nutr.*, 2016, **56**, S29.
100. K. Xiao, H. J. Zhang, L. J. Xuan, J. Zhang, Y. M. Xu and D. L. Bai, *Stud. Nat. Prod. Chem.*, 2008, **34**, 453.
101. J. A. Baur and D. A. Sinclair, *Nat. Rev. Drug Discovery*, 2006, **5**, 493.
102. M. Ganzera and S. Sturm, *J. Pharm. Biomed. Anal.*, 2018, **147**, 211.
103. F. Fratianni, M. Tucci, M. D. Palma, R. Pepe and F. Nazzaro, *Food Chem.*, 2007, **104**, 1282.
104. C. W. Beninger, G. L. Hosfield and M. J. Bassett, *J. Am. Soc. Hortic. Sci.*, 1999, **124**, 514.
105. C. Spanou, A. S. Veskoukis, T. Kerasioti, M. Kontou, A. Angelis, N. Aligiannis, A. L. Skaltsounis and D. Kouretas, *PLoS One*, 2012, **7**, e32214.
106. L. R. Silva, M. J. Pereira, J. Azevedo, R. F. Gonçalves, P. Valentão, P. G. de Pinho and P. B. Andrade, *Food Res. Int.*, 2013, **50**, 167.
107. S. J. Lee, J. J. Kim, H. I. Moon, J. K. Ahn, S. C. Chun, W. S. Jung, O. K. Lee and I. M. Chung, *J. Agric. Food Chem.*, 2008, **56**, 2751.
108. S. Neugart, S. Baldermann, B. Ngwene, J. Wesonga and M. Schreiner, *Food Res. Int.*, 2017, **100**, 411.
109. J. Pinela, M. A. Prieto, L. Barros, A. M. Carvalho, M. B. P. P. Oliveira, J. A. Saraiva and I. C. F. R. Ferreira, *Sep. Purif. Technol.*, 2018, **192**, 501.
110. L. Weisskopf, N. Tomasi, D. Santelia, E. Martinoia, N. B. Langlade, R. Tabacchi and E. Abou-Mansour, *New Phytol.*, 2006, **171**, 657.
111. L. Zhang, K. Wei, H. Cheng, L. Wang and C. Zhang, *Bot. Stud.*, 2016, **57**, 8.
112. T. Suzuki, N. Yamazaki, Y. Sada, I. Oguni and Y. Moriyasu, *Biosci., Biotechnol., Biochem.*, 2003, **67**, 2683.

CHAPTER 4

Role of Dietary Fiber in Legumes

V. BENÍTEZ*[a,b], R. M. ESTEBAN[a,b] AND M. A. MARTÍN-CABREJAS[a,b]

[a]Instituto de Investigación en Ciencias de la Alimentación, CIAL (CSIC-UAM), Universidad Autónoma de Madrid, Madrid, Spain; [b]Departamento de Química Agrícola y Bromatología, Facultad de Ciencias, Universidad Autónoma de Madrid, Madrid, Spain
*E-mail: vanesa.benitez@uam.es

4.1 What is Meant by Dietary Fiber?

Dietary fiber (DF) includes a complex and heterogeneous group of substances with different physical, chemical and physiological properties. This fact makes it difficult to define and to develop methodologies capable of efficiently reflecting its biological effects.[1] The term "*dietary fiber*" was first used by Hipsley in 1953 as the non-digestible constituents making up the plant cell wall, but this only represents part of what it is considered to be dietary fiber today.[2] Since then, diverse definitions have been proposed,[3] which include a wide range of substances, having greater physiological significance than previously expected. There is no universal definition of dietary fiber. However, it is generally accepted that a definition based on physiological aspects is necessary.[4]

Food Chemistry, Function and Analysis No. 8
Legumes: Nutritional Quality, Processing and Potential Health Benefits
Edited by María Ángeles Martín-Cabrejas

Published by the Royal Society of Chemistry, www.rsc.org

One of the latest definitions was proposed by The American Association of Cereal Chemists (AACC) that defines DF as "*the edible parts of plants or analogous carbohydrates that are resistant to digestion and absorption in the human small intestine with complete or partial fermentation in the large intestine. Dietary fiber includes polysaccharides, oligosaccharides, lignin, and associated plant substances. Dietary fibers promote beneficial physiological effects including laxation, and/or blood cholesterol attenuation, and/or blood glucose attenuation*".[5] This definition is analogous with that given by Codex Alimentarius in 2009,[6] which included non-digestible carbohydrate polymers with more than nine monomers that have beneficial functional effects in the human body. These non-digestible carbohydrates are included in the following groups: (a) those carbohydrate polymers present naturally in foods; (b) synthetic carbohydrates obtained by chemical synthesis which exhibit a demonstrated beneficial health effect; and (c) carbohydrates, obtained from raw plant material by chemical, physical and enzymatic methods, that also exhibit a demonstrated beneficial health effect. In addition, the Codex definition specifies that added fiber must be exhibited to have physiological benefits and places the decision to include indigestible oligosaccharides (from 3 to 9 monomeric units) as dietary fiber on national authorities.[2]

According to the definitions, dietary fiber includes non-starch polysaccharides, such as cellulose, hemicelluloses, pectins, gums, resistant oligosaccharides such as fructooligosaccharides and galactooligosaccharides, and other carbohydrates such as resistant starch (RS) and dextrins, as well as modified celluloses and synthesized carbohydrate polymers such as polydextrose. Moreover, some carbohydrates of animal origin, such as chitin, hyaluronan, and chondroitin sulfate, and lignin and other associated substances, which are joined to cellulose in the plant cell wall, are also considered in the definitions as dietary fibers.[2,4]

Some of the substances associated with polysaccharides in the cell walls are bioactive compounds such as phenolic compounds, carotenoids and phytosterols. These substances are relevant constituents of DF and deserve special attention since they exert profound effects on the physicochemical properties of DF and also influence its physiological properties in humans.[1,7] For example, phenolic compounds linked to dietary fiber are not absorbed in the small intestine, reaching the colon intact, where they are substrates for bacterial microflora, along with non-digestible carbohydrates and proteins. The fermentation produces metabolites and promotes an antioxidant environment that may provide health protection in the human body.[1,7,8] Literature indicates that the phenolic compounds associated with the plant cell walls constitute approximately 50% of total dietary antioxidants.[8] As a consequence of the substantial amount of phenolic compounds associated with the plant cell wall constituents, the concept of "*dietary fiber-antioxidants*" complex has appeared.[9]

Regarding DF classification, the most widely accepted approach is to classify fiber according to its solubility (soluble or insoluble dietary fiber), and/or its fermentability (partially or poorly fermented, and well-fermented) in an *in vitro* system, mimicking human enzymes.[3] In most cases, well-fermented fibers are soluble in water, while partially or poorly fermented fibers are insoluble. Cellulose, lignins and some hemicelluloses typically make up the bulk of insoluble dietary fiber (IDF), whereas pectin substances, glucans, inulin, gums, galactomannans and other non-starch polysaccharides make up soluble dietary fiber (SDF).[10] SDF and IDF exhibit different physiological effects. Thus, SDF is mainly related to metabolism of carbohydrates and lipids and some of its components present a potential prebiotic effect, whereas IDF is responsible for increasing intestinal regulation, water absorption and fecal bulk, and helps to reduce concentration and contact time of potential carcinogens with the intestinal mucosa.[10,11]

The recommended daily intake of fiber varies with the country, the ideal fiber intake ranging from 20 to 35 g per day. However, few countries report fiber intakes that follow this recommendation, the consumption of the greater part of the population in developed countries being about 11 g per day.[12] Thus, an increase in fiber intake in order to promote health and prevent diseases is an essential public health goal.[10]

4.2 Dietary Fiber in Legumes

Legumes have been reported as a good source of dietary fiber.[12] Table 4.1 summarizes the DF content of raw conventional and non-conventional legumes. The legume DF contents range from 15 to 55%, depending on the species. Moreover, each species, due to the existence of a huge number of varieties and cultivars, shows great heterogeneity in DF contents. In general, non-conventional legumes present higher levels of dietary fiber (20–55%) than conventional ones. Soybean, mucuna and dolichos stand out as the legumes with the highest DF contents (more than 40%), followed by mung bean, pole sitao, cowpea, jack bean (*Canavalia ensiformis*) and lupin with DF levels between 30 and 40%.[13–16] Nevertheless, lima bean, pigeon pea and some *Canavalia* spp show contents of DF slightly above 20%, similar or lower to that found in conventional pulses.[13,17] Regarding conventional legumes, in general, beans show the highest DF content, followed by chickpeas and peas, and the lowest content is found in lentils, although this order depends on the variety or cultivar.[16,18–25]

Dietary fiber of legumes, conventional and under-exploited ones, is mainly composed of insoluble dietary fiber, which represents between 62 and 99% of total dietary fiber (TDF) depending on species and varieties (Table 4.1). The lowest contribution of IDF is found in beans, since IDF represents between 66 and 80% of TDF.[16,18–23] But in the rest of legumes this fraction contributes more than 80 or 90% to TDF.[14–16,18–23,26] With respect to SDF, the contents of this fraction were remarkably different depending on the legume species, being much lower than IDF in all cases. The proportion of IDF and SDF is important for the physiological and technological properties of dietary fiber, in this sense, 30–50% of SDF and 50–70% of IDF is considered a well-balanced proportion to obtain the physiological effects of both fractions.[27] Thus, fiber destined for use as a food ingredient should

Table 4.1 Dietary fiber content of raw conventional and non-conventional legumes.

Legumes	Scientific name	IDF (%)	SDF (%)	TDF (%)	References
Beans	*Phaseolus vulgaris* L.	11.4–29.4	0.4–10.6	16.8–29.8	13, 16 and 18–25
Chickpea	*Cicer arietinum* L.	10.0–25.9	0.9–8.4	17.6–27.2	13, 19, 20, 24, 26 and 51
Cowpea	*Vigna unguiculata* (L.) Walp.	29.8–31.0	0.9–4.0	31.2–34.1	13–15
Dolichos	*Lablab purpureus* (L.) Sweet	39.9–44.0	2.1–4.0	42.0–48.0	14 and 15
Jack bean	*Canavalia ensiformis* (L) DC	31.7–42.5	1.5–3.6	33.2–45.6	14 and 15
Lentils	*Lens culinaris* L.	11.4–24.5	0.9–6.9	16.5–27.9	19, 21, 22, 24 and 26
Lima bean	*Phaseolus lunatus*	17.7	3.7	20.9	13
Lupin	*Lupinus albus* cv.multolupa	23.6	14.4	38.0	16
Maunaloa	*Canavalia cathartica*	22.6	1.0	23.6	17
Mucuna	*Stizolobium niveum*	39.8/41.0	2.9/4.0	42.7–45.0	14 and 15
Mung bean	*Vigna radiate* (L.) Wilczek	26.9	4.8	31.7	13
Peas	*Pisum sativum* L.	8.7–27.6	0.6–8.7	10.1–29.7	13, 21, 24, 34, 51 and 85
Pigeon pea	*Cajanus cajan*	19.4	2.4	21.8	13
Pole sitao	*Vigna unguiculata* subsp. Sesuipedalos L. Verde	29.5	5.5	35.0	13
Soybean	*Glycine max*	38.9–52.1	2.7–8.0	46.9–54.8	13 and 14
Sword bean	*Canavalia gladiata*	20.0–21.0	0.02–0.8	21.0–22.0	17

have a ratio of approximately 1 : 2 of soluble to insoluble fiber to be considered acceptable for this purpose.[10]

A knowledge of the cell wall polysaccharides and the bonds of sugars in polysaccharides, as well as the presence of non-cell wall polysaccharides which form part of dietary fiber, is also important in order to understand better the properties and functions of dietary fibers.[28] Dietary fiber of legumes is mainly composed of cell-wall polysaccharides, although non-cell wall components are also important in the dietary fiber of legumes. The main non-cell wall components of legume dietary fiber are RS and a group of non-digestible oligosaccharides, the raffinose family of oligosaccharides (RFOs).

IDF of legumes is mainly constituted by cellulose and hemicelluloses, although polyuronides and RS may be also present in legume IDF composition. Cellulose comprises linear chains of β-(1–4)-linked glucose associated by hydrogen bonds to form microfibrils. Regarding hemicelluloses, they include polysaccharides that can be solubilized by aqueous alkali after elimination of pectic and water soluble polysaccharides, which link cellulose microfibrils together. They contain backbones of glucose units with β-(1–4) glucosidic linkages, but unlike cellulose, they are smaller in size and are usually branched, and include a variety of sugars, mainly xylose and some galactose, mannose and arabinose.[28,29] Pectins are complex and heterogenous polysaccharides that act as bonding elements in all plant cell walls. In general, pectins are formed by the methylated ester of polygalacturonic acids, composed of chains of 300–1000 galacturonic acid units, linked to chains of neutral sugars such as L-arabinose, D-galactose and L-rhamnose joined with α-(1–4) linkages.[10,30]

The polysaccharide values found in the literature for IDF legumes are variable and depend on several factors such as soil, climate, collection time or botanical species and variety.[31] In general, IDF of chickpea and lentil is mainly constituted by glucose and arabinose, their main polysaccharides being cellulose and arabinans; other polysaccharides such as RS are also present, although in minor amounts.[26,32,33] Regarding peas, the main polysaccharides of IDF are cellulose and polyuronides, with appreciable amount of arabinans, whereas RS is not detected.[34] Similar composition is exhibited by beans, whose IDF also comprise glucose, uronic acids and arabinose as the main carbohydrate elements, however half of the glucose is of cellulosic origin and the other half comes from RS.[23] Regarding non-conventional legumes, glucose (47–59%), which may form cellulose, RS and xyloglucans, arabinose (14–24%) and uronic acids (10–22%), whose origin might be pectic polysaccharides, mainly constitute the bulk of IDF.[35,36]

With regard to SDF composition, uronic acids stand out as the main constituents of this fraction in beans, chickpeas and peas,[23,26,34] while arabinose is the main sugar component of SDF in lentil and cowpea (57 and 38% respectively), and they present a low content of uronic acids.[26,35] This low content of uronic acids is also found in other non-conventional legumes such as dolichos, mucuna and jack bean, glucose (dextrin origin) being the main component of SDF.[15]

Regarding non-cell wall polysaccharides, RS is found in raw and processed legumes, both conventional and non-conventional (Table 4.2). The content of RS found in the literature for raw legumes is variable, ranging from 1.5 in raw lentils to 37% in raw beans.[15,18,20,21,36–40] Processing, both cooking and germination, produce a decrease in the content of RS, except in two studies about the effect of cooking in which an increase of RS in lentils and chickpeas was shown after cooking.[39,40] RS can be defined as the fraction of starch and the starch degradation products that are not absorbed in the small intestine of healthy individuals and is supposed to have properties of both dietary fiber fractions, soluble and insoluble.[41] There are four types of RS. RS1, which consists of indigestible plant matrix surrounding starch granules that might be present in raw legumes. Likewise, RS2, which is a naturally occurring RS, for instance in high amylose corn or uncooked potato, and could also be part of the RS found in uncooked legumes. RS3 is the crystallized starch formed when foods containing starch are subjected to specific cooking and cooling processes, this would be the RS present in processed legumes. Lastly, RS4 is a synthetic starch made through chemical processes, and is not found in Nature.[42]

Table 4.2 Resistant starch content of raw and processed legumes (% dry weight).

Legumes	Scientific name	Raw legumes	Processed legumes	Type of process	References
Beans	*Phaseolus vulgaris* L.	3.7–37.0	2.3–7.4	Cooking	18, 20, 21, 38 and 39
Chickpea	*Cicer arietinum* L.	0.8–19.3	2.2–9.8	Cooking	20, 21 and 36–39
Cowpea	*Vigna unguiculata* (L.) Walp.	7.9	2.2	Germination	15
Dolichos	*Lablab purpureus* (L.) Sweet	2.9	1.4	Germination	15
Jack bean	*Canavalia ensiformis* (L.) DC	4.7	0.9	Germination	15
Lentils	*Lens culinaris* L.	1.5–19.9	2.5–10.2	Cooking	21, 37, 38 and 40
Mucuna	*Stizolobium niveum*	5.0	3.0	Germination	15
Peas	*Pisum sativum* L.	2.5	1.89/4.28	Cooking	21 and 38

On the other hand, the raffinose family of oligosaccharides or α-galactosides, a group of non-digestible oligosaccharides (NDO) recognized as dietary fiber (in some countries) and as a functional ingredient, is found in noticeable quantities in pulses.[2,28] The presence of RFOs in legumes provokes flatulence which is considered undesirable. In the past, studies have focused on building up methods to remove or reduce RFOs. However, currently, the literature shows that they have a prebiotic effect, promoting the growth of probiotic bacteria in human gut microbiota.[38]

The α-galactosides of most significance in legumes include raffinose, stachyose, and verbascose. Raffinose is constituted of one galactose monomer linked to a sucrose molecule *via* α-(1–6) glycosidic linkage and is the first compound in the RFOs. The second member of the series is the tetrasaccharide stachyose and the next one is the pentasaccharide verbascose, both are synthesized from raffinose by the addition of α-D-linked galactosyl units. Furthermore, other α-galactosides based on a cyclitol, such as ciceritol, have been detected in significant amounts in beans, chickpeas, lupin and lentils.[20,26,28]

The total content of these RFOs in chickpeas, lentils, peas and beans ranges from 2 to 10 g/100 g dry weight.[20,25,26,28,38] However, RFOs of non-conventional legumes such as dolichos, cowpea, soybean, mucuna and jack bean range from 2 to 4 g/100 g dry weight.[43,44] Diversity in the RFO contents among lines and cultivars of legumes has been exhibited; playing a vital role in this variability are factors such as seed maturity, environment and genetics.[28] Recently, the consumption of RFOs has been related to an increase in gut bifidobacteria and Ca absorption, in addition to increased colonic butyrate production and improved immune function.[38] Therefore, the processing methods that maintain the level of these oligosaccharides are important for increasing their consumption.

4.3 Effects of Processing on Legume Dietary Fiber

The presence of certain antinutritional factors limits the biological value and acceptance of legumes as food. This is the reason why seeds have to be processed before being consumed.[20] Moreover, processing is also necessary to improve the sensorial characteristics of legume seeds. Processing produces changes in the content and composition of dietary fiber and, therefore, in its physicochemical and functional properties, therefore the study of these modifications is highly recommended. The more common processes to which legumes are subjected are thermal treatments, such as cooking, dehydration or canning, germination and fermentation.[12]

In general, changes observed in dietary fiber content due to thermal processing are very complex and depend on the type of legume, the processing method and its duration, as well as the analytical method applied (Table 4.3).[18] Several studies have evaluated the effect of soaking and cooking on dietary fiber of legumes with mixed results.[18,33,39,40,45–47] Vasishtha & Srivastava[33] evaluated how dietary fiber in different types of chickpea genotypes changes after soaking and cooking. The results revealed that cellulose, hemicelluloses, lignin and pectins of seeds change drastically during soaking and cooking in all the genotypes studied, improving their concentration and, eventually, enhancing health benefits of chickpeas. Likewise, Wang *et al.*[39,40,47] found increases in TDF and IDF of chickpeas, beans, lentils and peas after cooking, and also in SDF of chickpeas and beans, while SDF of peas and lentils decreased.[40,47] The decrease in SDF may be due to the solubilisation of polysaccharide contents that compose the SDF fraction.[46]

However, Kutos *et al.*,[18] who studied the effect of soaking, cooking and canning on dietary fiber of beans, found decreases in IDF and TDF levels with all the processes studied, as well as SDF decreases after cooking of soda-soaked beans and canning. Nevertheless, soaking of beans with water produces an increase in the SDF content, and cooking without soaking also increases it. From a nutritional point of view, only thermally processed beans are relevant, because these legumes are never consumed as raw or soaked. Thus, soaking before cooking should be avoided to maintain a high SDF and a better ratio between soluble and insoluble fractions. Dueñas *et al.*[46] also found decreases in TDF, SDF and IDF of soaked lentils after cooking, whereas TDF and IDF of dark beans remain stable after cooking and SDF increases slightly.

The combined effect of cooking together with freeze-drying in conventional legumes was studied by Almeida Costa *et al.*[21] The SDF and IDF levels remained stable in the freeze-dried cooked legumes compared to the raw ones. Marconi *et al.*[48] evaluated the effect of boiling and microwave cooking on non-starch polysaccharides (NSP) content. It is interesting that despite the considerable differences between both heating methods, microwaving and boiling had similar effects on total NSP levels. Cooking procedures, both microwave and boiling, did not modify the total NSP of chickpeas and beans compared to raw legumes, but both processes produced a redistribution of the insoluble NSP to the soluble fraction, improving the SDF : IDF ratio.

Table 4.3 Effect of cooking, industrial dehydration and canning conditions on dietary fiber content of legumes.

Treatment	Conditions	Legumes		IDF	SDF	TDF	Reference
Cooking	S 18h water + C 60 min	Dark beans	*Phaseolus vulgaris* cv Tolosa	=	↑	=	46
	S 18h water + C 20 min	Lentils	*Lens culinaris* cv Rubia de la Armuña	↓	↓	↓	46
	S 18h water + C 40 min	Beans	*Phaseolus vulgaris* cv Pinto	↓	↑	↓	18
	S 18h baking soda + C 40 min	Beans	*Phaseolus vulgaris* cv Pinto	↓	↓	↓	18
	NS C 2h	Beans	*Phaseolus vulgaris* cv Pinto	↓	↑	↓	18
	S 18h water + C 110 min or C 55 min	Chickpeas, beans	*Cicer arietinum*, *Phaseolus vulgaris* cv Barlotto	↓	↑	=	48
	S 18h water + C 11 min or C 9 min microwave boiling	Chickpeas, beans	*Cicer arietinum*, *Phaseolus vulgaris* cv Barlotto	↓	↑	=	48
	S 8h water + C 15 min	Soya beans, mung beans, ground nuts	*Glycine hispida*, *Phaseolus aureus*, *Arachius hypogea*	↑, ↓, =	↑	=	45
	S 16h water + C 20 min (40 min chickpea) + freeze dried	Peas, beans, chickpeas, lentils	*Pisum sativum* cv Maria, *Phasealus vulgaris* cv IAC, *Cicer arietinum*, *Lens culinaris*	=	=	-	21
	S overnight + C 240 min	Chickpeas	*Cicer arietinum*	↑	↓	↑	19
	S 24h + C	Lentils, peas	*Lens culinaris*, *Pisum sativum*	↑	↓	↑	40 and 47
	S 24 h + C	Beans, chickpeas	*Phasealus vulgaris*, *Cicer arietinum*	↑	↑	↑	39
Industrial dehydration	S 16h water + C 20–30 min (beans and lentils) 70 min (chickpeas) + D	Chickpeas, beans, lentils	*Cicer arietinum*, *Phaseolus vulgaris*, *Lens culinaris*	↑	↑ (↓ lentils)	↑	20 and 26
Canning	Commercial can	Beans	*Phaseolus vulgaris* cv Pinto	↓	↓	↓	18

Likewise, Azizah and Zainon[45] found no changes in TDF content of mung bean, soya bean and ground nut after cooking and boiling, although SDF and IDF suffered modifications which improved the SDF : IDF ratio. These modifications might be due to a partial depolymerization and solubilisation of insoluble pectic compounds and hemicelluloses. Thermal processing and cooking may produce a softening of plant tissues, attributed in part to the degradation of pectic substances involved in the adhesion of cells in the middle lamella of the cell walls.[28]

Aldwairji *et al.*[49] studied the fiber content of boiled and canned legumes commonly consumed in the United Kingdom. TDF content of boiled legumes ranged from 3.6% in green beans to 11.2% in red kidney beans, whereas TDF values of canned legumes are between 2.7% (canned green beans) and 7.4% (canned chickpeas); legumes preserved by canning had significantly lower TDF content than boiled ones. Moreover, canning did not significantly modify the ratio of IDF to SDF compared to boiled legumes.[12]

On the other hand, the effects of an industrial dehydration process on dietary fiber of chickpeas, lentils and beans were evaluated by Aguilera *et al.*[20] and Martin-Cabrejas *et al.*[26] The authors studied an industrial dehydration process that involved three consecutive steps: soaking, cooking and dehydration. In general, soaking did not change the TDF content of legumes. However, soaking plus cooking produced a significant rise in TDF (10–17%), which was due to increases of IDF in lentils and chickpeas (31 and 19%, respectively) and increases in SDF in beans and chickpeas (33–56%, respectively). The dehydration stage showed further rises in both IDF and SDF fractions in cooked pulses, contributing to TDF increments of 12–27%. Therefore, the industrial dehydration process affected the content of DF in legumes, promoting significant increases in TDF, with improved ratios between SDF and IDF fractions in chickpeas and beans.[20]

Regarding α-galactosides, soaking of legumes reduced the content of RFOs, probably due to their solubilisation in the soaking solution, in addition, cooking after soaking treatment also helped in the solubilisation of these compounds, and lastly the industrial dehydration process caused the largest reduction in the level of RFOs. All these processes result in legumes provoking less flatulence.[20,26] After the evaluation of the industrial dehydration process, the authors indicated that these processes may be used to modify the dietary and functional characteristics of legumes in order to use them as food ingredients for their beneficial health effects, specifically to create special diets for specific population groups with digestive problems.

Germination is considered an appropriate technique to increase the nutritional value of legumes by decreasing the level of antinutritional factors.[16] Several studies have reported the impact of germination on the DF contents in legumes and the changes were dependent on the germination conditions and the legume studied (Table 4.4).[14–16,34,46,50] In general, the germination process increases IDF and SDF fractions, producing an increment in TDF content, as a consequence mainly of the increase in cellulosic glucose due

Table 4.4 Effect of germination and fermentation conditions on dietary fiber content of legumes.

Treatment	Conditions	Legume		IDF	SDF	TDF	Reference
Germination	Darkness 2, 4, 6 days 20 °C	Peas	*Pisum sativum*	↑ (↓ 2 days)	↑	↑ (= 2 days)	34
	Darkness 7 days 20 °C	Dark beans	*Phaseolus vulgaris* cv Tolosa	↑	↓	↑	46
	Darkness 7 days 20 °C	Lentils	*Lens culinaris* cv Rubia de la Armuña	↓	↓	↓	46
	Darkness 4 days 25 °C	Dolichos, cowpea, mucuna, jack bean	*Lablab purpureus*, *Vigna unguiculata*, *Stizolobium niveum*, *Canavalia ensiformis*	↑	↑ (↓ jack bean)	↑	14
	Darkness 2 days 28 °C	Lupin, soybeans	*Lupinus albus* cv multulupa, *Glycine max*	↑	=	↑	16
	Darkness 2 days 28 °C	Black beans	*Phaseolus vulgaris*	↓	↑	=	16
	12 h light 4 days 25 °C	Dolichos, cowpea, mucuna, jack bean	*Lablab purpureus*, *Vigna unguiculata*, *Stizolobium niveum*, *Canavalia esiformis*	↑ (↓ jack bean)	↑ (= jack bean)	↑ (= jack bean)	35
	24 h light 4 days 25 °C	Dolichos, cowpea, mucuna, jack bean	*Lablab purpureus*, *Vigna unguiculata*, *Stizolobium niveum*	↓ (↑ cowpea)	= (↑ jack bean)	↑ (↓ dolichos, mucuna)	14
	24h light 2, 4, 6 days 20 °C	Peas	*Pisum sativum*	= (↑ 6 days)	↑	↑	34
Fermentation	Natural fermentation	Beans	*Phaseolus vulgaris* cv carilla	=	↓	↓	23
	Lactic acid fermentation	Beans	*Phaseolus vulgaris* cv carilla	=	↓	↓	23
Fermentation + autoclaving	Natural fermentation + 121 °C 20 min	Beans	*Phaseolus vulgaris* cv carilla	=	↓	=	23
	Lactic acid fermentation + 121 °C 20 min	Beans	*Phaseolus vulgaris* cv carilla	=	↓	↓	23

to the metabolic reactions undergone by seeds during germination.[14,15] The modifications occurred during germination produced an improvement in the IDF : SDF ratio that might influence the physicochemical properties and physiological effects of DF. However, Dueñas *et al.*[46] found a reduction in the TDF level of lentils after germination, due to a decrease in both fractions, SDF and IDF. SDF decrease may be due to the degradation of high molecular weight SDF to smaller parts and, depending on the level of depolymerization, a part of the SDF could be incompletely recuperated by ethanolic precipitation in the fiber analysis.

Regarding RFOs, germination reduces the levels of α-galactosides in legumes although its effect depends on the kind of pulse. The considerable losses of oligosaccharides found during germination are related to the increase in the activity of α-galactosidase, which hydrolyses the α-1-6-galatosidic bonds, thereby causing a rise in the total soluble sugar content.[14,16]

Fermentation is one of the inexpensive and simple processing methods that can be used to reach the desired changes in seed composition and to improve palatability. It is applied all over the world and especially in developing countries. Both natural fermentation and fermentation with micro-organisms produced decreases of SDF in beans, TDF also suffered slight decreases, while IDF remained stable (Table 4.4). These changes provoked modifications in the SDF : IDF ratio. These results suggest that the different fermentation treatments might be used to modify the dietary and functional characteristics of fiber.[23]

4.4 Physicochemical Properties of Legume Dietary Fiber

The physical and chemical properties of fiber are of importance in predicting its functional behavior in food systems and its physiological properties as it passes through the gastrointestinal tract. These properties include particle size and bulk volume, surface area characteristics and porosity, solubility, viscosity and gel-forming ability, hydration properties, oil binding capacity, mineral and organic molecule binding capacity, flavor, color, and cation exchange capacity.[10,29]

4.4.1 Particle Size, Porosity, Surface Area Characteristics and Bulk Density

The particle size depends on the cell wall type presented in foods and on the degree of processing and it is related to transit time, fermentation or fecal excretion and to other physicochemical properties such as oil holding capacity or hydration properties.[29] In general, reduction in the particle size produces an increase in oil holding capacity and hydration properties, probably an increase in surface area and porosity, and modification of structure could be responsible for these changes. Moreover, the regiochemistry of the surface layer may also play a role in the adsorption or binding of some molecules.[29]

The bulk density (BD) is an indirect measurement of the porosity of a material and is important with regard to the packaging. It depends on the product structure, its particle size and their distribution, and is also correlated to other physicochemical properties.[35] The evaluation of bulk density of cotyledon fibers and hull from lentils, peas and chickpeas indicated that soluble fibers and hulls showed much higher densities than did insoluble fibers. In particular, lentils showed the highest density in insoluble fibers and hulls, while chickpea soluble fiber is the densest of the soluble fibers.[51] In addition, some physiological properties could be related to BD. Chau and Cheung[52] found a positive correlation between bulk density of insoluble fiber of three legumes and their cholesterol-lowering effect and also with the HDL:total cholesterol ratio.

4.4.2 Viscosity and Gel-forming Ability

Viscosity, or gel-forming capacity, is related to the fiber's ability to absorb water and form a gelatinous mass. This property is especially associated with the solubility of dietary fibers, such as pectins, gums and β-glucans,[4] and depends on several factors, such as fiber concentration, pH, temperature, ionic strength, shear conditions of processing and chain length or molecular weight of fiber.[29] The viscosity increases as the chain length and molecular weight of fiber increase. Thus, highly soluble and branched fibers with relatively short chain polymers usually have low viscosities.[29] Regarding legume fibers, evaluation of the viscosity of soluble fibers of peas, lentils and chickpeas indicates that they have low viscosity in solution.[51]

Gel formation increases the viscosity of the contents of the gastrointestinal tract, which results in a decrease in the rate of nutrient release, probably due to physical trapping of molecules within solid particles or hydrated networks, or simply slowing the rate of transport to the epithelium.[51] Moreover, viscosity may also explain the delayed gastric emptying often associated with the ingestion of fibers. Lastly, gels may also provide lubrication to stool.[4]

4.4.3 Hydration Properties

Hydration properties are relevant from both a technological and physiological point of view and can influence the success of inclusion of fiber-enriched ingredients into foods.[53] They may be evaluated by measuring water-holding, water absorption and swelling capacity. Water-holding capacity (WHC) is the quantity of water that is retained by 1 g of dry fibers under specified conditions of time soaked, temperature, and duration and speed of centrifugation. Nevertheless, water absorption is the kinetics of water uptake determined by a Baumann apparatus and provides detailed information about the fiber, specifically its substrate pore volume. Swelling (SW) can be measured by the bed volume technique, determined by swelling the fibers in water overnight, in a volumetric cylinder.[54]

The main factors influencing hydration properties are the polysaccharide composition of the plant cell wall and the intermolecular organization of the polysaccharides within the cell wall.[53] Thus, a study about hydration properties of pea hulls revealed that dietary fiber of pea showed low water absorption capacity (WAC) (3.2 g mL^{-1}) and swelling capacity (SWC) (5.2 mL g^{-1}) and medium WHC (7.4 g g^{-1}). These low values may be due to the presence of tightly packed polysaccharides (notably cellulose linked with pectins) in native pea hull.[55] However, when fibers were subjected to depectination, the hydration properties improved, probably due to the expansion of the tissues when pectins are removed. Therefore, extraction of specific classes of polysaccharides can affect hydration properties in a predictable way in pea hull fibers, providing modified pea hulls with hydration capacities much closer to the values of other commercial products such as wheat bran.[55]

Apart from polysaccharide composition and its intermolecular organization, the source of the dietary fiber and some physical properties such as particle size, porosity, ionic form, temperature, pH, type of ions in solution, ionic strength, stresses upon fibers and density are also important to comprehend the different behavior of samples during hydration.[53,54] Dalgetty and Baik[51] studied hydration properties of hull and cotyledon fibers of three legumes. Water holding capacity of insoluble fibers (10.1 mL g^{-1} in chickpeas to 13.4 mL g^{-1} in peas) was much higher than in hull fibers (3.6 mL g^{-1} in lentils to 6.2 mL g^{-1} in chickpeas), as was swelling capacity. Insoluble fibers can adsorb water like a sponge forming a hydrophilic matrix in which water is entrapped, wherein the water fills the interstices of the polysaccharide quasicrystalline to cause a considerable swelling.[56] Of the hulls, chickpea exhibited the highest swelling power and water holding, whereas lentils yielded the highest swelling power of insoluble fibers.

Dietary fibers with high hydration properties can be used as functional ingredients to avoid synaeresis and modify the viscosity and texture of some formulated foods.[27] Moreover, fibers with good hydration properties are able to improve the conditions of fecal bulk and consistency so as to prevent constipation.[56] Furthermore, a linear correlation was found between WHC of IDF of legumes and cholesterol-lowering effect and the ratio between HDL and total cholesterol.[52]

4.4.4 Oil and Organic Molecule Binding Capacity

Besides hydration properties, fibers possess the capacity to hold or bind oil. This property is measured as oil holding capacity (OHC), which is the quantity of oil maintained by the fibers after mixing, incubation with oil and centrifugation.[54] OHC is closely dependent on the bulk density of the material, but it may be also related to the surface properties and to the hydrophilic nature of the polysaccharides. Particles with minor bulk density usually have higher surface area, and consequently have greater capacity to bind or adsorb the lipid components.[56] The ability of dietary fibers to retain oil can be important in food applications, for example, for stabilization of foods with a high percentage of fat, and emulsions, or in preventing fat losses upon cooking.[27,57,58] Moreover, it is also interesting from a nutritional point of view, since the ability to bind or absorb fats and increase their excretion is related to plasma cholesterol reduction.[28]

The evaluation of oil holding capacity of cotyledon fibers and hulls of three legumes, peas, lentils, and chickpeas, showed that legume fibers have good values of OHC. However, there are differences due to the type of fiber. Thus, OHC of insoluble fibers (4.01 to 6.93 mL g^{-1}) is much higher than those of hulls and soluble fibers (0.89 to 1.76 mL g^{-1}).[51] Moreover, pea fibers may be utilized as fat binders and replacers in beef products thanks to their high oil and water binding capacities.[57,58]

Apart from oil and fat, fibers have the ability to sequester and even chemically bind other organic molecules, like bile acids, glucose, cholesterol, carcinogenic compounds or enzymes. This capacity has been suggested as a potential mechanism by which certain dietary fibers, especially those rich in uronic acids and phenolic compounds, may have a hypocholesterolemic or hypoglycemic effect. The oil adsorption of fibers may be influenced by the environmental conditions (duration of exposure, pH), the physical and chemical forms of fibers, and the nature of bile acids.[29]

4.4.5 Mineral Binding and Cation Exchange Capacity

Fiber has been suspected of impairing mineral absorption because charged polysaccharides (such as pectins through their carboxyl groups) and associated substances may bind metal ions. Charged polysaccharides do not have an effect on mineral and trace element absorption, since they release the binding minerals when fermented in the colon. However, associated substances such as phytates can have a negative effect.[29]

With regard to cation exchange capacity (CEC), fibers with high CEC could entrap, destabilize, and disintegrate a lipid emulsion, leading to a decrease in the diffusion and absorption of lipids including cholesterol, accounting for the cholesterol-lowering effect of fibers.[56] The evaluation of CEC of pea hull showed a relatively high capacity to exchange cations.[55] This capacity depends on the content of galacturonic acid and degree of methylation, so the depectination of pea hulls produce a loss of CEC. However, the de-esterification of pectins in alkaline conditions, which increase galacturonic acid concentration, resulted in an increase in CEC.

4.5 Health Benefits of Dietary Fiber

Health benefits of DF have been discussed in numerous publications, and from the initial observations of Burkitt *et al.* in 1974,[59,60] many studies have demonstrated that habitual intake of fiber-rich plant foods is associated with the prevention and management of various chronic diseases.[61] Among others, the beneficial effects of DF are related to improvements of the digestive system, cholesterol lowering, diabetes control, obesity reduction and weight loss, reduction of cardiovascular diseases and cancer, and improvement of growth and activity of beneficial intestinal bacterial.[4]

4.5.1 Digestive System

High DF intake promotes a stool weight increase and transit time decreases, which may contribute to the prevention of large bowel disorders and diseases such as constipation, diverticulitis and bowel cancers.[54] The laxative effects of non-absorbed carbohydrates are due to their osmotic effects, to the increasing bacterial mass as a consequence of DF fermentation, and also to the water-holding capacity of the remaining unfermented fiber. DF in mixed diets, legumes and whole grain products, are particularly effective promoters of normal laxation.[3,62]

4.5.2 Cancer Prevention

The consumption of legumes has been related to cancer prevention, due to their dietary fiber content and the presence of antioxidants among other components.[63] The etiology of cancer involves not only inherited factors, but also environmental (dietary factors).[4] Dietary fiber and antioxidant compounds are some of the dietary factors related to the reduction of the risk of developing cancer.[1] Pectins and pectic oligosaccharides have been shown to induce apoptosis in *in vitro* human colonic adenocarcinoma cells, and the anticancer properties of pectins might be related to their complex side chains.[64] Cellulose may also be involved in preventing colon cancer in humans, reducing the production of fecal mutagens and the concentration of fecal secondary bile acids.[3,65] Furthermore, some types of DF may have antioxidant roles.[3,66] Fuller *et al.*[3] report studies in rats in which lignin from cereal brans are partially metabolized to mammalian lignin, which acts as an antioxidant *in vitro.* This fact might add to the protection by lignin against some cancers, although further studies will be needed in humans. Moreover, several original works and reviews have described the inverse relationship between colon cancer and the consumption of DF with associated antioxidants, mainly phenolic compounds.[1,67,68] Furthermore, consumption of RS increases stool bulk and decreases fecal concentrations of potential carcinogens, such as secondary bile acids, which may have a potential role in preventing colorectal cancer.[3]

4.5.3 Reducing Glycemic Response

The reducing effect on postprandial blood glucose and insulin response have been extensively studied for some types of DF. As mentioned before, soluble fibers increases viscosity of the stomach content so viscous fibers have been associated with prolonged gastric emptying and attenuated speed of starch digestion and glucose absorption. These facts explain the slow release of glucose to the portal circulation, lowering the postprandial glycemic and insulinemic responses. The glycaemic index (GI) allows the classification of foods on the basis of the blood glucose response relative to a starchy food such as white bread, or a standard glucose solution. High-dietary fiber foods, such as legumes, are low-GI foods (GI < 55),[69] which have been shown to reduce postprandial blood glucose and insulin responses in non-diabetic and diabetic subjects.[63] Moreover, RS has been shown to significantly decrease postprandial glucose and insulin responses.[3,70] A study of Behall *et al.*[71] indicates that synergistic effects could exist when foods contain multiple fibers, concluding that soluble fibers are mainly those linked with blood glucose lowering effects, whereas there is some evidence that RS is also effective in glucose attenuation. Kendall *et al.*[61] report that the action of phenolic compounds that are present in low GI foods might be another possible mechanism by which these types of foods can improve glycaemic control.

4.5.4 Body Weight Management

Obesity, an important challenge in most parts of the world, is a risk factor for type 2 diabetes and coronary heart disease (CHD). Fiber-rich pulse foods, other plant foods and DF isolates may assist in weight management among children or adults by different mechanisms.[63,72] Pulses, due to their high fiber content and slowly digested carbohydrates, could be related to satiety and weight control.[63] The recent review of Browlee *et al.*[73] evaluates the current evidence on the impact of fiber-rich foods and isolated foods on body weight management. The authors indicate that there is evidence from observational studies that consistently demonstrates that habitual increased intake of fruits, vegetables and whole grains is associated with lower body weight increase over time; however the randomized, controlled trials based on increase of fruit/vegetable or wholegrain food intake alone show no impact on body weight. Therefore, there is a need for further, well-designed intervention studies in legumes.

4.5.5 Control of Arterial Blood Pressure

There are few studies on the possible effects of DF on arterial blood pressure. Several animal studies[60,74,75] indicate that the addition of fiber to the diet attenuates the onset of hypertension. Although the antihypertensive mechanisms of DF have not been elucidated, the effect of soluble fibers in modulating insulin metabolism, in the reduction of plasma cholesterol levels, as well as in reducing weight loss, could be implicated in blood pressure reduction. Human studies carried out with oat DF, soy protein, *Psyllium*, cocoa husk, or lupin flour enriched bread, suggest a relationship between dietary fiber and blood pressure control, although further studies are needed to demonstrate the antihypertensive effect of DF.[60]

4.5.6 Cholesterol Lowering

Although obesity and type 2 diabetes are risk factors for heart disease, diets rich in DF directly reduce heart disease risk by modifying serum lipid risk factors. Elevated serum total and low density lipoprotein (LDL)-cholesterol levels are considering the main risk factors for cardiovascular diseases. The role of some fibers in lowering cholesterol levels been extensively studied and is one of the most well documented health effects of DF.[3] Soluble fibers, such as beta-glucans, pectins, or guar gum, form a highly viscous layer in the small intestine which reduces the reabsorption of bile acids, so the synthesis of bile acids from cholesterol increases, thus reducing the circulating plasma cholesterol levels.[3,76]

Legumes have been shown to have hypocholesterolaemic effects, for example, kidney beans and groundnuts,[69] and studies on mixed legumes suggested that their consumption lowers LDL-cholesterol by partially interrupting the enterohepatic circulation of bile acids. Consumption of Pinto beans showed decreases in both LDL- and high density lipoprotein (HDL)-cholesterol without affecting serum triglycerides, very low density lipoprotein (VLDL)-cholesterol or glucose.[77]

4.5.7 Gut Microbiota and Prebiotic Effects

Other benefits of DF are colonic health effects related to fermentation of DF in the large intestine. Human gut microbiota plays an important role in normal digestive functions and also affects other functions in the human organism.[3] Gut microbiota can be affected by many factors, including diet. In this sense, some types of DF have been shown to act as prebiotic factors, which selectively enrich the gut microbiota in beneficial bacteria, with the subsequent benefits to human health. Fructans, RFOs, galacto-oligosaccharides, and resistant starch are some of the non-digestible carbohydrates that have been shown[3] to have a prebiotic effect in human subjects. As explained above, peas and kidney beans are a good source of RS, RFOs and these prebiotic compounds are also present in other kinds of legumes. Moreover, kidney beans and lentils are also a source of fructans.[78]

4.6 Food Applications of Legume Fibers

The inclusion of fiber in foods can modify texture, consistency, sensory characteristics and rheological behavior of the end products.[29] Therefore, the use of specific fibers in food products is determined by their physicochemical properties and by food processing conditions. Likewise, other factors such as fiber color and flavor must be considered due to their effect on the sensory characteristics of the new product.[28]

Currently, pea fibers, both hull fibers and cotyledon fibers, are the only legume fibers available in the market for food applications. An increasingly important application of these fibers is the fiber enrichment of gluten-free food products classically made from corn, rice and/or potato flour.[28] Pea fibers can be found under several trademarks in different countries. Some of them are Exafine® and Swelite® (Cosucra, Belgium); BCP Natural Pea Fiber (Best Cooking Pulses, Inc., Portage la Prairie, Canada), Vitacel® pea fiber (J. Rettenmaier, Germany); Exlite (Parrheim Foods, Saskatoon, Canada), and Pea Fiber IG (Roquette, France).

The investigations about the food applications for dietary fiber of legumes, hulls or fiber fractions, mainly focus on fiber enrichment, fat binding and retention during cooking, nutrient fortification, and texture modification in low-fat products.[28] Breads, baked goods or cereal based products have been traditionally the main vehicles for fiber enrichment, and pea commercial fibers have been the most widely studied among the legume fibers, although investigations with other legume fibers have been developed.[28] The literature indicates that the incorporation of pea dietary fiber in breads produces breads with acceptable quality characteristics and high-fiber content, depending on the quantity of the pea fiber added.[79–81] Likewise, the addition of 5% of hulls or 5% of cotyledon insoluble fiber or 3% of soluble fibers obtained from lentils, peas and chickpeas produce breads with the total fiber content increased and an improved moistness without significantly increasing crumb firmness during storage.[82]

The effect of the inclusion of roasted hulls from pinto beans, navy beans, soybeans and black-eyed peas, as well as several cereal brans, were compared in low-fat, high-fiber muffin formulations. The results showed that muffins with pulse fibers had the highest water holding capacity and the highest moisture content, but these fibers produced a reduction in the muffin volume.[28] On the other hand, high-fiber pasta containing pea hull fiber was introduced in the market decades ago. The evaluation of cooking properties and textural characteristics of spaghetti enriched with pea hull fiber showed the same characteristics as the control pasta if the quantity of pea fiber added is between 7.5 and 12.5%.[83]

Pea cotyledon fibers have also been investigated as an ingredient for retaining fat in ground beef during high temperature heating[57,58] and as a texture modifier in lower fat beef patties to improve eating quality.[84] Pea fibers were identified as the product with the most potential as a fat retainer in ground beef. Additionally, these fibers improve tenderness and cooking yield, and had no negative effect on juiciness or beef flavor intensity.

4.7 Conclusion

One of the main components of legumes is dietary fiber, whose beneficial effects are mainly due to its physico-chemical properties. In general, the insoluble dietary fiber is the main fraction, but the species and variety of legumes are decisive elements in the content of dietary fiber. Its main constituents are cellulose, hemicelluloses, pectins, lignin and other non-cell wall polysaccharides such as resistant starch and RFOs. The processing of legumes generally affects the dietary fiber contents and composition, showing modifications in the structure of the polysaccharides, which influence the integrity of the tissues causing changes in the physicochemical and functional properties of dietary fiber. These fiber properties are of importance in predicting its functional behavior in food systems and its physiological properties as it passes through the gastrointestinal tract. Currently, the food industry is showing an increasing interest in dietary fiber of legumes, pea fibers being the main fiber of legumes available in the market for food applications because of its excellent properties. Future work has to be done to extend the use of other dietary fiber sources in the consumer diet.

References

1. F. T. Macagnan, L. P. da Silva and L. H. Hecktheuer, *Food Res. Int.*, 2016, **85**, 144.
2. B. R. Hamaker and Y. E. Tuncil, *J. Mol. Biol.*, 2014, **426**, 3838.
3. S. Fuller, E. Beck, H. Salman and L. Tapsell, *Plant Foods Hum. Nutr.*, 2016, **71**, 1.
4. D. Mudgil and S. Barak, *Int. J. Biol. Macromol.*, 2013, **61**, 1.
5. J. W. DeVries, M. E. Camire, S. Cho, S. Craig, D. Gordon, J. M. Jones, *et al.*, *Cereal Foods World*, 2001, **46**, 112.
6. Food and Agriculture Organization of the United Nations and World Health Organization, *Nutr. Foods Spec. Diet Uses*, 2009, 83.
7. F. Saura-Calixto, *J. Agric. Food Chem.*, 2011, **59**, 43.
8. I. Goñi, M. E. Díaz-Rubio, J. Pérez-Jiménez and F. Saura-Calixto, *Food Res. Int.*, 2009, **42**, 840.
9. P. Vitaglione, A. Napolitano and V. Fogliano, *Trends Food Sci. Technol.*, 2008, **19**, 451.
10. Y. Maphosa and V. A. Jideani, *Food Rev. Int.*, 2016, **32**, 98.
11. M. Cruz-Requena, C. N. Aguilar-González, L. A. Prado-Barragan, M. G. Carneiro-da Cunha, M. T. dos Santos Correia, J. C. Contreras-Esquivel, *et al.*, *Emirates J. Food Agric.*, 2016, **28**, 522.
12. R. M. Esteban, E. Mollá and V. Benítez, *Dietary Fiber for the Prevention of Cardiovascular Disease: Fiber's Interaction between Gut Micoflora, Sugar Metabolism, Weight Control and Cardiovascular Health*, ed. R. A. Samaan, Academic Press Elsevier, London, 1st edn, 2017, ch. 7, pp. 121–146.
13. A. C. Mallillin, T. P. Trinidad, R. Raterta, K. Dagbay and A. S. Loyola, *Br. J. Nutr.*, 2008, **100**, 485.
14. M. A. Martín-Cabrejas, M. F. Díaz, Y. Aguilera, V. Benítez, E. Mollá and R. M. Esteban, *Food Chem.*, 2008, **107**, 1045.
15. V. Benítez, S. Cantera, Y. Aguilera, E. Mollá, R. M. Esteban and M. F. Díaz, *et al.*, *Food Res. Int.*, 2013, **50**, 64.
16. C. M. Donangelo, O. L. C. Trugo, N. M. F. Trugo and B. Eggumb, *Food Chem.*, 1995, **53**, 23.
17. K. R. Sridhar and S. Seena, *Food Chem.*, 2006, **99**, 267.
18. T. Kutoš, T. Golob, M. Kač and A. Plestenjak, *Food Chem.*, 2003, **80**, 231.

19. M. A. Perez-Hidalgo, E. Guerra-Hernández and B. García-Villanova, *J. Food Compos. Anal.*, 1997, **10**, 66.
20. Y. Aguilera, M. A. Martín-Cabrejas, V. Benítez, E. Mollá, F. J. López-Andréu and R. M. Esteban, *J. Food Compos. Anal.*, 2009, **22**, 678.
21. G. E. De Almeida Costa, K. Da Silva Queiroz-Monici, S. M. Pissini Machado Reis and A. C. De Oliveira, *Food Chem.*, 2006, **94**, 327.
22. M. Dueñas, T. Sarmento, Y. Aguilera, V. Benitez, E. Mollá, R. M. Esteban, *et al.*, *LWT--Food Sci. Technol.*, 2016, **66**, 72.
23. M. A. Martín-Cabrejas, B. Sanfiz, A. Vidal, E. Mollá, R. M. Esteban and F. J. López-Andréu, *J. Agric. Food Chem.*, 2004, **52**, 261.
24. Y. Chen, R. McGee, G. Vandemark, M. Brick and H. J. Thompson, *Nutrients*, 2016, **8**, 1.
25. A. E. Kleintop, D. Echeverria, L. A. Brick, H. J. Thompson and M. A. Brick, *J. Agric. Food Chem.*, 2013, **61**, 9719.
26. M. A. Martín-Cabrejas, Y. Aguilera, V. Benítez, E. Mollá, F. J. López-Andréu and R. M. Esteban, *J. Agric. Food Chem.*, 2006, **54**, 7652.
27. N. Grigelmo-Miguel and O. Martín-Belloso, *Lebensm.-Wiss. Technol.*, 1999, **32**, 503.
28. S. M. Tosh and S. Yada, *Food Res. Int.*, 2010, **43**, 450.
29. D. Dhingra, M. Michael, H. Rajput and R. T. Patil, *J. Food Sci. Technol.*, 2012, **49**, 255.
30. R. Chawla and G. R. Patil, *Compr. Rev. Food Sci. Food Saf.*, 2010, **9**, 178.
31. Z. U. Rehinan, M. Rashid and W. H. Shah, *Food Chem.*, 2004, **85**, 245.
32. H. Vasishtha, R. P. Srivastava and P. Verma, *J. Food Sci. Technol.*, 2014, **51**, 4090.
33. H. Vasishtha and R. P. Srivastava, *J. Food Sci. Technol.*, 2013, **50**, 579.
34. M. A. Martín-Cabrejas, N. Ariza, R. Esteban, E. Mollá, K. Waldron and F. J. López-Andréu, *J. Agric. Food Chem.*, 2003, **51**, 1254.
35. V. Benítez, S. Cantera, Y. Aguilera, E. Mollá, R. M. Esteban, M. F. Díaz, *et al.*, *Food Res. Int.*, 2013, **50**, 64.
36. L. Bravo, P. Siddhuraju and F. Saura-Calixto, *Food Chem.*, 1999, **64**, 185.
37. Y. Aguilera, R. M. Esteban, V. Benítez, E. Mollá and M. A. Martín-Cabrejas, *J. Agric. Food Chem.*, 2009, **57**, 10682.
38. Y. Brummer, M. Kaviani and S. M. Tosh, *Food Res. Int.*, 2015, **67**, 117.
39. N. Wang, D. W. Hatcher, R. T. Tyler, R. Toews and E. J. Gawalko, *Food Res. Int.*, 2010, **43**, 589.
40. N. Wang, D. W. Hatcher, R. Toews and E. J. Gawalko, *LWT-Food Sci. Technol.*, 2009, **42**, 842.
41. S. G. Sáyago-Ayerdi, J. Tovar, F. J. Blancas-Benítez and L. A. Bello-Pérez, *J. Food Nutr. Res.*, 2011, **50**, 1.
42. A. Satija and F. B. Hu, *Curr. Atheroscler. Rep.*, 2012, **14**, 505.
43. M. A. Martín-Cabrejas, M. F. Díaz, Y. Aguilera, V. Benítez, E. Mollá and R. M. Esteban, *Food Chem.*, 2008, **107**, 1045.
44. Y. A. Adebowale, A. Adeyemi and A. A. Oshodi, *Food Chem.*, 2005, **89**, 37.
45. A. H. Azizah and H. Zainon, *Malays. J. Nutr.*, 1997, 131.
46. M. Dueñas, T. Sarmento, Y. Aguilera, V. Benitez, E. Mollá, R. M. Esteban, *et al.*, *LWT--Food Sci. Technol.*, 2016, **66**, 72.

47. N. Wang, D. W. Hatcher and E. J. Gawalko, *Food Chem.*, 2008, **111**, 132.
48. E. Marconi, S. Ruggeri, M. Cappelloni, D. Leonardi and E. Carnovale, *J. Agric. Food Chem.*, 2000, **48**, 5986.
49. M. A. Aldwairji, J. Chu, V. J. Burley and C. Orfila, *J. Food Compos. Anal.*, 2014, **36**, 111.
50. D. M. Valdez-Anguiano, E. Herrera-Cabrera, G. Dávila-Ortiz, D. Oomah, A. Cardador-Martínez and C. Jiménez-Martínez, *Int. J. Res. Agric. Food Sci.*, 2015, **2**, 1.
51. D. D. Dalgetty and B. K. Baik, *Cereal Chem.*, 2003, **80**, 310.
52. C. F. Chau and P. C. K. Cheung, *Nutr. Res.*, 1999, **19**, 257.
53. V. Benítez, E. Mollá, M. A. Martín-Cabrejas, Y. Aguilera, F. J. López-Andréu and R. M. Esteban, *Food Chem.*, 2011, **127**, 501.
54. M. Elleuch, D. Bedigian, O. Roiseux, S. Besbes, C. Blecker and H. Attia, *Food Chem.*, 2011, **124**, 411.
55. R. M. Weightman, C. M. G. C. Renard and J. F. Thibautl, *Carbohydr. Polym.*, 1994, **24**, 139.
56. H. Y. Yeh, N. W. Su and M. H. Lee, *J. Agric. Food Chem.*, 2005, **53**, 4361.
57. E. T. Anderson and B. W. Berry, *Food Res. Int.*, 2001, **34**, 689.
58. E. T. Anderson and B. W. Berry, *J. Food Qual.*, 2001, **24**, 291.
59. D. P. Burkitt, A. P. Walker and N. S. Painter, *JAMA, J. Am. Med. Assoc.*, 1974, **229**, 1068.
60. A. Aleixandre and M. Miguel, *Food Funct.*, 2016, **7**, 1864.
61. C. W. C. Kendall, A. Esfahani and D. J. A. Jenkins, *Food Hydrocolloids*, 2010, **24**, 42.
62. J. A. Marlett, M. I. McBurney and J. L. Slavin, *J. Am. Diet. Assoc.*, 2002, **102**, 993.
63. C. Hall, C. Hillen and J. Garden Robinson, *Cereal Chem. J.*, 2017, **94**, 11.
64. W. G. T. Willats, J. P. Knox and J. D. Mikkelsen, *Trends Food Sci. Technol.*, 2006, **17**, 97.
65. B. Reddy, A. Engle, S. Katsifis, B. Simi, H. P. Bartram, P. Perrino, *et al.*, *Cancer Res.*, 1989, **49**, 4629.
66. A. Fardet, *Nutr. Res. Rev.*, 2010, **23**, 65.
67. D. Lizarraga, M. P. Vinardell, V. Noé, J. H. van Delft, G. Alcarraz-Vizán, S. G. van Breda, *et al.*, *J. Nutr.*, 2011, **141**, 1597.
68. A. E. Quirós-Sauceda, H. Palafox-Carlos, S. G. Sáyago-Ayerdi, J. F. Ayala-Zavala, L. A. Bello-Perez, E. Álvarez-Parrilla, *et al.*, *Food Funct.*, 2014, **5**, 1063.
69. T. P. Trinidad, A. C. Mallillin, A. S. Loyola, R. S. Sagum and R. R. Encabo, *Br. J. Nutr.*, 2010, **103**, 569.
70. J. H. Kwak, J. K. Paik, H. I. Kim, O. Y. Kim, D. Y. Shin, H. J. Kim, *et al.*, *Atherosclerosis*, 2012, **224**, 457.
71. K. M. Behall, D. J. Scholfield and J. G. Hallfrisch, *Nutr. Res.*, 2006, **26**, 644.
72. M. O. Weickert and A. F. H. Pfeiffer, *J. Nutr.*, 2008, **138**, 439.
73. I. A. Brownlee, P. I. Chater, J. P. Pearson and M. D. Wilcox, *Food Hydrocolloids*, 2017, **68**, 186.
74. Y. P. Lee, T. A. Mori, I. B. Puddey, S. Sipsas, T. R. Ackland, L. J. Beilin, *et al.*, *Am. J. Clin.Nutr.*, 2009, **89**, 766.

75. M. Galisteo, M. Sanchez, R. Vera, M. Gonzalez, A. Anguera, J. Duarte, *et al.*, *J. Nutr.*, 2005, **135**, 2399.
76. R. A. Othman, M. H. Moghadasian and P. J. H. Jones, *Nutr. Rev.*, 2011, **69**, 299.
77. J. W. Finley, J. B. Burrell and P. G. Reeves, *J. Nutr.*, 2007, **137**, 2391.
78. M. S. Dodevska, B. I. Djordjevic, S. S. Sobajic, I. D. Miletic, P. B. Djordjevic and V. S. Dimitrijevic-Sreckovic, *Food Chem.*, 2013, **141**, 1624.
79. F. W. Sosulski and K. K. Wu, *Cereal Chem.*, 1988, **65**, 186.
80. J. Wang, C. M. Rosell and C. Benedito de Barber, *Food Chem.*, 2002, **79**, 221.
81. C. Collar, E. Santos and C. M. Rosell, *J. Food Eng.*, 2007, **78**, 820.
82. D. D. Dalgetty and B. K. Baik, *Cereal Chem.*, 2006, **83**, 269.
83. C. M. Tudorică, V. Kuri and C. S. Brennan, *J. Agric. Food Chem.*, 2002, **50**, 347.
84. E. T. Anderson and B. W. Berry, *J. Food Sci.*, 2000, **65**, 805.
85. M. D. Stoughton-Ens, D. W. Hatcher, N. Wang and T. D. Warkentin, *Food Res. Int.*, 2010, **43**, 547.

CHAPTER 5

Legume Bioactive Peptides

L. MOJICA[a] AND E. GONZALEZ DE MEJIA*[b]

[a]Tecnología Alimentaria, Centro de Investigación y Asistencia en Tecnología y Diseño del Estado de Jalisco, A. C., CIATEJ, 44270, Guadalajara, Mexico; [b]Department of Food Science and Human Nutrition, University of Illinois at Urbana-Champaign, Urbana, IL, 61801, USA
*E-mail: edemejia@illinois.edu

5.1 Introduction

Legumes are dry seeds high in protein, dietary fiber, vitamins, and minerals, thus considered a low-cost nutrient source. The legume group includes soybean, common bean, peanuts, chickpeas, lentils, fava beans, cowpeas, and snow peas, among others.[1] Pulses are the seeds of leguminous plants that differ from legume oilseeds in their low-fat content. Pulses also have low glycemic index resulting in slower digestion. This group includes common beans, lentils, peas, chickpeas, fava bean and cowpeas.[2] Legumes in the diet are linked to a decreased risk of suffering type 2 diabetes, cancer, cardiovascular diseases and obesity.[1,3]

Legumes are also an inexpensive source of protein and peptides with biological potential. They are also environmentally suitable as a source of protein.[4] Bioactive peptides generated from food proteins by enzymatic systems or during food processing may exhibit different bioactive properties related to their low molecular weight. These includes properties such as antihypertensive, antioxidant, anti-inflammatory, immunomodulatory, opioid, metal chelating and hypocholesterolemic.[4–6]

Food Chemistry, Function and Analysis No. 8
Legumes: Nutritional Quality, Processing and Potential Health Benefits
Edited by María Ángeles Martín-Cabrejas

Published by the Royal Society of Chemistry, www.rsc.org

Different processing approaches have been reported to generate protein hydrolysates with a desired biological potential. Researches have used different enzymatic systems, hydrolysis conditions and enzyme/substrate proportions in order to produce protein hydrolysates with enhanced biological potential.[7] Furthermore, there are several *in silico* tools that help in predicting hydrolysis products of proteins using several proteases and hydrolysis conditions. In addition, the physicochemical properties and biological potential can be calculated using databases. *In silico* approaches have the advantage of low cost, fast analysis, and handling of the high amount of data in order to predict or discriminate among several compounds or molecules. However, further biochemical, *in vitro* or *in vivo* assays are needed in order to validate the activity of the bioactive compounds. The objective of this chapter was to review the literature regarding the biological potential of legume peptides generated by hydrolysis of the main storage proteins from selected legumes and evaluate their biological potential using *in silico* approaches.

5.2 Common Beans (*Phaseolus vulgaris*)

The common bean is a legume widely consumed around the world, representing an important source of protein in the diet of people from developing countries.[8] The common bean nutritional value is associated with its content of protein (15–35%), complex carbohydrates (50–60%), dietary fiber, low lipid content, vitamins, and minerals. Moreover, common beans are a rich source of phytochemicals such as phenolic compounds, saponins, and phytosterols among others.[8–11] Common bean proteins include phaseolin, lectins, protease inhibitors (Kunitz and Bowman-Birk) and α-amylase inhibitors. The consumption of peptides derived from legume protein has been associated with several health benefits. These potential benefits are mainly related to blocking molecular markers of chronic diseases such as type 2 diabetes, hypertension, inflammation, obesity, and cancer.[8,12,13] Bioactive peptides can be generated by processing (germination or fermentation) or using different enzyme systems. Enzyme systems include gastrointestinal proteases such as pepsin, chymotrypsin, and trypsin. Alternative sources of proteases are also from micro-organisms (alcalase, flavourzyme, thermolysin) and from plant sources (papain, bromelin).[9,12–16]

The biological health potential of bean peptides has been associated with their antioxidant properties.[8,12,13,17,18] This includes anti-inflammatory,[16,19] and hypoglycemic potential[9,14,20] as well as its relation to the metabolic syndrome,[16] colorectal cancer,[10] antihypertensive,[17,18] and anti-obesity potential.[3,21]

Table 5.1 shows the common bean main storage proteins as reported in the protein database UniProtKB, including phaseolin α and β type, arcelin, leucoagglutinating and erytroagglutinating phytohemagglutinins, α-amylase inhibitors and Bowman-Birk type protease inhibitors. In the same database, several sequenced *Phaseolus vulgaris* metabolism associated proteins are

Table 5.1 Main common bean seed storage proteins.[a]

Common bean proteins	Sequence	ID	Amino acid residues	Molecular mass [Da]
Phaseolin, alpha-type	MMRARVPLLL LGILFLASLS ASFATSLREE EESQDNPFYF NSDNSWNTLFKNQYGHIRVL QRFDQQSKRL QNLEDYRLVE FRSKPETLLL PQQADAELLL VVRSGSAILV LVKPD-DRREY FFLTQGDNPI FSDNQKIPAG TIFYLVNPDPKEDLRIIQLA MPVNNPQIHE FFLSSTEAQQ SYLQEFSKHI LEASFNSKFEEINRVLFEEE GQQEEGQQEG VIVNID-SEQI EELSKHAKSS SRKSHSKQDNTIGNEFGNLT ERTDNSLNVL ISSIEMKEGA LFV-PHYYSKA IVILVVNEGEAHVELVGPKG NKETLEFESY RAELSKDDVF VIPAAYPVAI KATSNVNFTGFGINANNNNR NLLAGKTDNV ISSIGRALDG KDVLGLTFSG SGEEVM-KLINKQSGSYFVDG HHHQQEQQKG SHQQEQQKGR KGAFVY	P07219	436	49 271
Phaseolin	RRVPLLLLGI LFLASLSASF AISLREHNES QDNPFYFSSD NSWQTLFKNQ YGHIRVLQRF DQRSERLQNL EDYRLVEFMS KPETLLLPQQ ADAEFLLVVR SGSALLALVK PGG-TIIYSLK QQDTLKIPAG TIFFLINPEN NEDLRIIKLA MTVNNPQIQD FFLSSTEAQQ SYLYGFSKHI LDASFNSPIE KINRLLFAEE GRQEGVIVNI GSDLIQELSR HAKSSSRKSL DHNSLDISNE WGNLTDIVYN SLDVLLTYVE IKEGGLFVPH YNSKAIVILV VEEG-VAKVEL VGPKREKESL ELETYRADVSEGDVFVIPAAFPFAIKAISNVNFTSFGINANNNY-RIFLTGKGGPTGKEDNIISAGINPDV LGLMFPGSGE DVQKLFNNQN LSHFVNGSYH KNAHPHEQEQ QKQQKGRKGA FVY	Q43617	423	47 514
Phaseolin, beta-type	MMRARVPLLL LGILFLASLS ASFATSLREE EESQDNPFYF NSDNSWNTLF KNQYGHIRVL QRFDQQSKRL QNLEDYRLVE FRSKPETLLL PQQADAELLL VVRSGSAILV LVKPD-DRREY FFLTSDNPIF SDHQKIPAGT IFYLVNPDPK EDLRIIQLAM PVNNPQIHEF FLSSTEAQQS YLQEFSKHIL EASFNSKFEE INRVLFEEEG QQEGVIVNID SEQIKELSKH AKSSSRKSLS KQDNTIGNEFGNLTERTDNSLNVLISSIEMEEGALFVPHYYSKAIVILV-VNEGEAHVELVGPKGNKETLE YESYRAELSK DDVFVIPAAY PVAIKATSNV NFTGF-GINAN NNNRNLLAGK TDNVISSIGR ALDGKDVLGL TFSGSGDEVM KLINKQSGSY FVDAHHHQQE QQKGRKGAFVY	P02853	421	47 565
Erythroagglutinating phytohemagglutinin	MASSNLLSLA LFLVLLTHAN SASQTSFSFQ RFNETNLILQ RDATVSSKGQ LRLTNVNDNG EPTLSSLGRA FYSAPIQIWD NTTGAVAASP TSFTFNIDVP NNSGPADGLA FVLLPVGSQP KDKGGLLGLF NNYKYDSNAH TVAVEFDTLY NVHWDPKPRH IGIDVNSIKS IKTTT-WDFVK GENAEVLITY DSSTKLLVAS LVYPSLKTSF IVSDTVDLKS VLPEWVIVGF TATTGITKGN VETNDILSWS FASKLSDGTT SEALNLANFA LNQIL	P05088	275	29 746

Leucoagglutinating- phytohemagglutinin	MASSKFFTVL FLVLLTHANS SNDIYFNFQR FNETNLILQR DASVSSSGQL RLTNL-NGNGE PRVGSLGRAF YSAPIQIWDN TTGTVASFAT SFTFNIQVPN NAGPADGLAF ALVPVGSQPK DKGGFLGLFD GSNSNFHTVA VEFDTLYNKD WDPTERHIGI DVN-SIRSIKT TRWDFVNGEN AEVLITYDSS TNLLVASLVY PSQKTSFIVS DTVDLKSVLP EWVSVGFSAT TGINKGNVET NDVLSWSFAS KLSDGTTSEG LNLANLVLNK IL	P05087	272	29 556
Arcelin-1	MASSKLLSLA LFLVLLTHAN SATETSFNFP NFHTDDKLIL QGNATISSKG QLQLTGVGSN ELPRVDSLGR AFYSDPIQIK DSNNVASFNT NFTFIIRAKN QSISAYGLAF ALVPVNSPPQ KKQEFLGIFN TNNPEPNART VAVVFNTFKN RIDFDKNFIK PYVNENCDFH KYN-GEKTDVQ ITYDSSNNDL RVFLHFTVSQ VKCSVSATVH LEKEVDEWVS VGFSATSGLT EDTTETHDVL SWSFSSKFRN KLSNILLNNI LNKIL	P19329	265	29 255
Alpha-amylase inhibitor 2	MASSNLLTLA LFLVLLTHAN SASDTSFNFY SFNETNLILQ GDATVSSKGYLQLHTVDSMC SAFYSAPIQI RDSTTGNVAS FDTNFTMNIT TQREANSVIGLDFALVPVQP KSKGHT-VTVQ FDTFRSRISI DVNNNDIKSV PWDEQDYDGQNAKVRITYNS STKVLAVSLS NPST-GKSNEV SARMEVEKEL DDWVRVGFSAISGVHEYSFE TRDVLSWSFS SKFSQHTTSE RSNILLNNIL	Q41114	240	26 597
Bowman-Birk type proteinase inhibitor 2	MMVLKVCLLL VFLAGVTTAR MDLNHLIGSN HHDSSDEPSE SSEPCCDICV CTASIPPICQ CTDVRLNSCH SACKSCMCTR SMPGKCRCLD TTDYCYKSCK SSGEDDD	P01060	107	11 637

[a]Protein sequences were obtained from UniProtKB; ID: unique and stable entry identifier from UniProtKB; amino acid nomenclature: C, cysteine; H, histidine; I, isoleucine; M, methionine; S, serine; V, valine; A, alanine; G, glycine; L, leucine; P, proline; T, threonine; F, phenylalanine; R, arginine; Y, tyrosine; W, tryptophan; D, aspartic acid; N, asparagine; E, glutamic acid; Q, glutamine; K, lysine. http://www.uniprot.org/.

also available. Common bean storage proteins in Table 5.1 ranged from 107 amino acids (11 637 Da) to 436 amino acids (49 271 Da). These proteins are the source of bioactive peptides that are released during protein hydrolysis, and each peptide sequence length and amino acid sequence depends on the protease specificity used to digest the protein.

Table 5.2 shows a list of potential peptides that could be generated from phaseolin common bean main storage protein by gastrointestinal (GI) proteases (pepsin, trypsin, and chymotrypsin). Simulated GI digestion was performed using the peptidecutter tool. Fifty-nine peptide sequences ranging from three to 25 amino acids are reported in Table 5.2. During digestion, digestive proteases hydrolyze proteins to generate peptide sequences. Pepsin generated 44% of peptides, trypsin 41%, and chymotrypsin 15%. Furthermore, GI proteases are able to generate small dipeptides and free amino acids during digestion. These peptide sequences present different potential biological activities according to the Biopep database.

Figure 5.1 shows the potential biological activities associated with peptides sequences in the common bean protein phaseolin. Dipeptidyl peptidase IV (DPP-IV), an enzyme associated with the treatment of type 2 diabetes presented the highest inhibition (%), followed by angiotensin-converting enzyme (ACE), an enzyme related to hypertension and antioxidative activity (%). However, a small number of peptide sequences associated with other metabolic pathways are also reported, for instance, glucose uptake stimulating peptides, renin inhibition peptides, anticancer peptides, and immunomodulating peptides among other biological activities.

The predicted biological potential due to bioactive peptides produced from common bean is mainly related to antidiabetes and antihypertensive potential. These results are in agreement with *in vitro* and *in vivo* research on common bean peptides focused on these diseases. However, the biological potential of common bean peptides needs more research in order to explore other bioactivities also reported.

5.3 Lentils (*Lens culinaris*)

Lentils are an important source of proteins (24.3 to 30.2%), starch, fiber and micronutrients in the human diet. They contain bioactive components, including phytochemicals such as phenolic compounds and protein derived bioactive peptides.[4,22–24] Albumins and globulins represent the highest proportion of proteins in lentils. Legumin and vicilin-like proteins belong to the globulins portion and represent around 70% of total protein.[24] The health benefits of lentils have been associated with their content of protein, fiber, and phenolic compounds. Lentil bioactive peptides have been associated with ACE inhibition, bile salts binding potential and antioxidant activity.[4,22,23,25–27]

Lentil storage protein sequences available on the UniProtKB database are limited (Table 5.3). Eight protein sequences from lentil storage proteins were identified on the database. Two convicilin proteins of 60 128 Da and 19 303

Table 5.2 *In silico* digestive enzymes peptide generation from common bean main storage protein phaseolin.[a]

Peptide sequence	Peptide length [aa]	Peptide mass [Da]	Cleaving enzyme	Peptide sequence	Peptide length [aa]	Peptide mass [Da]	Cleaving enzyme	Peptide sequence	Peptide length [aa]	Peptide mass [Da]	Cleaving enzyme
EEEGQQEEGQQEG-VIVNIDSEQIEE	25	2845.8	Pepsin	DQQSK	5	604.6	Trypsin	VGPK	4	399.4	Trypsin
VDGHHHQQEQQK	12	1470.5	Trypsin	ATSNVN	6	604.6	Pepsin	ESY	3	397.3	Chymo-trypsin
AMPVNNPQIHE	11	1249.4	Pepsin	SDNQK	5	590.5	Trypsin	EDL	3	375.3	Pepsin
VVNEGEAHVEL	11	1195.2	Pepsin	IPAGTI	6	570.6	Pepsin	INK	3	373.4	Trypsin
VIPAAYPVAIK	11	1141.4	Trypsin	SKPET	5	560.6	Pepsin	LQN	3	373.4	Pepsin
EEEESQDNP	9	1075.9	Pepsin	QSGSY	5	540.5	Chymotrypsin	VVR	3	372.4	Trypsin
GSHQQEQQK	9	1069.0	Trypsin	VPHY	4	514.5	Chymotrypsin	IIQ	3	372.4	Pepsin
TDNVISSIGR	10	1061.1	Trypsin	GHIR	4	481.5	Trypsin	ETL	3	361.3	Pepsin
SSTEAQQSY	9	999.9	Chymotrypsin	SHSK	4	457.4	Trypsin	HAK	3	354.4	Trypsin
GINANNNNR	9	986.0	Trypsin	VPLL	4	440.5	Pepsin	NSK	3	347.3	Pepsin
SGSGEEVMK	9	923.0	Pepsin	MMR	3	436.5	Trypsin	DDV	3	347.3	Pepsin
TQGDNPIF	8	890.9	Chymotrypsin	SSSR	4	435.4	Trypsin	DVL	3	345.3	Pepsin
QDNTIGNE	8	889.8	Pepsin	TDNS	4	435.3	Pepsin	AEL	3	331.3	Pepsin
ISSIEMK	7	806.9	Trypsin	SGSAI	5	433.4	Pepsin	GNK	3	317.3	Trypsin
PQQADAE	7	757.7	Pepsin	LDGK	4	431.4	Trypsin	GAF	3	293.3	Chymo-trypsin
VKPDDR	6	728.8	Trypsin	EDY	3	425.3	Chymotrypsin	ATS	3	277.2	Pepsin
NSDNSW	6	721.6	Chymotrypsin	NQY	3	423.4	Chymotrypsin	EGA	3	275.2	Pepsin
VNPDPK	6	668.7	Trypsin	LQR	3	415.4	Trypsin	AGK	3	274.3	Trypsin
HILEAS	6	668.7	Pepsin	AIVI	4	414.5	Pepsin	SAS	3	263.2	Pepsin
EEINR	5	659.6	Trypsin	TER	3	404.4	Trypsin				

[a]Predicted peptide sequences using peptidecutter ExPASy tool; amino acid nomenclature: C, cysteine; H, histidine; I, isoleucine; M, methionine; S, serine; V, valine; A, alanine; G, glycine; L, leucine; P, proline; T, threonine; F, phenylalanine; R, arginine; Y, tyrosine; W, tryptophan; D, aspartic acid; N, asparagine; E, glutamic acid; Q, glutamine; K, lysine. https://web.expasy.org/peptide_cutter/.

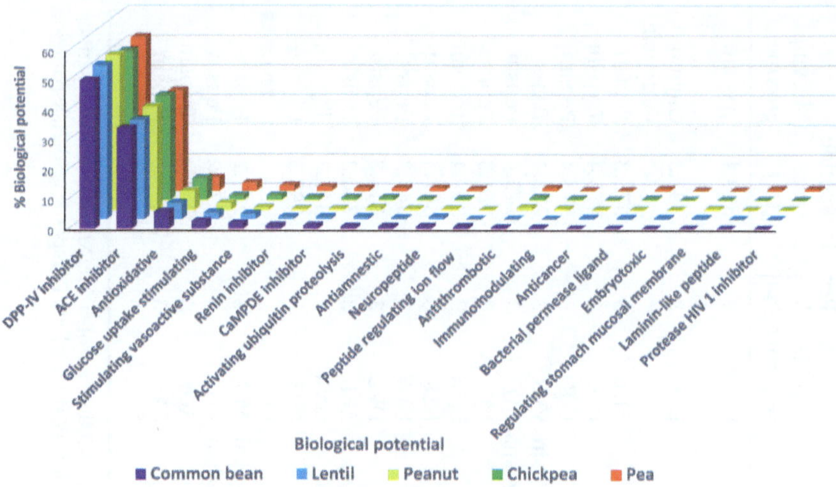

Figure 5.1 Common bean, lentil, peanut, chickpea and pea storage protein derived bioactive peptides predicted biological potential. Biological potential was obtained from the Biopep database. Percentage of biological potential was calculated from the occurrence frequency of bioactive fragments with a particular activity. http://www.uwm.edu.pl/biochemia/index.php/pl/biopep.

Da molecular weight, one vicilin type c protein of 10 331 Da and an albumin-2 (4180 Da), belonged to storage proteins not associated with defense-related proteins, such as lectins, Bowman-Birk inhibitor, and trypsin/chymotrypsin inhibitors. From convicilin (60 128 Da) simulated GI digestion (peptidecutter) it was possible to generate 71 peptide sequences with more than three amino acid residues and up to 22 amino acids.

Gastrointestinal enzymes pepsin and trypsin are more active in hydrolyzing the lentil protein than chymotrypsin (Table 5.4). Peptides generated according to their amino acid functional group sequence may exert different biological functions. Lentil convicilin peptides biological function is reported in Figure 5.1, according to the Biopep database. The main biological activity was associated with the inhibition of DPP-IV (52.0%), followed by the potential to block ACE (33.6%). However, most reported studies on lentil protein hydrolysates biological potential are linked to ACE inhibition and antioxidant activity. In this case, the antioxidative peptides only represent 5.5% of the biological potential of all of the protein sequences.

Other biological potential activities were identified for lentil convicilin derived peptides. They included stimulation of glucose uptake, vasoactive substance, renin inhibitor, calmodulin-dependent cyclic nucleotide phosphodiesterase (CaMPDE) inhibitor, activating ubiquitin proteolysis, antiamnestic, neuropeptide, anticancer, bacterial permease ligand, embryotoxic and regulation of stomach mucosal membrane.

Table 5.3 Main lentil seed storage proteins.[a]

Lentil proteins	Sequence	ID	Amino acid residues	Molecular mass [Da]
Convicilin	MATTIKSRFP LLLLLGIIFL AFVCVAYAND DEGSEPRVTG QRERGRQEGE KAEQSREQSP GQWRPSHGKE EDEEEKEQKE AQSGREKWER KEDEEKVVEE EEGEWRGSQR HGDPEERSRQ RHREEKTKRQ VEEQTEEKDR RYQHEGKEEE TSSESQERRN PFLFK- SNKFL TLFENENGHI RRLQRFDKRS DLFENLQNYRLVEYRAKPHS IFLPQHIDAE FIVVVLSGKA ILTVLSPNDR NSYNLERGDA IKSPAGATYY LVNPDDEEDL RVVD- FVISLN RPGKFEAFDL SANRRQYLRG FSKSVLEASL NTKYDTIEKV LLEEQENEPH QRRDRKGRPQ GQEKHAIVKV SREQIEELRR LAKSSSKKSL PSEFEPFNLR SQNP- KYSNKF GKFFEVTPEK KYPQLQDLDL LVSSVEINEG GLLLPHYNSR AIVVLLVNEG KGNLELVGFK NEQQEREDNK ERNNEVQRYE ARLSPGDVVI IPAGHPVSIS ASSNLN- LLGF GINAENNERN FLTGSDDN	Q9M3X8	518	60 128
Lectin	MASLQTQMIS FYLIFLSILL TTIFFFKVNS TETTSFSITK FSPDQKNLIF QGDGYTTKGK LTLTKAVKST VGRALYSTPI HIWDRDTGNV ANFVTSFTFV IDAPSSYNVA DEFTFFI- APV DTKPQTGGGY LGVFNSKEYD KTSQTVAVEF DTFYNAAWDP SNKERHIGID VNSIKSVNTK SWNLQNGERA NVVIAFNAATNVLTVTLTYP NSLEEENVTS YTLNEV- VPLK DVVPEWVRIG FSATTGAEFA AHEVHSWSFH SELGGTSSSK QAADA	P02870	275	30 352
Convicilin	GIIFLAFVCV AYANDDEGSE PRVPGQRERG RQEGEKAEQS REQSPGQWRP SHGKEED- EEE KEQKEAQPGR EKWERKEDEE KVVEEEGEWR GSQRHGDPEE RSRQRHREEK TKRQVEEQTE EKDRRYQHEG KEEETSSESQ ERRNPFLFKS NKFLTLFENE N	B0BCI7	161	19 303
Bowman-Birk inhibitor	MVLMNKKAIM KLALMLFLLG FTANVVDARF DSTSFITQVL SNGDDVKSACCDT- CLCTRSQ PPTCRCVDVR ESCHSACDKC VCAYSNPPQC QCYDTHKFCYKACHN- SEIEE LIKY	C7BUS0	114	12 784
Bowman-Birk type proteinase inhibitor	MVLMNKKAIM KLALMLFLLG FTANVVDARF DSTSFITQVL SNGDDVKSAC CDTCLCTRSQ PPTCRCVDVR ESCHSACDKC VCAYSNPPQC QCYDTHKFCY KACHNSEIEE	Q8W4Y8	110	12 266
Trypsin/chymotrypsin inhibitor	MVLMNKKAIM KLVLMLFLLG FTANVVNARF DSTSFITQVL SNGDDVKSAC CDT- CLCTRSQ PPTCRCVDVRESCHSACDKCVCAYSNPPQCQCYDTHNFCY KTCH	Q8W4Y1	104	11 608
Vicilin type C	MAATRLKSSF PLLMVLGIAF LASVCVSSRS DQDNPFIFES KRFQSFEKWK NGHIRLLQKF DQHSKILENL QNYRLLEYKS KPHTLFLP	Q40117	88	10 331
Albumin-2	TLTGYIANFS VLNXEAYLFI NDKYVLLDYA PGTXNDK	P86782	37	4180

[a]Protein sequences were obtained from UniProtKB; ID: unique and stable entry identifier from UniProtKB; amino acid nomenclature: C, cysteine; H, histidine; I, isoleucine; M, methionine; S, serine; V, valine; A, alanine; G, glycine; L, leucine; P, proline; T, threonine; F, phenylalanine; R, arginine; Y, tyrosine; W, tryptophan; D, aspartic acid; N, asparagine; E, glutamic acid; Q, glutamine; K, lysine. http://www.uniprot.org/.

Table 5.4 *In silico* digestive enzymes peptide generation from lentil main storage protein convicilin.[a]

Peptide sequence	Peptide length [aa]	Peptide mass [Da]	Cleaving enzyme	Peptide sequence	Peptide length [aa]	Peptide mass [Da]	Cleaving enzyme	Peptide sequence	Peptide length [aa]	Peptide mass [Da]	Cleaving enzyme
LSPGDVVIIPAGH-PVSISASSN	22	2117.3	Pepsin	EAQSGR	6	646.6	Trypsin	SSSK	4	407.4	Trypsin
EEETSSESQER	11	1310.2	Trypsin	EQIEE	5	646.6	Pepsin	EEK	3	404.4	Trypsin
EEQENEPHQR	10	1295.2	Trypsin	TGSDDN	6	607.5	End of sequence	EQK	3	403.4	Trypsin
QVEEQTEEK	9	1119.1	Trypsin	DTIEK	5	604.6	Trypsin	AIVV	4	400.5	Pepsin
VVEEEEGEW	9	1105.1	Chymotrypsin	QHEGK	5	597.6	Trypsin	NSY	3	382.3	Chymotrypsin
ANDDEGSEPR	10	1089.0	Trypsin	AEQSR	5	589.6	Trypsin	NPF	3	376.4	Chymotrypsin
GINAENNER	9	1016.0	Trypsin	QEGEK	5	589.6	Trypsin	NSR	3	375.3	Trypsin
VSSVEINEGG	10	990.0	Pepsin	SQNPK	5	572.6	Trypsin	EAR	3	374.3	Trypsin
ENENGHIR	8	967.9	Trypsin	NRPGK	5	570.6	Pepsin	NTK	3	361.3	Trypsin
VNPDDEED	8	931.8	Pepsin	HAIVK	5	566.7	Trypsin	VSR	3	360.4	Trypsin
EEDEEEK	7	906.8	Trypsin	VTGQR	5	559.6	Trypsin	SNK	3	347.3	Pepsin
GRPQGQEK	8	898.9	Trypsin	VCVAY	5	553.6	Chymotrypsin	SNK	3	347.0	Pepsin
HGDPEER	7	838.8	Trypsin	VNEGK	5	545.5	Trypsin	PLL	3	341.4	Pepsin
EQSPGQW	7	830.8	Chymotrypsin	SLPSE	5	531.5	Pepsin	SDL	3	333.3	Pepsin
PQHIDAE	7	808.8	Pepsin	YPQL	4	519.5	Pepsin	VVD	3	331.3	Pepsin
NEQQER	6	802.7	Trypsin	EDNK	4	504.4	Trypsin	LAK	3	330.4	Trypsin
AKPHSIF	7	798.9	Chymotrypsin	GDAIK	5	502.5	Trypsin	VIS	3	317.3	Pepsin
NNEVQR	6	758.7	Trypsin	GSQR	4	446.4	Trypsin	SVL	3	317.3	Pepsin
EVTPEK	6	701.7	Trypsin	SANR	4	446.4	Trypsin	AIL	3	315.4	Pepsin
LSPNDR	6	700.7	Trypsin	IVVV	4	428.5	Pepsin	EAS	3	305.2	Pepsin
RPSHGK	6	680.7	Trypsin	QNY	3	423.4	Chymotrypsin	GNL	3	302.3	Pepsin
SPAGATY	7	665.7	Chymotrypsin	LQR	3	415.4	Trypsin	GII	3	301.3	Pepsin
MATTIK	6	663.8	Trypsin	PHY	3	415.4	Chymotrypsin	SGK	3	290.3	Trypsin
EDEEK	5	648.6	Trypsin	VEY	3	409.4					

[a]Predicted peptide sequences using peptidecutter ExPASy tool; amino acid nomenclature: C, cysteine; H, histidine; I, isoleucine; M, methionine; S, serine; V, valine; A, alanine; G, glycine; L, leucine; P, proline; T, threonine; F, phenylalanine; R, arginine; Y, tyrosine; W, tryptophan; D, aspartic acid; N, asparagine; E, glutamic acid; Q, glutamine; K, lysine. https://web.expasy.org/peptide_cutter/.

5.4 Peanuts (*Arachis hypogaea*)

Peanuts are a rich source of oil (45–55%) and are a high nutritional value protein (24–36%). The peanut industry has focused on oil extraction, generating a high protein residue which can range from 47 to 55% total protein content.[28,29] Peanut industrial meal contains high amounts of proteins and essential amino acids and could be a source of functional ingredients. However, peanuts are considered a common food allergen and more important peanut allergenic proteins are Arah1 63.5 kDa and Arah2 17.0 kDa. Allergenic proteins can be reduced in food products by enzymatic hydrolysis; this process also increases protein solubility.[30] Peanut also contains less anti-nutritional compounds compared to other oilseeds, such as soybean. Peanut protein-derived bioactive peptides have been associated with antioxidant, ACE inhibitory, and antithrombotic potentials.[31,32] Table 5.5 shows the protein sequences of available peanuts storage proteins on UniProtKB, including conglutin-7, conglutin, profilin, arachin Ahy-3, arachin 25 kDa, arachin 21 kDa and Bowman-Birk type proteinase inhibitor B-II.

From *in silico* GI digestion of arachin Ahy-3, 73 peptide sequences were generated (Table 5.6). Peptide sizes ranged from 245.2 Da to 1675 Da. Pepsin and trypsin enzymes are more active in hydrolyzing peanut proteins compared to chymotrypsin. However, from the Biopep database it can be observed that arachin peanut protein is more likely to generate antidiabetes peptides (DPP-IV inhibitors, 52.2%), antihypertensive peptides (ACE, inhibitors 34.8%), followed by antioxidant activity (6.3%) (Figure 5.1).

Antithrombotic biological activity of peanut protein hydrolysates represents only 0.75% of the total biological activity of the peptides generated; however, Zang[31] reported the antithrombotic potential of peanut peptides. More research is needed to evaluate the antithrombotic biological potential of peanut proteins. Other peptide sequences presented lower biological potential associated with several types of diseases, representing 6% of total peanut peptide biological activity; these include glucose uptake stimulating, stimulating vasoactive substance, renin inhibitor, CaMPDE inhibitor, activating ubiquitin proteolysis, antiamnestic, neuropeptide, immunomodulating, and regulating stomach mucosal membrane.

5.5 Chickpea (*Cicer arietinum*)

Chickpea protein content ranges from 15% to 30%. It is also a good source of carbohydrates such as fiber, resistant starch and oligosaccharides, vitamins, minerals and other phytochemicals.[5,33–35] Chickpea is a good source of dietary protein with a balanced content of essential amino acids and low level of antinutritional factors.[5,36] Research on chickpea health benefits has linked the composition of this legume to the prevention of chronic diseases such as type 2 diabetes, digestive diseases, and cancer.[35]

Table 5.5 Main peanut seed storage proteins.[a]

Peanut proteins	Sequence	ID	Amino acid residues	Molecular mass [Da]
Arachin Ahy-3	MAKLLALSVC FCFLVLGASS VTFRQQGEEN ECQFQRLNAQ RPDNCIESEGGYIETWN- PNN QEFQCAGVAL SRFVLRRNAL RRPFYSNAPQ EIFIYQGSGYFGLIFPGCPG TFEEPIQGSE QFQRPSRHFQ GQDQSQRPLD THQKVHGFREGDLIAVPHGV AFWI- YNDQDT DVVAISVLHT NSLHNQLDQF PRRFNLAGKQEQEFLRYQQR SGRQSPK- GEE QEQEQENEGG NVFSGFSTEF LSHGFQVNEDIVRNLRGENE REEQGAIVTV KGGLSILVPP EWRQSYQQPG RGDKDFNNGIEETICTATVK MNIGKSTSAD IYN- PQAGSVR TVNELDLPIL NRLGLSAEYGSIHRDAMFVP HYNMNANSMI YALHG- GAHVQ VVDCNGNRVF DEELQEGQSLVVPQNFAVAA KSQSEHFLYV AFKTNSRASI SNLAGKNSYM WNLPEDVVANSYGLQYEQAR QLKNNNPFTF LVPPQDSQMI RTVA	Q647H2	484	54 569
Arachin 25 kDa	TDSEPGAVIL GFKASMTDVY AFPMTIGVRA HEPSDVMSKV RALEPSDMSKRAVAGM- KEFY VRAEGIHGTD KSPVITARMK PVHSIGSVRA HFIDGKPTPIKVRAFGPITE SML- DEHSMTI GRYEVIASRK YPSDYRHVFG METSALDRVYAIKGEHRAFS DVSPTHATIE MESPAAHDFL YGRAKVDEPM TIRVAFHSTCM	P04149	201	22 219
Arachin 21 kDa	GAPEIEYHDFDREYCHELMTYFRPIYFGEKGSLCHDVEHLDEMYRAFSKPLV- LERDGTESFEYLERPDGAEIVEIDKLGEMPFEYRFEDAEIDYECFAMESFGTD- KLPYGAAVTCHREYPKDAEFYGMDYCEVGPFEYDRELMFPYEGYHDESKLYAE- GRSTDCDFVRPELFADLE	P20780	176	20 863
Conglutin-7	MAKLTILVALALFLLAAHASARQQWELQGDRRCQSQLERANLRPCEQHLMQKIQRD- EDSYGRDPYSPSQDPYSPSQDPDR RDPYSPSPYDRRGAGSSQHQERCCNELNEF ENNQRCMCEA LQQIMENQSDRLQGRQQEQQ FKRELRNLPQQCGLRAPQRC DLEVESGGRDRY	Q6PSU2	172	20 114
Conglutin	MAKSTILVALLALVLVAHASAMRRERGRQGDSSSCERQVDRVNLKPCE- QHIMQRIMGEQEQYDSYDIRSTRSSDQQQRCC DELNEMENTQRCMCEALQQI- MENQCDRLQD RQMVQQFKRE LMNLPQQCNFRAPQRCDLDVSGGRC	Q647G9	145	16 920
Profilin	MSWQTYVDNHLLCEIEGDHLSSAAILGQDGGVWAQSSHFPQFKPEEITAIMNDFAEPG- SLAPTGLYLGGTKYMVIQGEPGAIIPGKKGPGGVTIEKTNQALIIGIYDKPMTPGQCN- MIVE RLGDYLIDTG L	Q9SQI9	131	14 051
Bowman-Birk type proteinase inhibitor B-II	AASDCCSACI CDRRAPPYFE CTCGDTFDHC PAACNKCVCT RSIPPQCRCT- DRTQGRCPLT PCA	P01067	63	6795

[a]Protein sequences were obtained from UniProtKB; ID: unique and stable entry identifier from UniProtKB; amino acid nomenclature: C, cysteine; H, histidine; I, isoleucine; M, methionine; S, serine; V, valine; A, alanine; G, glycine; L, leucine; P, proline; T, threonine; F, phenylalanine; R, arginine; Y, tyrosine; W, tryptophan; D, aspartic acid; N, asparagine; E, glutamic acid; Q, glutamine; K, lysine. http://www.uniprot.org/.

Table 5.6 *In silico* digestive enzymes peptide generation from peanut main storage protein arachin Ahy-3.[a]

Peptide sequence	Peptide length [aa]	Peptide mass [Da]	Cleaving enzyme	Peptide sequence	Peptide length [aa]	Peptide mass [Da]	Cleaving enzyme	Peptide sequence	Peptide length [aa]	Peptide mass [Da]	Cleaving enzyme
GEEQEQEQENEGGNV	15	1675.5	Pepsin	MNIGK	5	561.6	Trypsin	DEE	3	391.3	Pepsin
NAQRPDNCIESEGGY	15	1652.7	Chymotrypsin	VVPQN	5	555.6	Pepsin	NSY	3	382.3	Chymotrypsin
HGGAHVQVVDCNGNR	15	1562.6	Trypsin	QCAGVA	6	547.6	Pepsin	MAK	3	348.4	Pepsin
NNGIEETICTATVK	14	1492.6	Trypsin	IETW	4	547.6	Chymotrypsin	PIL	3	341.4	Pepsin
LVPPQDSQMIR	11	1283.5	Trypsin	QEGQS	5	547.5	Pepsin	MW	2	335.4	Chymotrypsin
NDQDTDVVAISV	12	1275.3	Pepsin	QEQE	4	532.5	Pepsin	DAM	3	335.3	Pepsin
EEQGAIVTVK	10	1073.2	Trypsin	PGCPGT	6	530.5	Pepsin	STE	3	335.3	Pepsin
QQGEENECQ	9	1064.0	Pepsin	GASSVT	6	520.5	Pepsin	EGD	3	319.2	Pepsin
NMNANSMIY	9	1057.2	Chymotrypsin	VPHY	4	514.5	Chymotrypsin	SGR	3	318.3	Trypsin
QGQDQSQRP	9	1043.0	Pepsin	HNQL	4	510.5	Pepsin	NAL	3	316.3	Pepsin
EEPIQGSEQ	9	1016.0	Pepsin	QGSGY	5	510.5	Chymotrypsin	VHG	3	311.3	Pepsin
PEDVVANSY	9	993.0	Chymotrypsin	EQAR	4	502.5	Trypsin	QY	2	309.3	Chymotrypsin
QVNEDIVR	8	972.0	Trypsin	ASISN	5	490.5	Pepsin	SVC	3	307.3	Pepsin
NPQAGSVR	8	827.8	Trypsin	TNSR	4	476.4	Trypsin	QR	2	302.3	Trypsin

Table 5.6 (*continued*)

Peptide sequence	Peptide length [aa]	Peptide mass [Da]	Cleaving enzyme	Peptide sequence	Peptide length [aa]	Peptide mass [Da]	Cleaving enzyme	Peptide sequence	Peptide length [aa]	Peptide mass [Da]	Cleaving enzyme
IAVPHGVA	8	762.9	Pepsin	SAEY	4	468.4	Chymo-trypsin	SHG	3	299.2	Pepsin
SNAPQEI	7	757.7	Pepsin	TVNE	4	461.4	Pepsin	IY	2	294.3	Chymotrypsin
STSADIY	7	755.8	Chymo-trypsin	AVAAK	5	458.5	Trypsin	LY	2	294.3	Chymotrypsin
LDTHQK	6	740.8	Trypsin	QSPK	4	458.5	Pepsin	TVA	3	289.3	end of sequence
LVPPEW	6	739.8	Chymo-trypsin	HTNS	4	457.4	Pepsin	NR	2	288.3	Trypsin
NPNNQE	6	714.6	Pepsin	NNNP	4	457.4	Pepsin	LR	2	287.3	Trypsin
QRPSR	5	642.7	Trypsin	GDKD	4	433.4	Trypsin	IF	2	278.3	Pepsin
GENER	5	603.5	Trypsin	QQR	3	430.4	Chymo-trypsin	AGK	3	274.3	Trypsin
SQSEH	5	586.5	Pepsin	RPF	3	418.4	Pepsin	GGL	3	245.3	Pepsin
QQPGR	5	584.6	Trypsin	DQF	3	408.4					
GSIHR	5	568.6	Trypsin	QSY	3	396.4					

[a]Predicted peptide sequences using peptidecutter ExPASy tool; amino acid nomenclature: C, cysteine; H, histidine; I, isoleucine; M, methionine; S, serine; V, valine; A, alanine; G, glycine; L, leucine; P, proline; T, threonine; F, phenylalanine; R, arginine; Y, tyrosine; W, tryptophan; D, aspartic acid; N, asparagine; E, glutamic acid; Q, glutamine; K, lysine. https://web.expasy.org/peptide_cutter/.

Chickpea protein could also be a source of protein hydrolysates and bioactive peptides.[6,34] Several investigations on chickpea protein hydrolysates have been published focusing on their antioxidant, antifungal activity, metal-chelating potential, ACE inhibition, and hypocholesterolemic potential.[5,34–36] Extensive research has been performed on chickpea protein as a source of bioactive peptides; however, only a few chickpea storage proteins have been sequenced and are available on the protein database. Table 5.7 shows four storage protein sequences available on the UniProtKB database. Legumin is one of the chickpeas main storage proteins, constituted by 496 amino acid residues and with a molecular weight of 56 251 Da. Only three other small chickpea non-metabolism related proteins were available from UniProtKB, including arietin, cicerarin, and cicerin.

From the *in silico* GI digestion of legumin protein, it was possible to identify 66 peptide sequences that were generated when pepsin, trypsin, or chymotrypsin were cut in the primary amino acid chain of the protein (Table 5.8). Peptide sequences ranged from 304.2 Da to 1648.5 Da, containing from 3 to 15 amino acid residues respectively. Regarding the biological potential of chickpea protein-derived peptides, Figure 5.1 shows that 50.2% of peptide sequences presented DPP-IV inhibition activity, followed by ACE inhibition (35.2%), and antioxidant activity (7.3%). Interestingly, glucose uptake stimulating peptides and stimulating vasoactive substance were also identified as part of the potential biological activity associated with chickpeas. Furthermore, chickpea protein could be used as a functional ingredient in food formulations, increasing food nutritional value and biological potential.

5.6 Pea (*Pisum sativum*)

Peas, like other legumes, are known as a source of protein (18–30%), dietary fiber, vitamins, and minerals.[37,38] Approximately 60% of pea proteins are considered globular proteins including legumin, vicilin, and convicilin.[37,39] Peas have been recognized for their health benefits including prevention of non-communicative diseases such as type 2 diabetes, cardiovascular disease, and cancer.[40] Vicilin and convicilin pea proteins are major allergens; however, during processing, these proteins can be inactivated. Besides generating bioactive peptides, other investigations have improved protein solubility and reduced antinutrient components using enzymatic hydrolysis.[39] Pea protein hydrolysates and bioactive peptides have been associated with ACE inhibition[33,41,42] and antioxidant activity.[5] Pea storage proteins have been widely studied. Eleven pea storage proteins are reported by the UniProtKB database, including different fractions of vicilin, convicilin and legumin, and albumins with a molecular weight ranging from 13 912 Da to 66 990 Da (Table 5.9).

Furthermore, from the *in silico* hydrolysis of provicilin storage protein, 57 protein sequences were generated ranging from 2 to 16 amino acid residues (Table 5.10). Around 50% of provicilin protein was more susceptible to be hydrolyzed by trypsin compared to pepsin (32%) and chymotrypsin (18%).

Table 5.7 Main chickpea seed storage proteins.[a]

Chickpea proteins	Sequence	ID	Amino acid residues	Molecular mass [Da]
Legumin	MAKLLALSLSFCFLLFGTCFALRDQPQQNECQLEHLNALKPDNRIKSEGGLIETWNPSNK QFA-CAGVALS RATLQPNSLL QTFLHQRSPE IFIQQGNGYF GMVFPGCVET FEEPRESEQGEGSKFSD-SHQKVNRFREGDIIAVPTGVVFWMFNDQDTPVIAVSLIDTSSFQNQLDQMPRR FYLAGNHEQE FLRYQQEGSE EEENEGGNIF SGFKRDFLEDALNVNRRIVNKLQGRNEDEEKGAIVKVKGGL-SITTPPEKEPRQKRGSRQEEDEDEDEKRQPHRHSRQDEDEDEKRQPHHHSRGGSKSQRDNG-FEETICTARLHQNIGSSSSPDIYNPQAGRIKTVTSFDLQALRFLKLSAEFGSLHKNAMFVPHYN-LNANSILYALKGRA RLLYALNCKGNSVFDGELEA GRALIVPQNF AIAAKSLSDR FSYVAFKTND RALINVCQKKLLQLLSIWKEMRPGSSSSTAPFHFLFHPAVTQTTKQQLDLVPNQYE	Q9SMJ4	496	56 251
Arietin	GVGYKVVVTT TAAADDDDVV	P83988	20	1995
Cicerarin	VKSTGRADDD LAVKTKYLPP	P83986	20	2174
Cicerin	ARCENFADSY RQPPISSSQT	P83987	20	2257

[a]Protein sequences were obtained from UniProtKB; ID: unique and stable entry identifier from UniProtKB; amino acid nomenclature: C, cysteine; H, histidine; I, isoleucine; M, methionine; S, serine; V, valine; A, alanine; G, glycine; L, leucine; P, proline; T, threonine; F, phenylalanine; R, arginine; Y, tyrosine; W, tryptophan; D, aspartic acid; N, asparagine; E, glutamic acid; Q, glutamine; K, lysine. http://www.uniprot.org/.

Table 5.8 *In silico* digestive enzymes peptide generation from chickpea main storage protein legumin.[a]

Peptide sequence	Peptide length [aa]	Peptide mass [Da]	Cleaving enzyme	Peptide sequence	Peptide length [aa]	Peptide mass [Da]	Cleaving enzyme	Peptide sequence	Peptide length [aa]	Peptide mass [Da]	Cleaving enzyme
QQEGSEEEENEGGNI	15	1648.5	Pepsin	DQMPR	5	645.7	Trypsin	TVTS	4	406.4	Pepsin
HQNIGSSSSPDIY	13	1404.4	Chymo-trypsin	NPQAGR	6	641.6	Trypsin	SIW	3	404.4	Chymo-trypsin
QEEDEDEDEK	10	1265.1	Trypsin	PGCVET	6	604.6	Pepsin	EPR	3	400.4	Trypsin
EGDIIAVPTGVV	12	1169.3	Pepsin	NPSNK	5	558.5	Trypsin	HSR	3	398.4	Trypsin
NDQDTPVIAVS	11	1158.2	Pepsin	IETW	4	547.6	Chymotrypsin	SQR	3	389.4	Trypsin
DQPQQNECQ	9	1089.1	Pepsin	QPHR	4	536.5	Trypsin	QNQ	3	388.3	Pepsin
QDEDEDEK	8	1006.9	Trypsin	EEPR	4	529.5	Trypsin	VNR	3	387.4	Trypsin
HPAVTQTTK	9	982.1	Trypsin	IDTSS	5	521.5	Pepsin	QQL	3	387.4	Pepsin
ESEQGEGSK	9	949.9	Pepsin	NANSI	5	517.5	Pepsin	GNSV	4	375.3	Pepsin
EETICTAR	8	922.0	Trypsin	VPHY	4	514.5	Chymotrypsin	LLQ	3	372.4	Pepsin
QPHHHSR	7	897.9	Trypsin	TNDR	4	504.5	Trypsin	NCK	3	363.4	Trypsin
SITTPPEK	8	871.9	Trypsin	NVNR	4	501.5	Trypsin	QGR	3	359.3	Trypsin
LINVCQK	7	817.0	Trypsin	ACAGVA	6	490.5	Pepsin	MAK	3	348.4	Pepsin
PGSSSSTAP	9	789.7	Pepsin	LSDR	4	489.5	Trypsin	SEGG	4	348.3	Pepsin
AGNHEQE	7	783.7	Pepsin	GAIVK	5	486.6	Trypsin	GGSK	4	347.3	Trypsin
IQQGNGY	7	778.8	Chymo-trypsin	IVNK	4	472.5	Pepsin	WM	2	335.4	Pepsin
NEDEEK	6	762.7	Trypsin	AIAAK	5	472.5	Trypsin	NAM	3	334.3	Pepsin
LKPDNR	6	741.8	Trypsin	GMVF	4	452.5	Pepsin	EDA	3	333.2	Pepsin
LVPNQY	6	732.8	Chymo-trypsin	SPEI	4	444.4	Pepsin	DGE	3	319.2	Pepsin
ATLQPNS	7	729.7	Pepsin	HQR	3	439.4	Trypsin	GSR	3	318.3	Trypsin
SDSHQK	6	700.7	Trypsin	EMR	3	434.5	Trypsin	SAE	3	305.2	Pepsin
LIVPQN	6	682.8	Pepsin	EAGR	4	431.4	Trypsin	DNG	3	304.2	Pepsin

[a]Predicted peptide sequences using peptidecutter ExPASy tool; amino acid nomenclature: C, cysteine; H, histidine; I, isoleucine; M, methionine; S, serine; V, valine; A, alanine; G, glycine; L, leucine; P, proline; T, threonine; F, phenylalanine; R, arginine; Y, tyrosine; W, tryptophan; D, aspartic acid; N, asparagine; E, glutamic acid; Q, glutamine; K, lysine. https://web.expasy.org/peptide_cutter/.

Table 5.9 Main pea seed storage proteins.[a]

Pea proteins	Sequence	ID	Amino acid residues	Molecular mass [Da]
Convicilin	MATTVKSRFP LLLFLGIIFL ASVCVTYANY DEGSETRVPG QRERGRQEGEKEEKRH- GEWR PSYEKEEHEE EKQKYRYQRE KKEQKEVQPG RERWEREEDEEQVEEEWRGS QRREDPEERA RLRHREERTK RDRRHQREGE EEERSSESQEHRNPFLFKSN KFLT- LFENEN GHIRRLQRFD KRSDLFENLQ NYRLVEYRAKPHTIFLPQHI DADLILVVLN GKAILTVLSP NDRNSYNLER GDTIKIPAGTTSYLVNQDDE EDLRVVDFVI PVNRP- GKFEA FGLSENKNQY LRGFSKNILEASLNTKYETI EKVLLEEQEK KPQQLRDRKR TQQGEERDAI IKVSREQIEELRKLAKSSSK KSLPSEFEPF NLRSHKPEYS NKFGKLFEIT PEKKYPQLQDLDILVSCVEI NKGALMLPHY NSRAIVVLLV NEGKGNLELL GLKN- EQQEREDRKERNNEVQ RYEARLSPGD VVIIPAGHPV AISASSNLNL LGFGINAKNNQRN- FLSGSDD NVISQIENPV KELTFPGSSQ EVNRLIKNQK QSHFASAEPEQKEEESQRKR SPLSSVLDSF Y	P13915	571	66 990
Legumin A2	MATKLLALSL SFCFLLLGGC FALREQPEQN ECQLERLNAL EPDNRIESEGGLIETWN- PNN KQFRCAGVAL SRATLQHNAL RRPYYSNAPQ EIFIQQGNGYFGMVFPGCPE TFEEPQESEQ GEGRRYRDRH QKVNRFREGD IIAVPTGIVFWMYNDQDTPV IAVSLT- DIRS SNNQLDQMPR RFYLAGNHEQ EFLRYQHQQGGKQEQENEGN NIFSGFKRDF LEDAFNVNRH IVDRLQGRNE DEEKGAIVKVKGGLSIISPP EKQARHQRGS RQEED- EDEDE ERQPRHQRGS RQEEEEDEDEERQPRHQRRR GEEEEEDKKE RRGSQKGKSR RQGDNGLEET VCTAKLRLNIGPSSSPDIYN PEAGRIKTVT SLDLPVLRWL KLSAEHGSLH KNAMFVPHYNLNANSIIYAL KGRARLQVVN CNGNTVFDGE LEAGRALTVP QNYAVAAK- SLSDRFSYVAFK TNDRAGIARL AGTSSVINNL PLDVVAATFN LQRNEARQLKSNNPFK- FLVP ARQSENRASA	P15838	520	59 270
Legumin A	MAKLLALSLS FCFLLLGGCF ALREQPQQNE CQLERLDALE PDNRIESEGGLIETWN- PNNK QFRCAGVALS RATLQRNALR RPYYSNAPQE IFIQQGNGYFGMVFPGCPET FEEPQESEQG EGRRYRDRHQ KVNRFREGDI IAVPTGIVFWMYNDQDTPVI AVSLTDIRSS NNQLDQMPRR FYLAGNHEQE FLQYQHQQGGKQEQENEGNN IFSGFKRDYL EDAFN- VNRHI VDRLQGRNED EEKGAIVKVKGGLSIISPPE KQARHQRGSR QEEDEDEEKQ PRH- QRGSRQE EEEDEDEERQPRHQRRRGEE EEEDKKERGG SQKGKSRRQG DNGLEETVCT AKLRLNIGPSSSPDIYNPEA GRIKTVTSLD LPVLRWLKLS AEHGSLHKNA MFVPHYNL- NANSIIYALKGR ARLQVVNCNG NTVFDGELEA GRALTVPQNY AVAAKSLSDRFSYVAFK- TND RAGIARLAGT SSVINNLPLD VVAATFNLQR NEARQLKSNNPFKFLVPARE SENRASA	P02857	517	58 805

Legumin J	MSKPFLSLLS LSLLLFASAC LATSSEFDRL NQCQLDSINA LEPDHRVESEAGLTETWNPN HPELKCAGVS LIRRTIDPNG LHLPSFSPSP QLIFIIQGKGVLGLSFPGCP ETYEEPRSSQ SRQESRQQQG DSHQKVRRFR KGDIIAIPSGIPYWTYNHGD EPLVAISLLD TSNIAN- QLDS TPRVFYLGGN PETEFPETQEEQQGRHRQKH SYPVGRRSGH HQQEEESEEQ NEGNSVLSGF SSEFLAQTFNTEEDTAKRLR SPRDERSQIV RVEGGLRIIK PKGKEEEEKE QSHSHSHREEKEEEEEEEED EEEKQRSEER KNGLEETICS AKIRENIADA ARADLYNPRA- GRISTANSLT LPVLRYLRLS AEYVRLYRNG IYAPHWNINA NSLLYVIRGEGRVRIVNCQG NTVFDNKVRK GQLVVVPQNF VVAEQAGEEE GLEYVVFKTNDRAAVSHVQQ VFRATP- SEVL ANAFGLRQRQ VTELKLSGNR GPLVHPRSQSQSH	P05692	503	56 895
Vicilin	MAATTMKASF PLLMLMGISF LASVCVSSRS DPQNPFIFKS NKFQTLFENENGHIRLLQKF DQRSKIFENL QNYRLLEYKS KPHTIFLPQH TDADYILVVLSGKAILTVLK PDDRNSFNLE RGDTIKLPAG TIAYLVNRDD NEELRVLDLAIPVNRPGQLQ SFLLSGNQNQ QNYLSGFSKN ILEASFNTDY EEIEKVLLEEHEKETQHRRS LKDKRQQSQE ENVIVKLSRG QIEELSKNAK STSKKSVSSESEPFNLRSRG PIYSNEFGKF FEITPEKNPQ LQDLDIFVNS VEIKEGSLLL- PHYNSRAIVI VTVNEGKGDF ELVGQRNENQ QEQRKEDDEE EEQGEEEINKQVQNYKA- KLS SGDVFVIPAG HPVAVKASSN LDLLGFGINA ENNQRNFLAGDEDNVISQIQ RPVKELAFPG SAQEVDRILE NQKQSHFADA QPQQRERGSRETRDRLSSV	P13918	459	52 231
Provicilin	MLLAIAFLAS VCVSSRSDQE NPFIFKSNRF QTLYENENGH IRLLQKFDKRSKIFENLQNY RLLEYKSKPH TLFLPQYTDADFILVVLSGK ATLTVLKSNDRNSFNLERGDAIKLPAGSIA YFANRDDNEE PRVLDLAIPV NKPGQLQSFLLSGTQNQKSS LSGFSKNILE AAFNTNYEEI EKVLLEQQEQ EPQHRRSLKDRRQEINEENVIVKVSRDQIE ELSKNAKSSS KKSVSSESGP FNLRSRNPIYSNKFGKFFEI TPEKNQQLQD LDIFVNSVDI KVGSLLLPNY NSRAIVIVT- VTEGKGDFELV GQRNENQGKE NDKEEEQEEETSKQVQLYRAKLSPGDVFVIPAGH- PVAINA SSDLNLIGLG INAENNERNF LAGEEDNVIS QVERPVKELAFPGSSHEVDR	P02854	410	46 385
Convicilin	MATTIKSRFP LLLLLGIIFL ASVVSVTYAN YDEGSEPRVP AQRERGRQEGEKEEKRHGEW RPSYEKEEDE EEGQRERGRQ EGEKEEKRHG EWGPSYEKQEDEEEKQKYRY QRE- KEDEEEK QKYQYQREKK EQKEVQPGRE RWEREEDEEQVDEEWRGSQR REDPEERARL RHREERTKRD RRHQREGEEE ERSSESQERRNPFLFKSNKF LTLFENENGH IRLLQRF- DKR SDLFENLQNY RLVEYRAKPHTIFLPQHIDA DLILVVLSGK AILTVLSPND RNSYN- LERGD TIKLPAGTTSYLVNQDDEED LRLVDLVIPV NGPGKFEAFD LAKNKNQYLR GFSKNILEASYNTRYETIEK VLLEEQEKDR KRRQQGEETD AIVKVS	P13919	386	46 396

(*continued*)

Table 5.9 (*continued*)

Pea proteins	Sequence	ID	Amino acid residues	Molecular mass [Da]
Legumin B	GNSVLSGFNV EFLAHSLNTK EDTAKRLRSP QDERGQIVKV EDGLHIISPELQEEEEQSHS QRKEEEEEEQ EQRHRKHSKK EDEDEDEEEE EEREQRHRKHSEKEEEDEDE PRSY-ETRRKW KKHTAEKKRE SHGQGEEEEE LEKEEEEEEEIQRQHSKGRK NGLEETICSA KIRENIARPS RGDLYNSGAG RISTVNSLTLPILRNLRLSA EYVLLYRNGI YAPHWNINAN SLLYVIRGEG RVRIVNSEGNKVFDDKVSLG QLVVVPQNFV VAQQAGNEEG FEYVVFKTND RAAVSHVNQVFRATPGEVLA NAFGLRHSQV AQIKSNGNRG PLVQPQSQ	P14594	338	38 990
Legumin K	IPYWTYNHGD EPLVAISLLD TSNIANQLDS TPRVFYLGGN PETEFPETQEEQQGRHRQKH SYPVGRRSGH HQQEEESEEQ NEGNSVLSGV SSEFLAQTFNTEEDTAKRLR SPRDERSQIV RVEGGLRIIN PKGKEEEEEK EQSHSHSHREEEEEEEEDEE KQRSEERKNG LEETICSAKI RENIADAAGA DLYNPRAGRIRTANSLTLPV LRYLRLSAEY VRLYRNGIYA PHWNINANSL LYVIRGEGRVRIVNFQGDAV FDNKVRKGQL VVVPQNFVVA EQAGEEEGLE YVVFKTN-DRAAVSHVQQVLR ATPAEVLANA FGLRQRQVTE LKLSGNRGPL VHPQSQSQSH	P05693	350	39 800
Albumin-1 A	MASVKLASLI VLFATLGMFL TKNVGAASCN GVCSPFEMPP CGTSACRCIP GLVVGYCRN PSGVFLRTND EHPNLCESDA DCRKKGSGNF CGHYPNPDIE YGWCFASKSE AEDFF-SKITP KDLLKSVSTA	P62926	130	13 912
Albumin-1 B	MASVKLASLM VLFATLGMFL TKNVGAASCN GVCSPFEMPPGSSACRCIPVGLVVGYCRH PSGVFLRTND EHPNLCESDA DCRKKGSGNF CGHYPNPDIEYGWCFASKSE AEDFF-SKITQ KDLLKSVSTA	P62927	130	13 970

[a]Protein sequences were obtained from UniProtKB; ID: unique and stable entry identifier from UniProtKB; amino acid nomenclature: C, cysteine; H, histidine; I, isoleucine; M, methionine; S, serine; V, valine; A, alanine; G, glycine; L, leucine; P, proline; T, threonine; F, phenylalanine; R, arginine; Y, tyrosine; W, tryptophan; D, aspartic acid; N, asparagine; E, glutamic acid; Q, glutamine; K, lysine. http://www.uniprot.org/.

Table 5.10 *In silico* digestive enzymes peptide generation from pea main storage protein provicilin.[a]

Peptide sequence	Peptide length [aa]	Peptide mass [Da]	Cleaving enzyme	Peptide sequence	Peptide length [aa]	Peptide mass [Da]	Cleaving enzyme	Peptide sequence	Peptide length [aa]	Peptide mass [Da]	Cleaving enzyme
AGEEDNVISQVERPVK	16	1769.9	Trypsin	EEIEK	5	646.6	Trypsin	NSR	3	375.3	Trypsin
VIPAGHPVAINASSD	15	1447.6	Pepsin	DQIEE	5	632.6	Pepsin	QVQ	3	373.4	Pepsin
QEINEENVIVK	11	1314.4	Trypsin	LSPGDV	6	586.6	Pepsin	NSF	3	366.3	Chymotrypsin
EQQEQEPQHR	10	1308.3	Trypsin	SKPHT	5	568.6	Pepsin	VSR	3	360.4	Trypsin
EEEQEEETSK	10	1237.2	Trypsin	NTNY	4	510.5	Chymo-trypsin	ANR	3	359.3	Trypsin
AIVIVTVTEGK	11	1129.3	Trypsin	NPIY	4	505.5	Chymo-trypsin	NIL	3	358.4	Pepsin
GINAENNER	9	1016.0	Trypsin	ENDK	4	504.4	Trypsin	LAF	3	349.4	Pepsin
PGSSHEVDR	9	983.0	end of sequence	GDAIK	5	502.5	Trypsin	SNK	3	347.3	Pepsin
ENENGHIR	8	967.9	Trypsin	SNDR	4	490.4	Trypsin	SLK	3	346.4	Trypsin
AIPVNKPGQ	9	923.1	Pepsin	VGQR	4	458.5	Trypsin	GDF	3	337.3	Chymotrypsin
DDNEEPR	7	873.8	Trypsin	LEY	3	423.4	Chymo-trypsin	NAK	3	331.3	Trypsin
ASVCVSSR	8	807.9	Trypsin	QNY	3	423.4	Chymo-trypsin	SSL	3	305.3	Pepsin
VNSVDIK	7	773.8	Trypsin	TDAD	4	420.3	Pepsin	ATL	3	303.3	Pepsin
SGTQNQK	7	761.8	Trypsin	SSSK	4	407.4	Trypsin	ER	2	303.3	Trypsin
SVSSESGP	8	748.7	Pepsin	PQY	3	406.4	Chymo-trypsin	SGK	3	290.3	Trypsin
EITPEK	6	715.8	Trypsin	PNY	3	392.4	Chymo-trypsin	DR	2	289.2	Trypsin
NENQGK	6	688.7	Trypsin	NQQ	3	388.3	Pepsin	EAA	3	289.2	Pepsin
SDQENP	6	688.6	Pepsin	LQK	3	387.4	Pepsin	LF	2	278.3	Chymotrypsin
PAGSIAY	7	677.7	Chymotryp-sin	SNR	3	375.3	Trypsin	AIA	3	273.3	Pepsin

[a]Predicted peptide sequences using peptidecutter ExPASy tool; amino acid nomenclature: C, cysteine; H, histidine; I, isoleucine; M, methionine; S, serine; V, valine; A, alanine; G, glycine; L, leucine; P, proline; T, threonine; F, phenylalanine; R, arginine; Y, tyrosine; W, tryptophan; D, aspartic acid; N, asparagine; E, glutamic acid; Q, glutamine; K, lysine. https://web.expasy.org/peptide_cutter/.

Regarding the biological functions of peptides generated from pea proteins, the higher proportion of total biological activity was for antidiabetic potential (51.6%), followed by antihypertensive (33.7%), and antioxidant (4.2%) activity, stimulation of glucose uptake (2.5%), and vasoactive substance (1.6%), renin inhibitor (1.3%), CaMPDE inhibitor (0.8%), activating ubiquitin proteolysis (0.8%) antiamnestic (0.8%), neuropeptide (0.4%), antithrombotic (0.8%), bacterial permease ligand (0.4%), laminin-like peptide (0.4), and protease HIV 1 inhibitor (0.4) properties. Pea protein is one of the most studied vegetable proteins. However, most research has been in relation to its technological and functional properties and further research on pea protein biological potential is needed in order to validate *in silico* results.

5.7 Concluding Remarks

Legumes are an inexpensive source of protein and complex carbohydrates. They are also rich in bioactive compounds such as phenols, phytosterols, and bioactive peptides generated by hydrolysis of their proteins. Several health benefits have been associated with the consumption of legumes, including antidiabetes, antihypertensive, anti-inflammatory, antiobesity, and anticancer properties, among others. Legume proteins have raised attention due to their potential as a techno-functional ingredient in food formulations and as a source of bioactive peptides. Common bean and pea storage proteins have been widely studied and their amino acid sequences have been reported on protein databases. However, information on lentil, chickpea, and peanut storage protein is less available. *In silico* approaches offers advantages for procuring fast and low-cost results. Legume protein hydrolysates generated bioactive sequences with high potential to inhibit dipeptidyl peptidase IV, an important molecular target in diabetes; however, most research on these legume proteins have been oriented to their antihypertensive and antioxidant properties. Further research is needed to complete the characterization of legume storage proteins. Novel technological applications and the biological potential of legume storage proteins related to health is an open field for research in both directions since legumes represent a low-cost protein alternative for human consumption.

References

1. D. Luna-Vital and E. González de Mejía, *Curr. Opin. Food Sci.*, 2018, **20**, 13.
2. CODEX ALIMENTARIUS, 2007, http://www.fao.org/docrep/010/a1391s/a1391s00.htm.
3. M. E. Oseguera Toledo, E. Gonzalez de Mejia, M. Sivaguru and S. L. Amaya-Llano, *J. Funct. Foods*, 2016, **27**, 160.
4. P. García-Mora, M. Martín-Martínez, M. Angeles Bonache, R. González-Múniz, E. Peñas, J. Frias and C. Martinez-Villaluenga, *Food Chem.*, 2017, **221**, 464.

5. A. M. Ghribi, A. Sila, R. Przybylski, N. Nedjar-Arroume, I. Makhlouf, C. Blecker, H. Attia, P. Dhulster, A. Bougatef and S. Besbes, *J. Funct. Foods*, 2015, **12**, 516.
6. C. Torres-Fuentes, M. Alaiz and J. Vioque, *Food Chem.*, 2011, **129**, 485.
7. L. Mojica and E. Gonzalez de Mejia, *Food Funct.*, 2016, **7**, 713.
8. J. Carrasco-Castilla, A. J. Hernández-Álvarez, C. Jiménez-Martínez, C. Jacinto-Hernández, M. Alaiz, J. Girón-Calle, J. Vioque and G. Dávila-Ortiz, *Food Chem.*, 2012, **135**, 1789.
9. T. de Souza Rocha, L. M. R. Hernandez, L. Mojica, M. H. Johnson, Y. K. Chang and E. González de Mejía, *Food Res. Int.*, 2015, **76**, 150.
10. D. A. Luna Vital, E. González de Mejía, V. P. Dia and G. Loarca-Piña, *Food Chem.*, 2014, **157**, 347.
11. D. A. Luna-Vital, L. Mojica, E. González de Mejía, S. Mendoza and G. Loarca-Piña, *Food Res. Int.*, 2015, **76**, 39.
12. R. J. S. de Castro, V. G. Cason and H. H. Sato, *Biocatal. Agric. Biotechnol.*, 2017, **10**, 291.
13. L. López-Barrios, M. Antunes-Ricardo and J. A. Gutiérrez-Uribe, *Food Chem.*, 2016, **203**, 417.
14. L. Mojica, E. Gonzalez de Mejia, M. Á. Granados-Silvestre and M. Menjivar, *J. Funct. Foods*, 2017, **31**, 274.
15. M. E. Oseguera-Toledo, E. González de Mejía, R. Reynoso-Camacho, A. Cardador-Martínez and S. L. Amaya-Llano, *Nutrafoods*, 2014, **13**, 147.
16. A. Jakubczyk, M. Karaś, U. Złotek and U. Szymanowska, *Food Res. Int.*, 2017, **100**, 489.
17. L. Mojica., D. A. Luna-Vital and E. González de Mejía, *J. Sci. Food Agric.*, 2017, **97**, 2401.
18. A. Valdez-Ortiz, C. I. Fuentes-Gutiérrez, L. J. Germán-Báez, R. Gutiérrez-Dorado and S. Medina-Godoy, *LWT–Food Sci. Technol.*, 2012, **46**, 91.
19. M. E. Oseguera-Toledo, E. G. de Mejia, V. P. Dia and S. L. Amaya-Llano, *Food Chem.*, 2011, **127**, 1175.
20. M. E. Oseguera-Toledo, E. Gonzalez de Mejia and S. L. Amaya-Llano, *Food Res. Int.*, 2015, **76**, 839.
21. Y.-Y. Ngoh, S. B. Choi and C.-Y. Gan, *J. Funct. Foods*, 2017, **33**, 67.
22. S. Bautista-Expósito, E. Peñas, M. Dueñas, J. M. Silván, J. Frias and C. Martínez-Villaluenga, *Food Chem.*, 2018, **257**, 341.
23. M. I. Torino, R. I. Limón, C. Martínez-Villaluenga, S. Mäkinen, A. Pihlanto, C. Vidal-Valverde and J. Frias, *Food Chem.*, 2013, **136**, 1030.
24. M. Jarpa-Parra, F. Bamdad, Y. Wang, Z. Tian, F. Temelli, J. Han and L. Chen, *LWT–Food Sci. Technol.*, 2014, **57**, 461.
25. C. Barbana and J. I. Boye, *Food Chem.*, 2011, **127**, 94.
26. J. I. Boye, S. Roufik, N. Pesta and C. Barbana, *LWT–Food Sci. Technol.*, 2010, **43**, 987.
27. P. Garcia-Mora, E. Peñas, J. Frias, R. Gomez and C. Martinez-Villaluenga, *Food Chem.*, 2015, **171**, 224.
28. L. Jiang, D. Hua, Z. Wang and S. Xu, *Food Bioprod. Process.*, 2010, **88**, 233.
29. X. Dong, M. Zhao, J. Shi, B. Yang, J. Li, D. Luo, G. Jiang and Y. Jiang, *Innovative Food Sci. Emerging Technol.*, 2011, **12**, 478.

30. P. Li, W. Zhang, X. Han, J. Liu, Y. Liu, M. A. A. Gasmalla and R. Yang, *J. Food Eng.*, 2017, **204**, 64.
31. S. B. Zhang, *Food Chem.*, 2016, **202**, 1.
32. R. Das and A. M. Kayastha, *Int. J. Biol. Macromol.*, 2018, **111**, 148.
33. C. Barbana and J. I. Boye, *Food Res. Int.*, 2010, **43**, 1642.
34. T. Zhang, B. Jiang, M. Miao, W. Mu and Y. Li, *Food Chem.*, 2012, **135**, 904.
35. C. Torres-Fuentes, M. del Mar Contreras, I. Recio, M. Alaiz and J. Vioque, *Food Chem.*, 2015, **180**, 194.
36. Y. Li, B. Jiang, T. Zhang, W. Mu and J. Liu, *Food Chem.*, 2008, **106**, 444.
37. Z. Ma, J. I. Boye and X. Hu, *Food Res. Int.*, 2017, **92**, 64.
38. Z. Ma, J. I. Boye and X. Hu, *LWT–Food Sci. Technol.*, 2018, **92**, 147.
39. M. Klost and S. Drusch, *Food Hydrocolloids*, 2018, DOI: 10.1016/j.foodhyd.2018.03.013.
40. R. Campos-Vega, G. Loarca-Piña and B. D. Oomah, *Food Res. Int.*, 2010, **43**, 461.
41. D. Chao, R. He, S. Jung and R. E. Aluko, *Food Res. Int.*, 2013, **54**, 1528–1534.
42. R. E. Aluko, A. T. Girgih, R. He, S. Malomo, H. Li, M. Offengenden and J. Wu, *Food Res. Int.*, 2015, **77**, 10.

CHAPTER 6

Melatonin

Y. AGUILERA*[a,b], M. REBOLLO-HERNANZ[a,b] AND M. A. MARTÍN-CABREJAS[a,b]

[a]Instituto de Investigación en Ciencias de la Alimentación, CIAL (CSIC-UAM), Universidad Autónoma de Madrid, Madrid, Spain; [b]Departamento de Química Agrícola y Bromatología, Facultad de Ciencias, Universidad Autónoma de Madrid, Madrid, Spain
*E-mail: yolanda.aguilera@uam.es

6.1 Melatonin as a Bioactive Molecule in Plants

Melatonin (*N*-acetyl-5-methoxytriptamine) is a neurohormone and bioactive antioxidant derived from tryptophan and synthesized from L-tryptophan *via* serotonin. The indoleamine was first isolated from bovine pineal gland by Lerner *et al.* in 1958.[1] Its common name comes from a dye present in cells of amphibians (melanophores), named melanin, which determines the color of their skin.

Although melatonin (MEL) was considered mainly as an animal neurohormone, for many decades, it has also been identified in plants, insects, crustaceans, fungi, macroalgae and bacteria. So far, it has been detected and quantified in the shoots, flowers, fruits, roots, seeds, and leaves of a large variety of species that are relevant to humans as foods and medicinal plants. In mammals, melatonin is secreted during darkness hours and decreases during the hours of daylight, acting as a biological modulator of

Food Chemistry, Function and Analysis No. 8
Legumes: Nutritional Quality, Processing and Potential Health Benefits
Edited by María Ángeles Martín-Cabrejas

Published by the Royal Society of Chemistry, www.rsc.org

circadian processes. Moreover, this molecule modulates several important physiological functions such as body temperature, sleep, retinal physiology, mood, anxiety, stimulation of the immune system, sexual behaviour, and reproduction.[2–4] Actually, melatonin supplements are used to treat jet lag and sleeping disorders.

Melatonin is considered as a potent naturally occurring antioxidant, showing free radical and reactive oxygen species (ROS) scavenger properties. Melatonin has been described to detoxify different reactive molecules, such as hydrogen peroxide, peroxynitrite anion, hydroxyl radicals, hypochlorous acid, and nitric oxide.[5] Recent research has reported that this phytohormone exhibits health-promoting benefits against some cancers,[6] especially leukemia,[7,8] suggesting that there is a connection between cancer incidence and physiological melatonin concentration; this molecule manifests a preventative role in tumour proliferation. In addition, MEL plays an important role in other diseases related to the cardiovascular system and Alzheimer's disease.[9]

For this reason, melatonin-rich food consumption could cause potential health influences by increasing the blood melatonin levels and improving the antioxidant status. The presence of melatonin has been reported in nuts and medical herbs[10] and in Mediterranean products such as grapes, olive oil, cereals, legumes and tomatoes.[11] In fermented foods, such as beer and wine, this indoleamine has also been detected, its levels increasing during alcoholic fermentation. An adequate intake of melatonin-rich foods should be expected to increase its very low circulating levels (~200 pg mL^{-1} at the maximum night peak and lower than 10 pg mL^{-1} during the day).

Melatonin is widely used as a food supplement in several countries, including Europe, to compensate for the physiologically diminished activity of the pineal gland. However, this indoleamine is classified as a drug and its recommended dose is fixed at between 0.5 and 5 mg per day. Various authorities have distributed guidance on regulating dietary supplements, including melatonin. The European Food Safety Authority (EFSA) only approves melatonin intake for the treatment of sleeping disorders and jet lag. EFSA published the Scientific Opinion on the substantiation of a health claim about melatonin and improved sleep efficiency (ID 1698, 1780, 4080) pursuant to Article 13(1) of Regulation (EC) No. 1924/20061 in 2010.[10]

Therefore, the search for other natural sources through the consumption of enriched-melatonin foods is of great interest. New perspectives for dietetic melatonin as a natural ingredient of food intake have opened up because of the presence of this molecule in plants and processed food.

6.1.1 Structure and Physicochemical Properties of Melatonin and Its Metabolites

Melatonin is an amphiphilic molecule, since it can freely cross cell membranes and distribute to any lipid or aqueous compartment including the cytosol, membranes, mitochondria and nucleus of cells.[12] It has a low

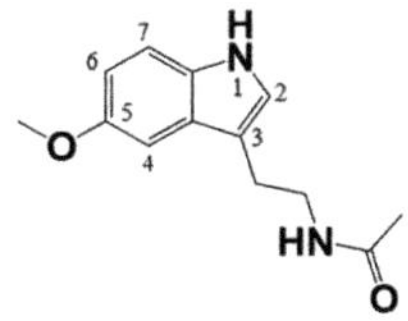

Figure 6.1 The structure of melatonin. The numbers identify the positions of the atoms in the indole heterocycle.

molecular weight (232.38 g mol^{-1}) and its solubility is higher in ethanol (8 mg mL^{-1}) than in water (0.1 mg mL^{-1}). This indoleamine shows an indole heterocycle and two side chains, namely, a 5-methoxy group and 3-amide group (Figure 6.1). Hypothetically, to form melatonin isomers (MIs), either one of these two side chains can be potentially relocated to any one of the seven positions in the indole ring. In order to distinguish among 42 potential MIs formed, it has been proposed that the methoxy group will be designated as side chain M (M) and the 3-(*N*-acetylaminoethyl) group is named as side chain A (A). Therefore, a MI can be designated according to the positions of side chains (M and A) at the seven positions on the indole ring.[12]

The indole heterocycle is the reactive centre, with high resonance stability and electroreactivity responsible for its free radical scavenger properties. However, side chains linked to the C5 and C3 positions (methoxy and aminoactyl side groups, respectively), also contribute significantly to the free radical scavenging capacity and they limit the prooxidative actions of melatonin.[13] On the other hand, suitable sites for hydroxyl radical or nitrogen radical reactions are positions 2, 3, 4, 6 and 7 of the indole heterocycle, since the activation energy barriers are quite low (from 1.1 to 5.9 kcal mol^{-1}) and therefore, the addition of $OH^{\bullet}$ or $NO^{\bullet}$ to these carbon atoms should occur quite rapidly. According to Aguilera *et al.* (2016),[14] melatonin presents potent scavenging properties because it requires a short time to scavenge the DPPH radical, reaching a steady state within 10–25 min. Therefore, in a specific situation less potent antioxidants might have the advantage of working more efficiently and quickly in some biological reactions, while powerful antioxidants are not always fast acting.

N1-acetyl-N2-formyl-5-methoxykynuramine (AFMK) and N[1]-acetyl-5-methoxykynuramine (AMK) are metabolites of melatonin that exhibit protective effects against radical species. AFMK is able to efficiently scavenge $OH^{\bullet}$ radicals and AMK is a good free radical scavenger, especially of a broad variety of reactive oxygen species (ROS) and reactive nitrogen species (RNS), but rather ineffective for scavenging $OOH^{\bullet}$. These compounds are produced by the interaction of melatonin with the ROS and RNS or by enzymatic or pseudoenzymatic pathways. The relative efficiency of these metabolites, with respect to melatonin, was influenced by the polarity of the environment, showing AMK to react faster than melatonin within all the studied free radicals in aqueous solutions.[15]

Other compounds belong to the melatonin family including 6-hydroxymelatonin and cyclic 3-hydroxymelatonin. The scavenging mechanism of 6-hydroxymelatonin may be similar to melatonin due to the molecules having similar structures. In some *in vitro* studies, it was reported that the antioxidant capacity of 6-hydroxymelatonin is even more potent than its parent molecule melatonin.[16] Cyclic 3-hydroxymelatonin seems to be a non-enzymatic compound produced by the reaction of melatonin with reactive species. To date, there is no evidence for the existence of oxidized melatonin metabolites such as AFMK and cyclic 3-hydroxymelatonin in plants.

6.1.2 Biosynthesis and Mechanism of Action

Melatonin is structurally related to the main and most abundant auxin present naturally in plants, indole-3-acetic acid (IAA). IAA is a tryptophan derived compound for which transporters, receptors, and transcription factors mediate its role in plant morphogenesis.

In mammals, melatonin is synthesized during the night-time in the pineal gland, and also in other organs, such as retina, lymphocytes, gastrointestinal tract, and bone marrow cells. Once synthesized, melatonin is released into the bloodstream reaching other body fluids, including saliva, urine, cerebrospinal fluid, bile, amniotic fluid and semen.[17] In mammals, melatonin acts in part *via* three membrane receptors, two of them are G-protein coupled receptors (MT1 and MT2), widely distributed in the central nervous system and in the peripheral tissues, while the third belongs to the family of quinone reductases (MT3), and no signalling pathways are associated with this enzyme.[18] In fact, melatonin is characterized by its small size, amphipathic nature and its active uptake mechanism; it also acts by non-receptor mediated means/actions, such as its antioxidant capacity.[13,19–22]

In vertebrates, the precursor of melatonin is the amino acid L-tryptophan, through the intermediate serotonin (5-HT, 5-hydroxytryptamine). In animals, the biosynthesis of melatonin proceeds in two steps including L-amino acid decarboxylase (AADC) and tryptophan 5-hydroxylase (TPH) in which TPH acts as the rate-limiting enzyme.[23] Serotonin is first converted into *N*-acetyl serotonin (acetylation step) followed by transformation of this compound into melatonin, due to the methyl group transferring from *S*-adenosylmethionine to the 5-hydroxy position of *N*-acetylserotonin to yield MEL. There are two enzymes involved in the biosynthetic reactions, serotonin-*N*-acetyl transferase (SNAcT), and hydroxyindole-*O*-methyl transferase (HIOMT).

The biosynthetic pathway of MEL in plants and in mammals seems to be similar.[24] In plants, MEL biosynthesis also occurs from L-tryptophan (TRP), which is transformed to serotonin; different pathways may synthesize this compound.[25] The first reaction converts L-tryptophan to tryptamine, which is catalysed by tryptophan decarboxylase (TDC) and then, the tryptamine is transformed to serotonin by the tryptamine 5-hydroxylase (T5H).[18,23]

Afterwards, serotonin is converted into *N*-acetylserotonin and then, into melatonin, catalysed by the enzymes SNAcT, acetylserotonin *N*-methyltransferase (AcSNMT) as a methyl donor, and HIOMT.[26] Nevertheless, some studies have reported some cohesion between plants and mammals because all of them can produce 5-hydroxytryptophan as an intermediate compound in the biosynthesis of melatonin.[25]

In higher plants, TRP is the precursor of some molecules such as IAA and tryptamine. IAA could be synthesized *via* indole-3-pyruvic acid (IPA) and indole-3-acetaldehyde (IAAld), *via* indole-3-acetonitrile (IAN) and finally, *via* indole-3-acetamide (IAM). In this sense, IAA is considered an intermediate product suggesting that tryptamine could be also transformed into IAA by tryptamine deaminase (TDA). Therefore, serotonin and melatonin biosynthesis would be directly connected *via* this intermediate (Figure 6.2).[26]

Melatonin biosynthesis pathways seem to be complex in plants; therefore, there are necessary additional *in vivo* assays to characterize orthologs for the biosynthetic pathway genes and to identify melatonin biosynthetic pathway mutants in plants.

6.1.3 Role of Melatonin in Higher Plants

Several studies have reported that melatonin may possess multiple roles in plants, including regulator of circadian rhythm, regulator and modulator of growth and reproductive development (root and shoot development, fruit development, increases in biomass...), controlling plant diseases and insect pests, protective role against abiotic and biotic stressors, and antioxidant for free radicals. Evidence also suggests that melatonin shows the capacity to increase crop production.[25,27–30]

6.1.3.1 Melatonin as a Biological Circadian Cycle Regulator

The role of melatonin as a regulator of the circadian rhythm has been described in animals, highest levels being reported at night and lowest levels during the day (hormone of darkness).[3,31,32] However, in plants, the circadian rhythm is related to a diversity of biological processes such as protein stability, metabolic regulation..., the photosynthetic process, and may also affect flowering and seed yield in crops.[33,34] An oscillating behaviour of melatonin has been observed in 12 h light:12 h dark cycles, showing low levels during the light period and highest levels in darkness in *Chenopodium rubrum* plants[31] as well as in microalga Ulva sp (16h light:8h dark). These decreases suggest that melatonin biosynthesis is induced by oxidative stress in response to solar radiation.[35,36] However, additional studies have found no significant differences in melatonin levels with respect to the light/dark cycle (*Pharbitis nil* and *Solanum lycopersicum*) or maximum levels of melatonin in the light phase of the light/dark cycle (*Eichhornia crassipes*).[31] The circadian rhythm of melatonin production seems to occur in plants, indicating that

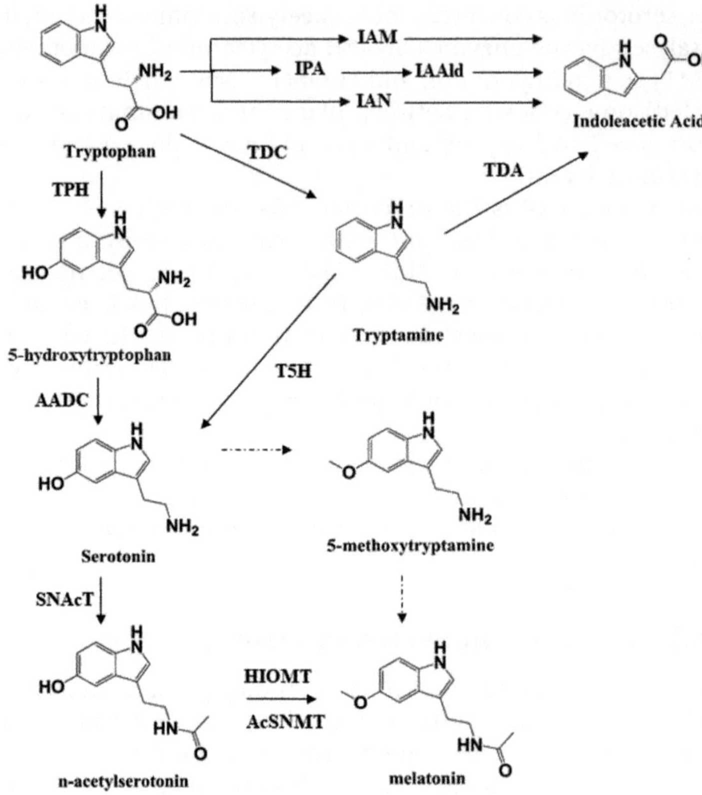

Figure 6.2 Biosynthetic pathway of melatonin (enzymes involved: TPH tryptophan 5-hydroxylase, AADC aromatic L-amino decarboxylase, SNAcT serotonin *N*-acetyltransferase, HIOMT hydroxyindole-*O*-methyltransferase, AcSNMT acetylserotonin *N*-methyltransferase, TDC tryptophan decarboxylase, T5H tryptamine 5-hydroxylase, TDA tryptamine desaminase). The basic tryptophan-dependent pathways of indoleacetic acid biosynthesis are shown with intermediates numbered: (1) indole-3-pyruvic acid (IPA) and indole-3-acetaldehyde (IAAld), (2) indole-3-acetamide (IAM), and (3) indole-3-acetonitrile (IAN).

photoperiodic change in melatonin levels might be species dependent in plants and act as a stress-protecting agent against abiotic factors, providing an adaptive mechanism to ensure reproduction.[32,37]

6.1.3.2 Melatonin as a Plant Growth Regulator

The enhancement of plant biomass, particularly root biomass, by melatonin has been reported in the literature, where it has been shown to demonstrate similar behaviour to IAA.[38–41] Arnao and Hernandez-Ruiz described that the optimal growth-promoting activity was reported at 10 μM melatonin.[40]

Melatonin also promotes lateral and adventitious roots probably because of the melatonin antioxidant capacity.[24] However, melatonin could act as an inhibitor at high concentrations, since it inhibited root growth.[42]

6.1.3.3 Melatonin as a Biotic and Abiotic Antistressor

Significant increments in melatonin concentration have been observed due to abiotic and biotic stresses. Several studies have described the role of melatonin in reducing abiotic and biotic stresses including drought, salinity–alkalinity, light, high and low ambient temperatures, UV radiation, chemical pollutants and fungal/bacterial infections.[43–48] The damaging effects induced by abiotic stresses on plants lead to oxidative stress, due to generation of free radicals. These can be relieved by melatonin *via* antioxidant capacity, regulating the expression of antioxidant enzymes and thus inhibiting damage to the plant tissues.[49]

6.1.3.4 Melatonin as an Antioxidant

The ability of melatonin to protect against free radical destruction as an antioxidant has been confirmed. Melatonin reduces oxidative stress as a direct radical scavenger, stimulating synthesis of other antioxidant and antioxidant enzymes, protecting against antioxidative enzyme damage, increasing the efficiency of the mitochondrial electron transport chain and generating secondary metabolites with high scavenger activities (AFMK and AMK).[22] These effects have been described in the literature in several plants protecting them against biotic or abiotic stress and thus, from generated radicals such as reactive oxygen, nitrogen and chlorine species (ROS, RNS and RCS, respectively).[50,51]

6.2 Dietary Sources of Melatonin

Melatonin has been identified in several varieties of plants as well as in their different parts, such as fruits, leaves, roots and seeds. In addition, significant amounts of melatonin have been found in common beverages, seeds, cereals, legumes, vegetables, edible oils, nuts fruits and medicinal herbs, showing higher melatonin levels than those determined in animals, indicating their possible health contribution as bioactive compounds.[5,52]

6.2.1 Analysis of Melatonin in Food

To validate a suitable method for the analysis of melatonin in food analysis presents some difficulties due to its low contents that range from mg to pg per gram, its amphipathic characteristics and its antioxidant character, which make it a potent reactive molecule. For this reason, it is necessary to set up suitable methods that offer sensitivity and specificity. In general, sampling of various food plant samples and their subsequent extraction, analysis, and evaluation present some analytical problems. Extraction methodologies are a critical step in the melatonin quantification, due to interferences from the matrix composition, so that a prior purification step is generally required. Additionally, melatonin extracts are subject to oxidation because of their light sensitivity. Melatonin extraction from plant foods was completed using liquid–liquid extraction with a polar solvent, such as ethanol, methanol, chloroform or ethylacetate, followed by a solid phase extraction.[40,53,54]

Determination of melatonin in plant foods has been carried out using several techniques, including chromatographic and immunological methods. High performance liquid chromatography (HPLC) with ultraviolet (UV), electrochemical (HPLC-EC) or fluorimetric (HPLC-F) detection (λ_{exc} = 280 nm; λ_{em} = 345 nm), and GC/HPLC-MS are the main chromatographic techniques applied to analyze melatonin, with the latter showing the highest resolution and sensitivity.[40,55–57] Recently, ultra-performance liquid chromatography-tandem mass spectrometry (UPLC-MS/MS), and ultra-performance liquid chromatography-high-resolution mass spectrometry (UPLC-HR-MS) have been proposed by Fernández-Pachón *et al.*[58] as innovative techniques to identify melatonin in a food matrix. With respect to immunological assays, enzyme-linked immunosorbent assay (ELISA) and radioimmunoassay (RIA) are used but the melatonin levels could be over-estimated due to reaction of antibodies and enzymes with other components presents in the same matrix, suggesting lower precision.[14,59,60]

6.2.2 Melatonin in Plant-based Food

6.2.2.1 Melatonin in Plants

The first time that melatonin was identified outside the animal kingdom was in *Lingulodinium polyedrum*, a dinoflagellate alga.[61] Thereafter melatonin was detected in all organisms including bacteria, algae, fungi, plants and insects. At present, melatonin has been found in more than 100 plant species belonging to at least 19 different families, including both monocotyledons and dicotyledons, of angiosperm plants.[29] Knowledge about the presence of melatonin, excluding in these flowering plants, is rather scarce. According to the findings, the levels of melatonin in plants are higher than those found in animals, because of the absence of tryptophan availability and thus, the lack of the shikimic acid pathway, or/and also the immobility of plants and hence their inability to avoid environmental hazards (unlike animals).[62]

A number of investigations have reported differences in melatonin concentration among plants, not only because of techniques of extraction and quantification but also because of species and varieties of the same plants, light regime, type of tissue analyzed and environmental conditions (abiotic and biotic stresses).[44,62] Under stress conditions, plants have to improve their defense system and detoxify a variety of reactive species, so they are obligated to increase enzymatic and non-enzymatic mechanisms, including melatonin activity. Some studies have focused on the effects of exogenous melatonin on plant adaptation to stress due to its capacity for upregulating physiological parameters, scavenging reactive species, stimulating the production of antioxidant compounds, improving the activity of some stress-related enzymes and altering the expression of some stress-response genes.[63]

The available scientific data on the concentration and distribution of melatonin within vegetative organs of plants has been scarce until now, however, several studies have found that seeds and leaves show the highest levels, while fruits present the lowest levels. Melatonin concentration ranges from being undetectable to levels of 5 ng g^{-1} of tissue.[52,64] The reasons for these variations among tissues could be due to the lack of central vacuoles and water in seed cells, intrinsic genetic variability of the plants or the protective function that melatonin exerts in seeds (lipid rich tissues protect the embryo from oxidative stress).[65,66] In addition to producing melatonin, plants are also able to take up melatonin from the soil.

It has been consistently reported that melatonin is an amphiphilic compound, it can freely pass through all cellular compartments and distribute to nucleus, cytosol and mitochondria, with the highest recorded concentration being found in mitochondria.[67] Some evidence indicates that both mitochondria and chloroplasts synthesize melatonin, since serotonin *N*-acetyltransferase has been observed in chloroplasts and arylalkylamine *N*-acetyltransferase (AANAT) in mitochondria. The evidence makes sense since the mitochondria and chloroplasts are the main sources of free radical production, and therefore, they require strong protection from oxidative stress.[68]

6.2.2.2 *Melatonin in Food and Processed Food*

Traditional dietary habits from Mediterranean countries, rich in vegetables and fruits, are associated with a low incidence of chronic diseases, such as reduced risks of cancer, and neurodegenerative and cardiovascular diseases, common to developed countries, and improve longevity.[69,70] Beneficial compounds present in several vegetables and fruits seem to be responsible for the healthy properties of the Mediterranean diet, these include minerals, vitamins, phenolic compounds and many biologically active molecules.[71] Over the past few years, melatonin, a new molecule, has been characterized and identified in numerous foodstuffs, including beverages, with antioxidant properties that have a potentially detoxifying effect on reactive species. The unique melatonin precursor obtained from the diet is tryptophan, which has similar circadian cycle regulatory properties.[11] Therefore, foods rich in tryptophan (pumpkin, squash, whole oats, wheat bran, wheat germ, brown rice...) could be related with different melatonin levels.

Melatonin content has been quantified in a wide range of edible plant-based foods and beverages and these foods are regarded as promising nutraceuticals. A representative group of foods and beverages in which levels of melatonin have been evaluated is presented below.

Cereals: In the literature, noticeably different levels of melatonin have been reported in crops, including rice, corn, oat, wheat, and barley, ranging from 1.4 to 2034 ng g^{-1}.[56,72–74] Setyaningsih *et al.*[75] reported differences among rice samples analyzed, where pigmented grains, such as black rice, black glutinous rice and red rice showed high levels of melatonin (>100 ng g^{-1}), and the highest contents were found in red rice (212 ng g^{-1}). Correlation analysis have indicated that the samples with higher levels of melatonin contain a higher concentration of vitamin B. Evidence indicates that vitamin B3 could prevent the breakdown of the melatonin precursor and vitamin B6 is involved in melatonin synthesis.[76]

Vegetables: Melatonin has also been discovered in widely consumed common vegetables including tomato, pepper, mushroom, cucumber, beetroot, onion, garlic, carrot and cabbage, its concentration in these vegetables varies from several nanograms to several thousand nanograms per gram (0.01–12 900 ng g^{-1}). The foods that showed relatively high melatonin concentration are tomatoes and mushrooms, *i.e.*, 66.6 ng g^{-1} fresh weight (FW) tomato (*Lycopersicon esculentum* cv. Bonda) and 12 900 ng g^{-1} FW Basidiomycota (*Lactarius deliciosus*) quantified by LC-MS/LC-FD and HPLC, respectively.[55,77] The concentration of melatonin in tomatoes varies according to tissue analyzed developmental stage, the content increasing from the immature green stage to the red stage.[78,79] Melatonin as well as serotonin have not been detected in potato.[80] In contrast, this compound has been identified in nuts such as walnut and pistachio, reaching 233000 ng g^{-1} dry weight (DW) in the case of *Pistacia vera* L. cv. Ahmad Aghaei.[81,82]

Fruits: The concentration of melatonin is relatively low in most of the fruits analyzed, except in grape skin, *Vitis vinifera* L. cv Malbec, and cranberry *Vaccinium macrocarpon* Ait., reaching 158.9 ng g^{-1} DW and 96 μg g^{-1} DW, respectively.[35] Melatonin has also been detected in different varieties of cherries and strawberries in low concentration (0.01–20 ng g^{-1} FW).[55,83,84] Banana is also a commonly consumed fruit, however the levels of melatonin are around 0.66 ng g^{-1} DW, with the maximum concentration being found in the peel and at the over-ripe stage.[80] Several studies have also reported melatonin in grapes, and unlike in other foods, evidence indicates that this indoleamine decreases with ripening.[85] It has been detected in the exocarp of different Italian and French wine grapes, with flesh and seeds exhibiting different concentrations probably due to many endogenous and exogenous factors such as day/night fluctuations, agrochemical treatments, geographic origin, phenological stages, pathogen infections...[64]

Beverages: Melatonin has been widely described in fermented alcoholic drinks such as beers and wines, in fermented non-alcoholic drinks like fruit juices and also, in hot beverages like cacao and coffee.[58,86–88] Additionally, increases in the levels of melatonin in these beverages have been attributed to processes such as roasting and fermentation, due to the synthesis of this neurohormone. *Saccharomyces cerevisiae* produces not only melatonin, but also other compounds, such as biogenic amines, during grape fermentation.[89] Melatonin isomers are produced during the process of wine making, showing the highest levels close to the end of the fermentation process.[86] Stege *et al.*[90] showed that the melatonin concentration in red wine was higher than in white. Furthermore, melatonin has been widely described in different commercial beers,[11,91] ranging from 51.8 to 169.7 pg mL^{-1}, and showing its highest levels in beers with the highest alcoholic content, due to its solubility in alcohol.[92] Melatonin isomers have also been detected in these beverages at higher concentrations (14.3 ng mL^{-1}). In addition, fermented fruit juices exhibited greater melatonin levels than the fresh fruits (orange and pomegranate).[58,93]

Other non-alcoholic drinks, such as coffee and cocoa usually contain high concentrations of melatonin. Significant levels of melatonin have been found in the green (6800 ng g^{-1} DW) and roasted beans (9600 ng g^{-1} DW) of *Arabica coffea* species.[87,88] On the other hand, green and black teas did not contain melatonin, unlike medicine herbal infusions, which showed higher melatonin concentrations.[94]

Herbal plants: Considerable amounts of melatonin have been identified in several of Chinese medicinal herbs; many of them having levels higher than 1000 ng g^{-1} DW, such as Sangye (leaf of *Morus alba* L.), with 1510 ng g^{-1}.[95,96] Herrera *et al.*[97] studied nineteen herbal infusions from commercial medicinal plants, chamomile being the infusion with the highest melatonin content while the lowest was found in St. John's wort infusion. The highest level of melatonin, 6.0 ng mL^{-1} was found in chamomile, which corresponded to 1.5 μg per cup of chamomile infusion (250 mL). These results are in agreement with those reported by Ansari *et al.*[98] who found outstanding levels of melatonin in heated and boiled plant flowers (*Tanacetum parthenium*, *Tripleurospermum disciforme* and *Viola odorata*). Interestingly, the use of these herbal teas rich in melatonin are associated with their antioxidative power, traditionally used to slow the aging process or to treat oxidative stress diseases.

6.2.2.3 *Melatonin Bioavailability*

Previously, the bioavailability of melatonin from plant food consumption has been demonstrated, with reported increments in physiological melatonin blood levels after consumption of enriched melatonin foods.[63] Oral melatonin is generally absorbed from the gastrointestinal tract, requiring 45 min, and then, is metabolized in the liver. The melatonin released to the blood circulation is in the range 3–76%, and it is found to be short-lived in the blood, about 20–40 min. There are no identified morphophysiological barriers to melatonin, which has access to all tissues, being able to act directly, as an antioxidant, or indirectly, through receptors MT1 and MT2.[64] <Then, melatonin is excreted mainly as soluble metabolites in the urine, essentially as 6-sulfatoxymelatonin (aMT6s).[63]

The time of day a melatonin-rich food is consumed could be significant in defining the response, since the sensitivity of the melatonin receptor for its ligand may vary over a 24-hour period. The first study *in vivo* was done by Hattori *et al.*,[72] who reported serum melatonin increments after foodstuffs consumption (from 20 to 34 pg mL^{-1} after 1.5 h), where chicks were subjected to a diet containing melatonin rich plant products (3.5 ng g^{-1}).[72] Additionally, rats were also fed with 3 g of walnuts (*Juglans regia* L), containing 10.5 ng of melatonin. Increases of serum melatonin concentrations and antioxidant capacity, measured by trolox equivalent antioxidant capacity (TEAC) and ferric reducing ability of plasma (FRAP), were observed for the walnut-fed animals.[53] Therefore, the melatonin present in plasma might provide the ability to detoxify toxic free radicals.[64]

Several Japanese studies correlate the consumption of vegetables by humans with high urinary levels of aMT6s.[99,100] Evidence indicates that 1288 ng melatonin per day from six selected vegetables, for 65 days, increased urinary aMT6s concentration slightly in the melatonin-rich group, *versus* decreases in a control group, showing that a greater consumption of vegetables can raise circulatory melatonin concentrations. In recent studies, Sae-Teaw[101] and colleagues focus on the effects of fresh fruits and fruit juices consumption and levels of melatonin in human blood, reporting the highest value at 120 min after orange juice intake (151 pg mL^{-1}) and enhanced antioxidant capacity, FRAP and ORAC (oxygen radical absorbance capacity). However, the melatonin levels found in fermented orange juice were higher than the levels in fresh fruits.[58] Significant increases in urine aMT6s concentrations were also reported after Montmorency cherries (*P. cerasus*) (whole fruit and juice) consumption in a randomized, double-blind, placebo-controlled, crossover design study (125–220%, $p < 0.05$).[102,103] These intakes show higher circulating levels of melatonin, and therefore, higher bioavailability in these fruits.

With respect to fermented beverages intakes, results obtained from studies with healthy volunteers are also relevant, where the administration of red wine and beer resulted in high levels of melatonin in human serum.[92,104] These results indicate that the ethanol formed during alcoholic fermentation could be a stressor for *Saccharomyces* to synthesize melatonin.[89] Maldonado *et al.*[92] have described that after drinking beer, melatonin and total antioxidant status in human serum increased in seven healthy volunteers, concluding that moderate beer consumption can preserve from oxidative stress.[92]

6.3 Melatonin and Legumes

Although melatonin was originally thought to be produced exclusively in animals, it has been found in a considerable number of plant foods, including legumes and seeds. There literature about melatonin and isomers quantification in legumes is scarce because of its low levels in raw seeds, unlike other fruits and vegetables. The melatonin concentration determined in different legumes has been summarized in Table 6.1.

Table 6.1 Melatonin concentration in legumes.

Common name	Scientific name	Organ/tissue plant	Processing	Melatonin content ($ng\ g^{-1}$ or $pg\ g^{-1}$)	Method of analysis	Reference
Lupin	*Lupinus albus* L.	Cotyledon	Germination	3.83 $ng\ g^{-1}$ FW	HPLC-FD	73
		Coat	Germination	37.50 $ng\ g^{-1}$ FW	HPLC-FD	73
		Flour	Germination	0.53 $ng\ g^{-1}$ DW	HPLC-FD	73
		Root	Germination	14.00 $ng\ g^{-1}$ FW	HPLC-FD	73
		Leaf	Germination	11.00 $ng\ g^{-1}$ FW	HPLC-FD	73
		Hypocotyl	Germination	1.10 $ng\ g^{-1}$ FW	HPLC-FD	73
Alfalfa	*Medicago sativum*	Seed	Raw	16 $ng\ g^{-1}$	HPLC-ECD	108
Lentils	*Lens culinaris* L	Seed	Raw	0.4 $ng\ g^{-1}$ DW	HPLC-MS/MS	105
		Seed	Germination	2.5 $ng\ g^{-1}$ DW	HPLC-MS/MS	105
Kidney beans	*Phaseolus vulgaris* L	Seed	Raw	1.0 $ng\ g^{-1}$ DW	HPLC-MS/MS	105
		Seed	Germination	9.4 $ng\ g^{-1}$ DW	HPLC-MS/MS	105
Alfalfa	*Medicago sativum*	Seed	Raw	0.05 $ng\ g^{-1}$ DW	ELISA	14
		Seed	Germination	0.13 $ng\ g^{-1}$ DW	ELISA	14
Lentil	*Lens sculenta* L	Seed	Raw	0.07 $ng\ g^{-1}$ DW	ELISA	14
		Seed	Germination	0.22 $ng\ g^{-1}$ DW	ELISA	14
Mung bean	*Vigna radiata* L	Seed		0.11 $ng\ g^{-1}$ DW	ELISA	14
		Seed	Germination	0.17 $ng\ g^{-1}$ DW	ELISA	14
Lentil	*Lens culinaris* L	Seed	Germination + exogenous Mel, 20 μM	1089.8 $ng\ g^{-1}$ DW	HPLC-MS/MS	54
Kidney beans	*Phaseolus vulgaris* L.	Seed	Germination + exogenous Mel, 20 μM	529.1 $ng\ g^{-1}$ DW	HPLC-MS/MS	54
Lentil	*Lens culinaris* L	Coat	Germination + exogenous Mel, 20 μM	897.2 $ng\ g^{-1}$ DW	HPLC-MS/MS	54
		Endosperm	Germination + exogenous Mel, 20 μM	150.3 $ng\ g^{-1}$ DW	HPLC-MS/MS	54
		Radicle	Germination + exogenous Mel, 20 μM	58.9 $ng\ g^{-1}$ DW	HPLC-MS/MS	54
Lentil	Not specified	Seed	Germination	0.92 $ng\ g^{-1}$ DW	RIA	107

(*continued*)

Table 6.1 (*continued*)

Common name	Scientific name	Organ/tissue plant	Processing	Melatonin content (ng g^{-1} or pg g^{-1})	Method of analysis	Reference
Soya bean	*Glycine max*	Seed	Germination	1.89 ng g^{-1} DW	RIA	107
Vetch	*Vicia sativa* L	Seed	Germination	1.91 ng g^{-1} DW	RIA	107
Mung bean	*Vigna radiata* L	Seed	Germination + hydro primed seeds with Mel, 50 μM	1.391 ng g^{-1} DW at 25 °C	HPLC-MS/MS	116
		Seed	Germination + hydro primed seeds with water (25 °C)	0.130 ng g^{-1} DW	HPLC-MS/MS	116
		Seed	Germination (25 °C)	0.237 ng g^{-1} DW	HPLC-MS/MS	116
		Seed	Germination (5 °C)	0.180 ng g^{-1} DW	HPLC-MS/MS	116
		Seed	Germination + rewarmed (25 °C) after chilling	0.090 ng g^{-1} DW	HPLC-MS/MS	116
Lupin	*Lupinus albus* L.	Seed	Germination + $ZnSO_4$ treatment	80 ng g^{-1} FW	HPLC-FD	113
		Seed	Germination control (without $ZnSO_4$)	6.7 ng g^{-1} FW	HPLC-FD	116
		Seed	Germination + water restriction (50 mL)	56.8 ng g^{-1} FW	HPLC-FD	116
		Seed	Germination control (150 mL)	13.9 ng g^{-1} FW	HPLC-FD	116
		Seed	Germination + oxygen restriction (0% oxygen)	10 ng g^{-1} FW	HPLC-FD	116
		Seed	Germination control (100% oxygen)	6.2 ng g^{-1} FW	HPLC-FD	116
		Seed	Germination + acid medium (pH 3.5)	11 ng g^{-1} FW	HPLC-FD	116
		Seed	Germination control (pH 7)	7 ng g^{-1} FW	HPLC-FD	116
		Seed	Germination + cold stress (6 °C)	16 ng g^{-1} FW	HPLC-FD	116
		Seed	Germination control (25 °C)	8.5 ng g^{-1} FW	HPLC-FD	116
		Seed	Germination + exogenous Mel (10 μM)	180 ng g^{-1} FW	HPLC-FD	116
		Seed	Germination control (without exogenous Mel)	7.7 ng g^{-1} FW	HPLC-FD	116

Gran cao	*Glycyrrhiza uralensis*	Root	Germination (7 months)	34 000 ng g^{-1} FW	HPLC-UV/VIS-PAD	110
		Leaf	Germination (7 months)	200 ng g^{-1} FW	HPLC-UV/VIS-PAD	110
		Seed	Germination (7 months) under red light	20 000 ng g^{-1} FW	HPLC-UV/VIS-PAD	110
		Seed	Germination (7 months) under blue light	12 000 ng g^{-1} FW	HPLC-UV/VIS-PAD	110
		Seed	Germination (7 months) under white light	9000 ng g^{-1} FW	HPLC-UV/VIS-PAD	110
		Seed	Germination (7 months) under high UV-B irradiation	70 000 ng g^{-1} FW	HPLC-UV/VIS-PAD	110
		Seed	Germination (7 months) under low UV-B irradiation	38 000 ng g^{-1} FW	HPLC-UV/VIS-PAD	110
		Seed	Germination (7 months) control (without UV-B irradiation)	11 000 ng g^{-1} FW	HPLC-UV/VIS-PAD	110

6.3.1 Melatonin in Raw Legumes

As reported Aguilera *et al.*,[105] two legume seeds were examined for the presence of melatonin, kidney beans and lentils, with kidney beans exhibiting higher melatonin content (1.0 ng g^{-1} DW) than lentils (0.4 ng g^{-1} DW).[105] Because of the difficulties in the quantification of melatonin, due to its low content and amphipathic characteristic, it was necessary to use HPLC-ESI-MS/MS triple quadrupole and the external standard method, using pure melatonin, to achieve the highest resolution and sensitivity. This methodology was more sensible than that used by Hernandez-Ruiz and Arnao[106] in lupines, where melatonin was quantified by HPLC (Jasco Co., Tokyo, Japan) with fluorescence detection. In contrast, the results shown by Zielinski *et al.*[107] and Manchester *et al.*[108] using RIA and HPLC-ECD, respectively, for its quantification, detected values between 0.45 and 16 ng g^{-1} DW, in soya bean, vetch, lentil and alfalfa. The high melatonin levels for alfalfa seeds (16 ng g^{-1} DW) were determined using RIA; these concentrations could be over-estimated because of reaction of enzymes and antibodies with other compounds present in the seed, which might indicate lower precision.

6.3.1.1 *Melatonin and Environmental Stresses in Legumes*

Several studies have described the protective role of melatonin against environmental stresses, such as drought, light (UV-B radiation), heavy metals, low temperatures and salinity–alkalinity.[109–112] However, only some of them have been evaluated in legumes (*Lupinus albus* L., *Glycyrrhiza uralensis*, *Vigna radiate* and soybean).

Some researchers have described the influence of light on melatonin concentration, but it has not been elucidated how different quality of light (red, blue and white) and response to UV-B irradiation (high and low UV-B) could affect the concentration of melatonin in the root tissues of *Glycyrrhiza uralensis*. Afreen *et al.*[110] have reported melatonin variations, which depend on the wavelength of the light spectrum, showing higher levels of melatonin after 3 months with red light (20 μg g^{-1} FW), followed by white and blue light. Moreover, they have also found that the increases in melatonin were correlated to the intensity of UV exposure (the greater the intensity of UV-B irradiation, the higher the melatonin concentration, 70μ/g FW), thus, this compound might display a relevant role in oxidative damage (from radicals as a response to UV irradiation).

Artificially, adverse situations plants (cold stress, drought, pH, and anaerobic) have been induced to study the response of lupin and the changes in endogenous melatonin concentration. Arnao and Hernandez-Ruíz[113] used NaCl, H_2O_2, and $ZnSO_4$ and natural environmental conditions to simulate stress conditions to the lupin. The results showed that melatonin biosynthesis is stimulated by stress situations, since totally stress-producing treatments caused an increase in the root melatonin concentration, reaching the highest levels (12-fold increment) when the plant was stressed with $ZnSO_4$ (from 6.7 to 80 ng g^{-1} FW).

Occasionally, in some crops, artificial coating of seeds with external materials is used to improve handling, protection and germination enhancement.[114] This practice of covering seeds has been used by Wei *et al.*,[115] adding different concentrations of melatonin with seed-coating-reagent to soybean seeds. The use of melatonin improved seed production, soybean tolerance to drought, and salt stresses, simulated with water restrictions and 1% (w/v) NaCl. These increments might be due to enhancement of antioxidative actions, carbohydrate metabolism and photosynthesis process. Thus, the application of melatonin to the seeds of *Vigna radiate*, has shown improvements in the defense mechanism of plants at low temperatures.[116]

6.3.1.2 Effect of Legume Germination on Melatonin and Plant Tissues

Researchers have reported that the germination process of legumes leads to a significant increase in melatonin levels and thus, the development of plant biomass, particularly root biomass, as well as antioxidant activity.[38,40] Recent results have shown that light germination treatment (photoperiodic cycle, 16 h light:8 h dark) and dark germination conditions (24 h dark) resulted in similar melatonin concentrations being reached in lupin plants, the roots being the organ that contained the highest levels of melatonin, followed by leaves, hypocotyls and cotyledons. The germination brought about a pronounced increase in melatonin content (3-fold in the case of roots), that reached the highest levels at 18 days under light treatment.[73] On the other hand, Aguilera *et al.*[105] found different results for the levels of melatonin in kidney beans (*Phaseolus vulgaris* L.) when different germination conditions were used, 12 h light/12 h dark *vs.* 24 h dark, showing higher concentrations under dark conditions. Germination under 24 h dark exhibited the highest melatonin values after 6 days of germination in kidney bean legumes (9 ng g^{-1} DW), while the concentration of phenolic compounds decreased. In addition, the antioxidant activity in lentil and bean samples improved compared to the initial seeds. These results are more in accordance with those published by Posmyk *et al.*[111] and Posmyk *et al.*,[117] who obtained higher melatonin levels in darkness in cucumber and red cabbage germinated seeds, as those of Zieliski *et al.*[107] who found increments of melatonin at 7 days of germination in soybean, vetch and lentil seeds.

Some studies focus on the effects of exogenous melatonin dilutions on roots and cotyledons to demonstrate the ability of this molecule to stimulate the regeneration and the expansion of these plants.[40,106] The presence of melatonin provokes the appearance of root primordia from pericicle cells, changing the distribution of lateral and adventitious roots and the length and number of lateral and adventitious roots in lupins. Additionally, these authors have also determined that melatonin stimulates the expansion of cotyledons of lupins, similar to IAA behavior. From the results, it might be concluded that melatonin has a physiological effect on adventitious and lateral roots (growth promoter and rooting promotor) as well as on etiolated cotyledons in lupins.

The application of exogenous melatonin (enriched watering, 20 μM) on the germination of kidney beans *(Phaseolus vulgaris* L.) and lentils (*Lens culinaris* L.) have been evaluated by Aguilera *et al.*,[54] reaching melatonin levels of 1090 ng g^{-1} and 529 ng g^{-1} in lentil and kidney bean sprouts, respectively. In addition, the study showed the development of radicles, in contrast to conventional germination, was significantly increased when enriched water was used.[54] Similar results have been obtained after sprout tissues evaluation (seed coat, endosperm, and radicle) during the germination process. As time progressed, the levels of melatonin increased in all sprout tissues, the seed coats exhibited the highest levels (897.2 ng g^{-1} DW), and afterward, the melatonin level decreased sharply (eighth day).

6.3.2 Melatonin Legume Intake

The oral consumption of melatonin-rich foods is reported to increase the serum melatonin concentration. The determination of melatonin in plasma and of its metabolites in urine samples (aMT6s) were evaluated in young Sprague Dawley rats after kidney bean sprouts administration.[118] The intake of germinated kidney bean, after 90 min, increased the plasmatic melatonin levels (16%), as well as the level of aMT6s in urine, which indicated a noticeable increment (4-fold). However, antioxidant capacity levels and phenolic compounds did not exhibit any significant difference.

Until now, the results for melatonin levels in plasma after the intake of melatonin-rich foods in animals and humans, has been variable, since several factors might be involved. The amount of melatonin intake and the vehicle are probably the main factors that affect the melatonin levels in plasma. In this sense, the intake period (from days to weeks), the number of doses per day, and the time of blood collection also have an influence on the levels of melatonin measured by analysis.

To conclude, the consumption of plant foods containing melatonin is recommended to moderate the melatonin plasmatic levels, because they decrease with age and during specific periods (fasting, oxidative stress, energy restriction...). Therefore, the consumption of germinated legumes might be a novel way to improve not only dietary melatonin, but also antioxidant defense, as demonstrated with other melatonin rich foods.

References

1. A. B. Lerner, J. D. Case, Y. Takahashi, T. H. Lee and W. Mori, *J. Am. Chem. Soc.*, 1958, **80**, 2587.
2. J. Veenstra-VanderWeele, G. M. Anderson and E. H. Cook Jr, *Eur. J. Pharmacol.*, 2000, **410**, 165.
3. R. J. Reiter, *Experientia*, 1993, **49**, 654.

4. A. Galano, D. X. Tan and R. J. Reiter, *J. Pineal Res.*, 2011, **51**, 1.
5. R. J. Reiter and D. Tan, *Ann. N. Y. Acad. Sci.*, 2002, **957**, 341.
6. C. Bartsch, H. Bartsch and M. Karasek, *Neuroendocrinol. Lett.*, 2002, **23**, 30.
7. C. Vijayalaxmi, R. Thomas Jr, R. J. Reiter and T. S. Herman, *J. Clin. Oncol.*, 2002, **20**, 2575.
8. T. Ravindra, N. K. Lakshmi and Y. R. Ahuja, *Indian J. Med. Sci.*, 2006, **60**, 523.
9. R. Hardeland, *J. Pineal Res.*, 2013, **55**, 325.
10. X. Meng, Y. Li, S. Li, Y. Zhou, R. Gan, D. Xu and H. Li, *Nutrients*, 2017, 9.
11. M. Iriti, E. M. Varoni and S. Vitalini, *J. Pineal Res.*, 2010, **49**, 101.
12. D. Tan, R. Hardeland, L. C. Manchester, S. Rosales-Corral, A. Coto-Montes, J. A. Boga and R. J. Reiter, *J. Pineal Res.*, 2012, **53**, 113.
13. D. X. Tan, R. J. Reiter, L. C. Manchester, M. T. Yan, M. El-Sawi, R. M. Sainz, J. C. Mayo, R. Kohen, M. Allegra and R. Hardeland, *Curr. Top. Med. Chem.*, 2002, **2**, 181.
14. Y. Aguilera, T. Herrera, V. Benítez, S. M. Arribas, A. L. López De Pablo, R. M. Esteban and M. A. Martín-Cabrejas, *Food Chem.*, 2015, **170**, 203.
15. A. Galano, D. X. Tan and R. J. Reiter, *J. Pineal Res.*, 2013, **54**, 245.
16. H. Seeger, A. O. Mueck and T. H. Lippert, *Br. J. Clin. Pharmacol.*, 1997, **44**, 283.
17. D. Tan, L. C. Manchester, R. J. Reiter, W. Qi, M. A. Hanes and N. J. Farley, *Life Sci.*, 1999, **65**, 2523.
18. D. Tan, L. C. Manchester, L. Fuentes-Broto, S. D. Paredes and R. J. Reiter, *Obes. Rev.*, 2011, **12**, 167.
19. R. J. Reiter, S. D. Paredes, A. Korkmaz, L. C. Manchester and D. X. Tan, *Adv. Med. Sci.*, 2008, **53**, 119.
20. P. A. Witt-Enderby, J. Bennett, M. J. Jarzynka, S. Firestine and M. A. Melan, *Life Sci.*, 2003, **72**, 2183.
21. K. Back, D. Tan and R. J. Reiter, *J. Pineal Res.*, 2016, **61**, 426.
22. R. J. Reiter, D. Tan and A. Galano, *Front. Physiol.*, 2014, 5.
23. V. Bajwa, S. J. Murch and P. K. Saxena, in *Production of Biomass and Bioactive Compounds Using Bioreactor Technology*, ed. K. Y. Paek, H. Murthy, J. J. Zhong, Springer, 2014, p. 445.
24. M. B. Arnao and J. Hernández-Ruiz, *Plant Signaling Behav.*, 2006, **1**, 88.
25. S. Park, K. Lee, Y. - Kim and K. Back, *J. Pineal Res.*, 2012, **52**, 211.
26. S. J. Murch and P. K. Saxena, *In Vitro Cell. Dev. Biol.: Plant*, 2002, **38**, 531.
27. R. J. Reiter, D. Tan, Z. Zhou, M. H. C. Cruz, L. Fuentes-Broto and A. Galano, *Molecules*, 2015, **20**, 7396.
28. J. Liu, W. Wang, L. Wang and Y. Sun, *Plant Growth Regul.*, 2015, **77**, 317.
29. S. D. Paredes, A. Korkmaz, L. C. Manchester, D. - Tan and R. J. Reiter, *J. Exp. Bot.*, 2009, **60**, 57.
30. W. J. Park, *J. Plant Biol.*, 2011, **54**, 143.
31. J. Kolár, I. Machácková, J. Eder, E. Prinsen, W. Van Dongen, H. Van Onckelen and H. Illnerová, *Phytochemistry*, 1997, **44**, 1407.
32. D. L. Van Tassel, N. Roberts, A. Lewy and S. D. O'Neill, *J. Pineal Res.*, 2001, **31**, 8.

33. A. N. Dodd, N. Salathia, A. Hall, E. Kévei, R. Tóth, F. Nagy, J. M. Hibberd, A. J. Millar and A. A. R. Webb, *Science*, 2005, **309**, 630.
34. K. A. Atkins and A. N. Dodd, *Curr. Opin. Plant Biol.*, 2014, **21**, 43.
35. H. E. Boccalandro, C. V. González, D. A. Wunderlin and M. F. Silva, *J. Pineal Res.*, 2011, **51**, 226.
36. B. Zuo, X. Zheng, P. He, L. Wang, Q. Lei, C. Feng, J. Zhou, Q. Li, Z. Han and J. Kong, *J. Pineal Res.*, 2014, **57**, 408.
37. M. B. Arnao and J. Hernández-Ruiz, *J. Pineal Res.*, 2015, **59**, 133.
38. V. N. Sarropoulou, I. N. Therios and K. N. Dimassi-Theriou, *J. Pineal Res.*, 2012, **52**, 38.
39. S. J. Murch, S. S. B. Campbell and P. K. Saxena, *In Vitro Cell. Dev. Biol.: Plant*, 2001, **37**, 786.
40. M. B. Arnao and J. Hernández-Ruiz, *J. Pineal Res.*, 2007, **42**, 147.
41. Q. Chen, W. Qi, R. J. Reiter, W. Wei and B. Wang, *J. Plant Physiol.*, 2009, **166**, 324.
42. J. Hernández-Ruiz, A. Cano and M. B. Arnao, *Planta*, 2004, **220**, 140.
43. M. B. Arnao and J. Hernández-Ruiz, *J. Pineal Res.*, 2009, **46**, 295.
44. C. Wang, L. Yin, X. Shi, H. Xiao, K. Kang, X. Liu, J. Zhan and W. Huang, *J. Food Sci.*, 2016, **81**, 958.
45. C. Li, P. Wang, Z. Wei, D. Liang, C. Liu, L. Yin, D. Jia, M. Fu and F. Ma, *J. Pineal Res.*, 2012, **53**, 298.
46. V. S. Bajwa, M. R. Shukla, S. M. Sherif, S. J. Murch and P. K. Saxena, *J. Pineal Res.*, 2014, **56**, 238.
47. N. Zhang, B. Zhao, H. Zhang, S. Weeda, C. Yang, Z. Yang, S. Ren and Y. Guo, *J. Pineal Res.*, 2013, **54**, 15.
48. N. Zhang, Q. Sun, H. Zhang, Y. Cao, S. Weeda, S. Ren and Y. Guo, *J. Exp. Bot.*, 2015, **66**, 647.
49. D. Tan, L. C. Manchester, R. J. Reiter and B. F. Plummer, *Neurosignals*, 1999, **8**, 70.
50. L. Yin, P. Wang, M. Li, X. Ke, C. Li, D. Liang, S. Wu, X. Ma, C. Li, Y. Zou and F. Ma, *J. Pineal Res.*, 2013, **54**, 426.
51. S. Vitalini, C. Gardana, P. Simonetti, G. Fico and M. Iriti, *J. Pineal Res.*, 2013, **54**, 322.
52. R. Hardeland, S. R. Pandi-Perumal and D. P. Cardinali, *Int. J. Biochem. Cell Biol.*, 2006, **38**, 313.
53. R. J. Reiter, L. C. Manchester and D. Tan, *Nutrition*, 2005, **21**, 920.
54. Y. Aguilera, T. Herrera, R. Liébana, M. Rebollo-Hernanz, C. Sanchez-Puelles and M. A. Martín-Cabrejas, *J. Agric. Food Chem.*, 2015, **63**, 7967.
55. M. Stürtz, A. B. Cerezo, E. Cantos-Villar and M. C. Garcia-Parrilla, *Food Chem.*, 2011, **127**, 1329.
56. D. Tan, B. M. Zanghi, L. C. Manchester and R. J. Reiter, *J. Pineal Res.*, 2014, **57**, 213.
57. I. Reinholds, I. Pugajeva, V. Radenkovs, J. Rjabova and V. Bartkevics, *J. Chromatogr. Sci.*, 2016, **54**, 977.

58. M. S. Fernández-Pachõn, S. Medina, G. Herrero-Martín, I. Cerrillo, G. Berná, B. Escudero-Lõpez, F. Ferreres, F. Martín, M. C. García-Parrilla and A. Gil-Izquierdo, *J. Pineal Res.*, 2014, **56**, 31.
59. C. de la Puerta, M. P. Carrascosa-Salmoral, P. P. García-Luna, P. J. Lardone, J. L. Herrera, R. Fernández-Montesinos, J. M. Guerrero and D. Pozo, *Food Chem.*, 2007, **104**, 609.
60. M. T. Kollmann, M. Locher, F. Hirche, K. Eder, H. H. D. Meyer and R. M. Bruckmaier, *Domest. Anim. Endocrinol.*, 2008, **34**, 14.
61. B. Pöggeler, I. Balzer, R. Hardeland and A. Lerchl, *Naturwissenschaften*, 1991, **78**, 268.
62. D. Tan, R. Hardeland, L. C. Manchester, A. Korkmaz, S. Ma, S. Rosales-Corral and R. J. Reiter, *J. Exp. Bot.*, 2012, **63**, 577.
63. G. A. Ravishankar and A. Ramakrishna, *Serotonin and Melatonin: Their Functional Role in Plants, Food, Phytomedicine, and Human Health*, CRC Press, Boca Raton, 2016.
64. R. J. Reiter, D. X. Tan, S. Burkhardt and L. C. Manchester, *Nutr. Rev.*, 2001, **59**, 286.
65. L. C. Manchester, D. Tan, R. J. Reiter, W. Park, K. Monis and W. Qi, *Life Sci.*, 2000, **67**, 3023.
66. M. Iriti and F. Faoro, *Med. Hypotheses*, 2006, **67**, 833.
67. D. Acuña-Castroviejo, G. Escames, J. León, A. Carazo and H. Khaldy, *Adv. Exp. Med. Biol.*, 2003, **527**, 549.
68. D. Tan, L. C. Manchester, X. Liu, S. A. Rosales-Corral, D. Acuna-Castroviejo and R. J. Reiter, *J. Pineal Res.*, 2013, **54**, 127.
69. C. La Vecchia and C. Bosetti, *Public Health Nutr.*, 2006, **9**, 1077.
70. M. Iriti and S. Vitalini, *Pol. J. Food Nutr. Sci.*, 2012, **62**, 71.
71. A. Speciale, J. Chirafisi, A. Saija and F. Cimino, *Curr. Mol. Med.*, 2011, **11**, 770.
72. A. Hattori, H. Migitaka, M. Iigo, M. Itoh, K. Yamamoto, R. Ohtani-Kaneko, M. Hara, T. Suzuki and R. J. Reiter, *Biochem. Mol. Biol. Int.*, 1995, **35**, 627.
73. J. Hernández-Ruiz and M. B. Arnao, *J. Agric. Food Chem.*, 2008, **56**, 10567.
74. F. S. Hosseinian, W. Li and T. Beta, *Food Chem.*, 2008, **109**, 916.
75. W. Setyaningsih, I. E. Saputro, G. F. Barbero, M. Palma and C. García Barroso, *J. Agric. Food Chem.*, 2015, **63**, 1107.
76. A. Fardet, *Nutr. Res. Rev.*, 2010, **23**, 65.
77. B. Muszynska and K. Sulkowska-Ziaja, *Food Chem.*, 2012, **132**, 455.
78. M. Okazaki and H. Ezura, *J. Pineal Res.*, 2009, **46**, 338.
79. Q. Sun, N. Zhang, J. Wang, H. Zhang, D. Li, J. Shi, R. Li, S. Weeda, B. Zhao, S. Ren and Y. Guo, *J. Exp. Bot.*, 2015, **66**, 657.
80. F. A. Badria, *J. Med. Food*, 2002, **5**, 153.
81. M. I. Tapia, J. R. Sánchez-Morgado, J. García-Parra, R. Ramírez, T. Hernández and D. González-Gómez, *J. Food Compos. Anal.*, 2013, **31**, 232.
82. E. Oladi, M. Mohamadi, T. Shamspur and A. Mostafavi, *Spectrochim. Acta, Part A*, 2014, **132**, 326.

83. M. Garrido, J. Espino, D. González-Gómez, M. Lozano, C. Barriga, S. D. Paredes and A. B. Rodríguez, *Exp. Gerontol.*, 2012, **47**, 573.
84. Y. Zhao, D. Tan, Q. Lei, H. Chen, L. Wang, Q. Li, Y. Gao and J. Kong, *J. Pineal Res.*, 2013, **55**, 79.
85. S. Vitalini, C. Gardana, A. Zanzotto, G. Fico, F. Faoro, P. Simonetti and M. Iriti, *J. Pineal Res.*, 2011, **51**, 278.
86. M. I. Rodriguez-Naranjo, A. Gil-Izquierdo, A. M. Troncoso, E. Cantos-Villar and M. C. Garcia-Parrilla, *Food Chem.*, 2011, **126**, 1608.
87. A. Ramakrishna, P. Giridhar, K. U. Sankar and G. A. Ravishankar, *Acta Physiol. Plant.*, 2012, **34**, 393.
88. A. Ramakrishna, P. Giridhar, K. U. Sankar and G. A. Ravishankar, *J. Pineal Res.*, 2012, **52**, 470.
89. M. I. Rodriguez-Naranjo, J. L. Ordóñez, R. M. Callejón, E. Cantos-Villar and M. C. Garcia-Parrilla, *Food Chem. Toxicol.*, 2013, **57**, 140.
90. P. W. Stege, L. L. Sombra, G. Messina, L. D. Martinez and M. F. Silva, *Electrophoresis*, 2010, **31**, 2242.
91. H. Garcia-Moreno, J. R. Calvo and M. D. Maldonado, *J. Pineal Res.*, 2013, **55**, 26.
92. M. D. Maldonado, H. Moreno and J. R. Calvo, *Clin. Nutr.*, 2009, **28**, 188.
93. P. Mena, Á. Gil-Izquierdo, D. A. Moreno, N. Martí and C. García-Viguera, *LWT--Food Sci. Technol.*, 2012, **47**, 13.
94. T. Kocadagli, C. Yilmaz and V. Gökmen, *Food Chem.*, 2014, **153**, 151.
95. G. Chen, Y. Huo, D. Tan, Z. Liang, W. Zhang and Y. Zhang, *Life Sci.*, 2003, **73**, 19.
96. T. Padumanonda, J. Johns, A. Sangkasat and S. Tiyaworanant, *Daru, J. Pharm. Sci.*, 2014, 22.
97. T. Herrera, Y. Aguilera, M. Rebollo-Hernanz, E. Bravo, V. Benítez, N. Martínez-Sáez, S. M. Arribas, M. D. del Castillo and M. A. Martín-Cabrejas, *LWT--Food Sci. Technol.*, 2018, **89**, 65.
98. M. Ansari, R. Kh, N. Yasa, S. Vardasbi, S. M. Naimi and A. Nowrouzi, *Daru, J. Pharm. Sci.*, 2010, **18**, 173.
99. C. Nagata, T. Matsubara, H. Fujita, Y. Nagao, C. Shibuya, Y. Kashiki and H. Shimizu, *Cancer Epidemiol., Biomarkers Prev.*, 2005, **14**, 2877.
100. S. Oba, K. Nakamura, Y. Sahashi, A. Hattori and C. Nagata, *J. Pineal Res.*, 2008, **45**, 17.
101. M. Sae-Teaw, J. Johns, N. P. Johns and S. Subongkot, *J. Pineal Res.*, 2013, **55**, 58.
102. S. Burkhardt, D. Xian Tan, L. C. Manchester, R. Hardeland and R. J. Reiter, *J. Agric. Food Chem.*, 2001, **49**, 4898.
103. M. Garrido, S. D. Paredes, J. Cubero, M. Lozano, A. F. Toribio-Delgado, J. L. Muñoz, R. J. Reiter, C. Barriga and A. B. Rodríguez, *J. Gerontol., Ser. A*, 2010, **65**, 909.
104. M. Iriti and E. M. Varoni, *J. Sci. Food Agric.*, 2015, **95**, 2355.
105. Y. Aguilera, R. Liébana, T. Herrera, M. Rebollo-Hernanz, C. Sanchez-Puelles, V. Benítez and M. A. Martín-Cabrejas, *J. Agric. Food Chem.*, 2014, **62**, 10736.

106. J. Hernández-Ruiz and M. B. Arnao, *Plant Growth Regul.*, 2008, **55**, 29.
107. H. Zielinski, B. Lewczuk, B. Przybylska-Gornowicz and H. Kozłowska, *Biologically Active Phytochemicals in Food: Analysis, Metabolism, Bioavailability and Function*, ed. W. Pfannhauser, G. R. Fenwick and S. Khokhar, Royal Society of Chemistry, Cambridge, U.K., 2001, p. 110.
108. L. C. Manchester, D. Tan, R. J. Reiter, W. Park, K. Monis and W. Qi, *Life Sci.*, 2000, **67**, 3023.
109. X. Lei, R. Zhu, G. Zhang and Y. Dai, *J. Pineal Res.*, 2004, **36**, 126.
110. F. Afreen, S. M. A. Zobayed and T. Kozai, *J. Pineal Res.*, 2006, **41**, 108.
111. M. M. Posmyk, H. Kuran, K. Marciniak and K. M. Janas, *J. Pineal Res.*, 2008, **45**, 24.
112. C. Li, P. Wang, Z. Wei, D. Liang, C. Liu, L. Yin, D. Jia, M. Fu and F. Ma, *J. Pineal Res.*, 2012, **53**, 298.
113. M. B. Arnao and J. Hernández-Ruiz, *J. Pineal Res.*, 2013, **55**, 149.
114. S. Pedrini, D. J. Merritt, J. Stevens and K. Dixon, *Trends Plant Sci.*, 2017, **22**, 106.
115. W. Wei, Q. Li, Y. Chu, R. J. Reiter, X. Yu, D. Zhu, W. Zhang, B. Ma, Q. Lin, J. Zhang and S. Chen, *J. Exp. Bot.*, 2015, **66**, 695.
116. K. Szafranska, R. Szewczyk and K. M. Janas, *Cent. Eur. J. Biol.*, 2014, **9**, 1117.
117. M. M. Posmyk, M. Balabusta, M. Wieczorek, E. Sliwinska and K. M. Janas, *J. Pineal Res.*, 2009, **46**, 214.
118. Y. Aguilera, M. Rebollo-Hernanz, T. Herrera, L. T. Cayuelas, P. Rodríguez-Rodríguez, Á. L. L. De Pablo, S. M. Arribas and M. A. Martin-Cabrejas, *Food Funct.*, 2016, **7**, 1438.

CHAPTER 7

Non-nutritional Factors: Lectins, Phytic Acid, Proteases Inhibitors, Allergens

C. CUADRADO*[a], K. TAKACS[b], E. E. SZABÓ[b] AND M. M. PEDROSA[a]

[a]Departamento de Tecnología de Alimentos, SGIT-INIA, Ctra la Coruña km 8.5, Madrid, 28040, Spain; [b]National Agricultural Research and Innovation Centre, Food Science Research Institute (NARIC FSRI), Herman Ottó út 15., Budapest, H-1022, Hungary
*E-mail: cuadrado@inia.es

7.1 Introduction

Nowadays, a balanced diet is considered essential for nutrition as well as for the treatment and prevention of disorders, mainly when such disorders are caused by excessive, insufficient, or unbalanced dietary intake.[1,2] Currently, the establishment of an optimum human diet is one of the most controversial subjects of discussion. A balanced diet should include all the necessary foods to preserve optimal nutrition. The dietary recommendations established by the FAO-WHO Group of Experts[3] are: protein should provide less than 15% of total calories, 60% from carbohydrates (3% of fibre) and 25% from lipids, 75% being derived from plant foods and 25% from animal foods. A plant-based diet can diminish the risk of several chronic diseases according to many epidemiological reports.[1,4] The legume family is comprised of

Food Chemistry, Function and Analysis No. 8
Legumes: Nutritional Quality, Processing and Potential Health Benefits
Edited by María Ángeles Martín-Cabrejas

Published by the Royal Society of Chemistry, www.rsc.org

a large number of species.[5] Their justified nutritional interest is due to the presence of adequate proportions of starch, protein, fibre, microelements and vitamins.[6] However, plant foods contain some antinutrient compounds, which can reduce nutrient assimilation, can sometimes cause adverse physiological effects or can even be toxic. Nevertheless, there is a great interest in the health value of legume consumption and the beneficial effects of antinutrients on cancer and coronary diseases prevention.[7] Seeds of legumes are rich in bioactive compounds with variable biochemistry. These compounds can be proteins (lectins, protease inhibitors, α-amylases), glycosides (vicine and convicine, α-galactosides), tannins, alkaloids or saponins.[8] Therefore, their extraction and quantification methodology are highly specific.[9–11] These bioactive compounds have diverse physiological effects, some being involved as energy reserve sources or in plant defence mechanisms against environmental conditions or predators.[12]

Existing data shows that their concentration, biochemical structure, exposure time and interaction with other dietary components influence the balance between deleterious and beneficial effects.[13] Thus, they may be considered as antinutrients and/or pronutrients with negative and/or positive effects on health, respectively.[7,14] Therefore, it is relevant to determine the amount and type of these compounds in the food and their influence on human health. A relevant challenge for the future is an understanding of the action mode of these bioactive compounds on the organism, as was emphasised at the 4th International Workshop on Antinutritional Factors in Legume Seeds and Oilseeds.[15] Foods are processed by thermal and non-thermal methods to increase their safety, quality, preservation, and suitability for product applications. The degree of processing influences the solubility and digestibility of nutrients and other related parameters, and consequently, it can affect the non-nutritive factors. Hence, the potential beneficial applications of these compounds, their synergistic effects and the effect of processing on such components is now the focus of scientific interest.[1,4]

This contribution is aims to review some of the main non-nutritive or bioactive compounds found in legumes such as lectins, phytates, protease inhibitors and allergens and the influence of thermal and non-thermal processing treatments on these non-nutritive compounds in legumes.

7.2 Lectins

Lectins or haemagglutinins, are glycoproteins competent to reversibly attach to particular glycoproteins and sugars on the cell surface in the gut epithelium, thus hindering nutrient absorption. This binding is manifested by *in vitro* agglutination of erythrocytes of several species.[16] Lectins are highly specific and particular types have singular interactions on agglutination, blood groups, digestion and toxicity. Dietary lectins have variable antinutritional effects, being toxic, interfering with the hormone balance and depleting nutrient reserves, which can lead to acute growth inhibition or even death. Purified lectins from soybeans or beans induced enlargement of the small intestine of rats, caused damage to the small intestine wall, impaired growth and stimulated pancreas hypertrophy in short-term experiments.[17] An interference with intestinal glycose transportation and a decrease in activities of maltase and invertase were also reported.[18] Bean lectins, at higher dose, induced depletion of body and skeletal muscle.[19,20] However, lectins can have positive actions by stimulating gut metabolism, amending obesity and restraining tumour growth.[21] Reduced tumour growth rate, in a dose-dependent way, was achieved by the inclusion of purified bean phytohemagglutinin in the diet of mice, which suggests a competition for nutrients between the gut wall experiencing hyperplasia and the developing tumour.[22] In rats with obesity, a decrease in insulin levels caused by lectins was related to the reduction of lipid accumulation by inclusion of raw kidney bean in the diet. Nevertheless, even at high levels no losses of body or muscle protein occurred, as with normal rats, which suggests a possible use of lectins in obesity treatment.[23] Lectins can be measured by their haemagglutinating activity and two haemagglutination assay methods are commonly used: native rat blood cells and trypsinized blood cells with increased sensitivity of the test. However, ELISA immunological methods, when possible, are being used due to their superior specificity.[24,25] Table 7.1 shows the measurements of lectin levels of several Spanish legumes by using both methodologies. The highest haemagglutination value, given in haemagglutination units (HU), was found in *Lens culinaris* (500.0 HU g kg^{-1}). A competitive indirect ELISA was carried out using standard lectin of soybean (SBA), lentil (LCA) and beans (PHA), and *Phaseolus vulgaris* var. Oracada, Curruquilla and Processor showed the highest values (12 and 25 g kg^{-1}) of all the legumes analysed.

Table 7.1 Content of lectin determined by haemagglutination assay and ELISA (g kg^{-1}) in different varieties of Spanish legumes.[a]

Sample	HU[b]	ELISA lectin
Glycine max var. Ostrumi	40.00 ± 0.00	4.80 ± 0.10
Lens culinaris var. Magda	500.00 ± 1.67	8.00 ± 0.27
Phaseolus vulgaris var. Oracada	40.00 ± 0.00	12.07 ± 0.73
Phaseolus vulgaris var. Corcal	3.00 ± 0.00	1.53 ± 0.43
Phaseolus vulgaris var. Gernikesa	24.09 ± 0.50	5.50 ± 0.50
Phaseolus vulgaris var. Tolosana	17.01 ± 1.50	6.49 ± 1.50
Phaseolus vulgaris var. Almonga	0.32 ± 0.00	0.67 ± 0.00
Phaseolus vulgaris var. Curruquilla	4.00 ± 0.00	12.87 ± 0.79
Phaseolus vulgaris var. Processor	20.41 ± 0.03	25.25 ± 1.90
Pisum sativum var. Chicarrón	0.32 ± 0.00	N.A.[c]
Pisum sativum var. Cartouche	0.24 ± 0.08	N.A.
Pisum sativum var. Luna	0.63 ± 0.00	N.A.
Vicia ervilia var. Villanueva	0.48 ± 0.16	N.A.
Vicia ervilia var. Taranto	0.63 ± 0.00	N.A.
Lathyrus cicera cv. ZL-02	1.28 ± 0.00	N.A.
Lathyrus cicera cv. ZL-41	0.48 ± 0.16	N.A.

[a]*Ph. vulgaris* var. Processor and Pinto are included in each assay as standards.
[b]HU (haemagglutination unit) is defined as the amount of material in the last dilution at which 50% of the cells were agglutinated.
[c]N. A. not analysed.

Malting is a special type of germination that includes a germination period (24–48 h) that modifies the chemical composition of the seeds. Trugo *et al.*[26] studied the influence of malting for 24 and 48 h on lectin composition in soybean and black bean. The results showed that the malting process significantly reduced the PHA content of black bean. After 24 h of malting the original PHA content (0.166 mg per 100 mg) was reduced by 46% and a decrease of 76% was reached when malting continued for up to 2 days. Worthy of note is the parallel decrease in the concentration of lectin estimated by haemagglutination assay. The influence of malting process on lectin content of soybean showed a comparable trend even though there were not significant differences between raw and malted samples. Therefore, the influence of malting on the lectin concentration could be attributed to the germination carried out in that process. However, the results show a more rapid disappearance of haemagglutinating activity after 2 days of malting and therefore, it cannot be ruled out that other factors involved in the malting process could have an additional effect on the lectin content.

The influence of germination on the lectin content has been analysed in germinated seeds of *L. culinaris* under different settings of watering, time and light.[27] The seed's lectin was quantified by competitive ELISA and assayed by haemagglutination. SDS-PAGE and immunoblotting were used to test the lectin modifications during germination. Regardless of the conditions of the germination, lectin content and haemagglutinating activity of the seeds was not changed during the first 3 days. Nonetheless, after 6 days of germination, lectin concentration was significantly higher; particularly when germination was carried out in the light and with daily watering (Figure 7.1). The SDS-PAGE and immunoblotting results confirm that the lectin was not broken down during the first 6 days of germination although other storage-proteins were degraded by proteolysis.

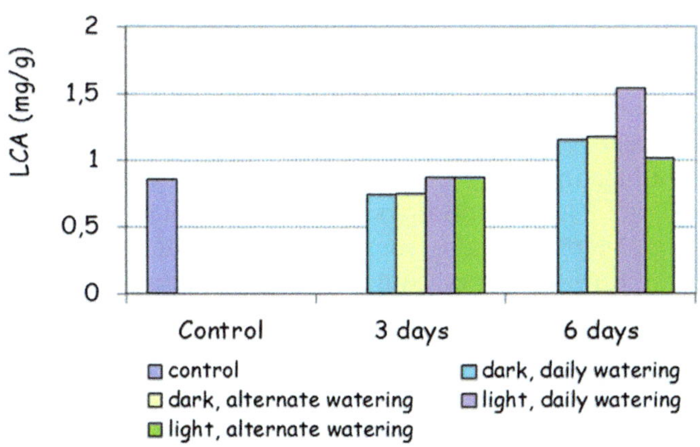

Figure 7.1 Effect of germination upon lectin concentration (LCA) of lectin seeds measured by indirect ELISA.

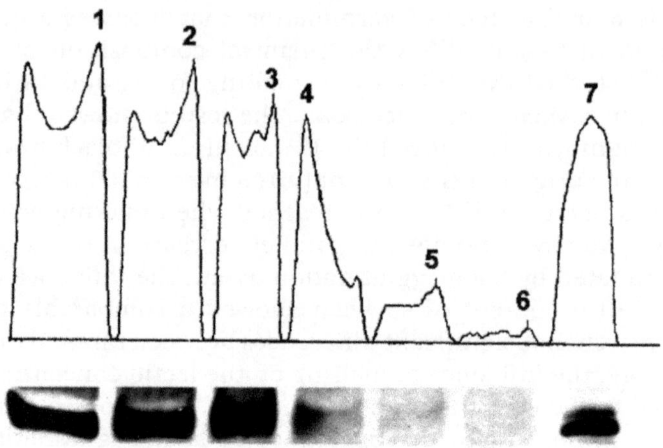

Figure 7.2 Densitometric measurement of the protein band corresponding to the β-chain (18k Da) of lectin lentil (LCA) on the basis of immunoblotting profiles of raw and fermented lentil samples after SDS-PAGE separation. The proteins were reacted on nitro-cellulose membrane with anti-LCA rabbit IgG. Raw lentil, 100% relative LCA content (peak 1): fermented lentil samples: 0 h, 77% relative LCA content (peak 2); 24 h, 58% relative LCA content (peak 3); 48 h, 37% relative LCA content (peak 4); 72 h, 13% relative LCA content (peak 5); 96 h, 3% relative LCA content (peak 6); and lentil lectin standard, LCA (peak 7).

Fermentation is a procedure used for conservation and production of different foods because of their high nutritive value and organoleptic characteristics. The consequences of natural fermentation on the lectin in the seeds of *L. culinaris* cultivar Magda 20 have been studied.[28] Lentil flour suspensions were allowed to ferment naturally at different concentrations of lentil flour (79, 150 and 221 g L^{-1}) and temperatures (28, 35 and 42 °C). Samples were taken at daily intervals along the fermentation (0, 24, 48, 72, and 96 h) and lentil lectin activity was determined by haemagglutination assay. In all the batches, there was a quick decay of the haemagglutination activity with the progress of fermentation. In all the conditions, the major decrease was between 0 h and 24 h of fermentation and the maximum reduction was achieved at 72 h and 96 h (98% and 97.8%, respectively), at the fermentation settings of 79 g L^{-1} and 42 °C. The lentil lectin changes were also monitored by SDS-PAGE and immunoblotting (Figure 7.2). The findings indicated that the lectin practically disappeared after 72 h of natural fermentation under the optimal conditions of flour concentration and temperature and confirmed those obtained by ELISA.

7.3 Phytic Acid

Phytic acid or myo-inositol-(1,2,3,4,5,6) hexakisphosphate, and its salts (called phytates), are the principal phosphorous storage molecules in legume seeds[29] and are formed during seed maturation representing

60–90% of their total phosphorous amount.[30] In general, legumes contain higher amounts of phytates than cereals and nuts and they are stored as globoid crystals in the aleurone cell layer of the seeds embryo.[31] Different authors[32,33] reported that the total inositol phosphates (IP) content (% dry weight basis) present in legumes is lower than 2%, although the climate conditions, and soil type in which seeds were grown and the variety all affect the IP content.[33] Muzquiz *et al.*[12] reported the total IP content in some Spanish legumes. The lower IP amount corresponds to *Lupinus mariae-josephi* (0.3%), and the higher content corresponds to *Glycine max* var. Ostrumi (2.0%). In *Vicia* genus, the content ranged from 0.6 to 0.9% in *V. narbonensis* and *V. faba*, respectively. Chickpea IP content ranged from 0.6 to 0.7% in different varieties, and for pea the amount ranged from 0.4% (var. Luna) to 0.6% (var. Cartouche). *L. culinaris* contained around a 1.2% of IP. Muzquiz *et al.*[34] studied the IP content in 19 Spanish dry bean varieties grown in four different locations; the total IP contents showed significant differences among varieties. Oboh *et al.*[35] studied the IP content in different underutilised legumes grown in Nigeria, and their content ranged from 0.15% in African yam bean to 0.47% in red lima bean. Analysis of inositol phosphates by HPLC methodology allows the determination of not only phytate (IP6) content but also the different inositol phosphates molecules called myo-inositol tri-, tetra-, and penta-phosphates (IP3, IP4 and IP5).[33] The main IP form in legumes is IP6, followed by IP5 and some legume seeds did not present the lower forms (IP4–IP1).[12,33] According to different authors the relative proportion of IP5 in soybeans ranged from 7.6 to 17.1% of the total IP content, from 3 to 12.1% in lupines, from 6.8 to 8.5% in peas, from 5.9 to 23.9% in chickpeas and, from 7.5 to 24.0% in common beans.[12,34,36] In lentils and faba beans the IP5 content represented around 25% of the total IP. In relation to IP4, the lowest relative proportion was present in *Pisum sativum* var. Iceberg (0.8%), and the highest amount correspond to *Cicer arietinum* var. Athenas (9.4%) and *L. culinaris* var. Magda (9.6%).

The bioavailability of legume minerals is reduced due to the presence of myo-inositol phosphates that make them unavailable for digestion. Excessive IP6 in the diet can negatively affect the mineral balance due to the insoluble complexes that it forms with essential mineral (Cu^{2+}, Zn^{2+}, Fe^{3+} and Ca^{2+}). Iron and zinc deficiencies are reported as highly prevalent nutritional problems in the world.[37] Phytate is also able to complex with proteins, starch and digestive enzymes such as lipase, α-amylase, pepsin, trypsin and chymotrypsin, and this complex can impair the solubility, digestibility and absorption of nutrients.[12,38] The smaller molecules (IP3–IP4) have less antinutritional effects than IP6 and IP5 due to these smaller molecules having a lower ability to complex with inorganic cations, proteins or enzymes. Although, Sandberg *et al.*[39] reported that, IP3 and IP4 in the presence of IP6 in breads also negatively affect iron absorption, so it is of great interest to determine the content of all the IP forms (IP3–IP6). There is no recommended daily intake of phytate, but depending on plant consumption in the diet and the type of food processing that takes place, the daily intake of inositol phosphates can reach 4500 mg. On average, daily intake of IP6 has been estimated at around 2000–2600 mg for vegetarian diets, whereas this intake only reached 150–1400 mg of IP6 for mixed (plants and animals) diets.[38] In a Mediterranean based diet, for individuals who consumed around 1–2 g of phytates per day no adverse effect on mineral bioavailability was reported.[40]

However, studies from the 1980s to present have shown that consumption of inositol phosphates does not appear to have only negative effects on human health. There are some healthy effects associated with inositol phosphates, such as lowering of cholesterol and triglycerides, reduction in kidney stone formation (quelating Ca^{2+}), reduction in the risk of a variety of cancers, improvement in the management of diabetes mellitus (regulating pancreatic β-cells) and strengthening of the immune system (enhancing natural killer cells).[12,38,40] Phytic acid is also well known as a natural antioxidative agent due to its ability to chelate Fe^{3+}, avoiding the formation of free radicals than can cause cell injury or cell death. Some diseases such as Alzheimer's and Parkinson's, cancer, arthritis and cirrhosis have been related to the oxidant activity of free radicals.[31,40]

As for other bioactive compounds, the positive or negative effect of IP depends on the amount present in the foods, time of consumption and how they interact with other dietary compounds. The content of some components might be positively or negatively affected by type of food, the cooking method and time, the addition of species, acidulants, antioxidants such as ascorbic acid, among other factors. Traditionally, thermal treatments (boiling, cooking, microwave or extrusion) have been applied to improve not only the nutritional values of pulses but also to reduce, or possibly eliminate, the adverse effects of antinutrients.

7.3.1 Effect of Non-thermal Processing on Inositol Phosphates Content

Soaking and germination are the most common non-thermal treatments used to process pulses, increasing their nutritional value. Soaking in a water solution at a certain temperature (from 4 to 80 °C) for a period of time (from 15 minutes to days) can also be used as a pre-treatment to cooking legume grains since it reduces the cooking time. The impact of the soaking process depends on time, temperature and pH of the soaking water. Fredrikson *et al.*[41] reported a decrease in IP6 of about 85% after pea soaking for 10 h at pH 7.5 and 45 °C. Soaking for 9 h at 30 °C significantly reduced the total IP of white lima bean and cream pigeon pea by about 26–45%, and soaking African yam bean and jack bean for 12 h reduced the IP content by 45–63%.[42] A pre-existing phytase located in the bean endosperm activated during soaking could be responsible for this reduction. As phytate is a water-soluble molecule, discarding the soaking water can reduce significantly the IP6 content.[37,43] Soaking lupine seeds in salted water (0.2%) overnight reduced the phytic acid content by 12% compared to raw samples.[44]

Germinated legumes showed improved palatability and nutritive value. Traditionally, pulse germination is carried out by soaking the seeds in tap water, discarding this water and letting the seeds sprout for a variable time (24–96 h). During this soaking period some biochemical modifications, such as the activation of phytases, amylases and proteases, allow the reduction of antinutritional factors.[37] In general, the reduction of inositol phosphates increased with the germination time, although some authors did not find a significant reduction of phytates with germination, probably due to a lower seed:water ratio.[45] The effect of different environmental conditions (light and watering) on phytic acid content of *L. culinaris* was studied.[46] Total IP decreased under all different conditions assayed from the third germination day; a strong IP6 reduction was observed after this period, reaching 80% after 6 days of germination in the dark and with watering on alternate days. Germination produces a slight increase of the IP3 and IP4 content, so, phytate reduction can improve mineral uptake after 6 days of germination. The phytase activity naturally present in plants hydrolyses phytate during germination since phosphorous is needed for the growth of the new plant. In addition, the smaller forms of IP might be dephosphorylated by phytase and phosphatases, which do not used IP6 as a substrate.[38] In germinated faba beans up to 9 days, a pronounced increase in phytase activity from 4 days of germination onwards with a concurrent reduction in IP6 content (3–14%) was reported.[47] The IP content of some African legumes after 4 days of germination was reduced by between 34 and 85%.[42] Aguilera *et al.*[48] reported the changes in IP content during up to 4 days germination in nonconventional legumes (cowpea, jack bean, dolichos and mucuna). The process caused a significant reduction in the IP3–IP6 contents in all seeds except for dolichos. The highest reduction corresponded to cowpea (18%), followed by mucuna and jack bean (15%). Therefore, germination improves the mineral bioavailability of different legumes compared to the raw ones. Wang *et al.*[49] observed that IP6 content decreased in mung beans by 76.0% and in germinated soybeans by 57.5% with an increase in the Zn and Fe bioavailability values. Trugo *et al.*[26] reported the effect of malting on IP content of black bean, soybean, chickpea and barley. Phytic acid was the main IP form and 2 days of malting produced reductions of around 25% and 15% in the sum of IP5 and IP6 in black bean and chickpea, respectively, with the concomitant formation of higher amounts of the smaller forms (IP3 and IP4), suggesting a better mineral bioavailability for these seeds.

Many legumes are included in the formulation of a variety of fermented foods, mainly soybean derivatives (such as soy sauce, miso, tempeh, *etc.*). The extent of IP6 reduction during the fermentation depends on the raw material, the fermentation conditions and the type of microorganism used. Natural fermentation of lentils var. Magda produced a significant reduction of total IP after 96 h of lactic acid fermentation because of the decrease of IP6 and IP5, the main compounds found in the raw lentil. The greatest modifications were observed in the first 24 h. Authors conclude that although the endogenous phytase would be active during the fermentation process, the observed reduction is mainly due to the bacterial phytases.[50] Natural fermentation of lupine seeds (*L. albus*) for 2 days reduced significantly (18%) the phytic acid content. Fermentation of different pulses (cowpea, kidney bean, pea) with *Saccharomyces cerevisiae* (3 g per 100 g of seeds) for 24 h at room temperature was more effective in reducing IP6 content (66.8–68.9%) than other treatments such as soaking (42.8–48.9%) or roasting (35.2–40.1%),[51] probably due to the combined action of both endogenous and yeast phytase.

7.3.2 Effect of Thermal Processing on Inositol Phosphates Content

Cooking, with or without pressure, is the most common method for pulse processing. Taking into account that phytate is thermally stable, a significant reduction in this molecule by cooking treatment is not expected to take place. However, as a soaking pre-treatment is usually applied prior to boiling seeds, discarding the soaking water help to reduce phytate content. Not all the studies found the same effect of heat treatments. Some authors reported a significant phytate reduction in both cooked and autoclaved chickpeas, although other authors did not find a significant reduction in phytate content after traditional, microwave and pressure cooking of chickpeas.[45] Chen[52] reported a reduction in IP6 content in different dry beans that ranged from 11.2% to 29.8% in cooked mung beans and pink beans, respectively. Oboh *et al.*[42] described a higher decrease in the IP amount in some Nigerian legumes when seeds were cooking under alkaline conditions. Pedrosa *et al.*[53] reported changes in the inositol phosphates (IP3–IP6) content in industrial canned beans. This process (soaking and boiling with pressure) reduced significantly the phytic acid content (around 50%), whereas it increased three times the contents of IP5 and IP4 in comparison to the raw samples, and IP3 was only detected in the canned samples. The IP reduction was probably due to leaching during the soaking and cooking processes or by activation of endogenous phytases during the soaking period. Different authors reported a similar or slightly higher reduction for other cooked beans, chickpeas, lentils and peas.[54,55] Autoclaving different legumes (cowpea, kidney bean and pea) at 121 °C for 20 min was very effective in reducing the IP6 content by about 65.0–70.4%, probably due to the formation of insoluble complexes between phytic acid and other compounds during the heat treatment.[51] Martín-Cabrejas *et al.*[54] studied the effect of combined treatment (soaking, cooking and industrial dehydration) on the inositol phosphates content in two chickpea varieties, lentils and two bean varieties. A general decrease of total IP content (44%) was observed in the combined treatment and the higher reduction was found in lentils. Chickpea varieties reduced slightly their IP content levels in comparison to raw seeds. The white bean variety showed higher IP reduction (14%) than the pink-mottle cream variety (6%). Thus, the dehydrated lentils would have a better mineral bioavailability in comparison to the other legumes studied.

DIC® (Instant Controlled Pressure Drop Treatment) is a processing technique that combines steam pressure (up to 8 bars) with heat (up to 170 °C). The effect of this process on the inositol phosphates content (IP3–IP6) was evaluated in lupine, chickpea, soybean, lentil, and roasted peanut.[11] DIC treatment produced a reduction in IP content but not all seeds were affected to the same extent. For example, the 3 bar, 3 min treatment decreased the total IP content by about 12% in peanut, 13% in soybean, 27% in chickpea, 37% in lentil and 41% in lupine. In soybean and peanut an increase in the content of IP4 and IP5 was observed, probably due to phytic acid dephosphorylation; however, lupine, chickpea and lentil showed a reduction of all inositol phosphate forms. Haddad *et al.*[56] reported a 55% reduction in IP6 content in *L. albus* and 60% in *L. mutabilis* by DIC treatment at 7 bar and 7 minutes.

Extrusion, a high-temperature, short time procedure, is widely used by the food industry to produce different types of foods. It is well known that the effectiveness of extrusion processing in changing nutritional and phytochemical composition depends on the parameters used, mainly temperature and moisture.[57] Alonso *et al.*[58] reported a reduction of 20.5% in extruded pea flours. Varela *et al.*[59] studied the effect of extrusion in different blends of pulses/corn starch (80/20), and they observed that the reduction of total IP was around 8% in kidney bean blends, 8.5% in chickpea blends and 35.8% in faba bean blends after extrusion at 150 °C. Morales *et al.*[60] also described an overall reduction in the total IP content (14–50%) in different extruded lentil flours blends at 160 °C. However, Lombardi-Boccia *et al.*[61] reported that legume phytate content was not affected by extrusion probably due to the mild extrusion conditions used.

The extent of inositol phosphate hydrolysis through food processing depends on the type of pulse used, the treatment and the amount of inositol phosphates present in the seeds. None of the treatments applied to legumes completely remove the inositol phosphates from the legume-based foods, but the reduction obtained after processing can improve the mineral bioavailability, and at the same time the remaining amount of inositol phosphates present in the food could provide health benefits for individuals.

7.4 Protease Inhibitors

Protease inhibitors (PI) are widely distributed in legumes. They are considered as antinutritional compounds hampering protein digestion, due to their ability to irreversibly inhibit, if not appropriately inactivated, the action of digestive enzymes. Nevertheless, after inactivation, protein inhibitors can in fact play a beneficial nutritional role, because of their high content of sulphur-containing amino acids in comparison to other seed proteins. Protease inhibitors are concerned in the proteolytic self-regulation process before and during the seed germination; they have a regulatory function by inhibiting endogenous proteases. Protease inhibitors also support defence mechanisms of the seed against insects and micro-organisms (bacteria, fungi). This natural role is being exploited in the development of insect resistant transgenic plants.[62]

Among legumes, protease inhibitors of soybean seeds (*Glycine max*) were first isolated by Kunitz and Bowman in 1945 and 1946, respectively. Both inhibitors, also described as serine protease inhibitors, or serpins, contain a small amount or no carbohydrates; they are water soluble proteins (albumin) and make up 0.2 to 2% of total soluble proteins of legumes.[12] Soybean Kunitz trypsin inhibitor's (KTI) parameters are the followings: 20.1–21.5 kDa, globular structure, monomeric protein containing 181 amino acid residues in a single polypeptide chain crosslinked by two disulphide bridges, 6% α-helix, 40–60% β-sheet (Figure 7.3).

These inhibitors are one-headed, namely one molecule of inhibitor inactivates one molecule of trypsin.[63] KTI inhibits trypsin of various species, *e.g.*, human, bovine, avian and fish.[64,65] Trypsin inhibitors are almost always capable of binding to lysine and arginine residues in trypsin. They are a competitive inhibitors, binding to the active sites of trypsin resulting in the hydrolysis of peptide bonds between amino acids of the reactive site of the inhibitor or the substrate. The soybean Kunitz trypsin inhibitor is an allergenic protein highly resistant to thermal and chemical denaturation.[66] Kunitz-type PIs were discovered to operate as inducers of IgE-mediated allergic reactions.[67]

Soybean Bowman–Birk inhibitor's (BBI) parameters are the followings: 8 kDa, globular monomeric molecule, containing 71 amino acid residues, 0% α-helix, 60% β-sheet (Figure 7.4). They are double-headed inhibitors due to their ability to inhibit simultaneously, but independently, two, either the same or different, enzymes, thus trypsin/trypsin, trypsin/chymotrypsin.[68] BBI inhibits human, bovine, porcine and avian trypsin and chymotrypsin, at a 1 : 1 enzyme : inhibitor ratio.[69] It has been found that the double headed BBIs have reduced affinity to bind proteases when one site is already engaged. Peanut (*Arachis hypogaea*) inhibitor has been found to show activity against chymotrypsin when it is previously engaged with trypsin and *vice versa*.[70]

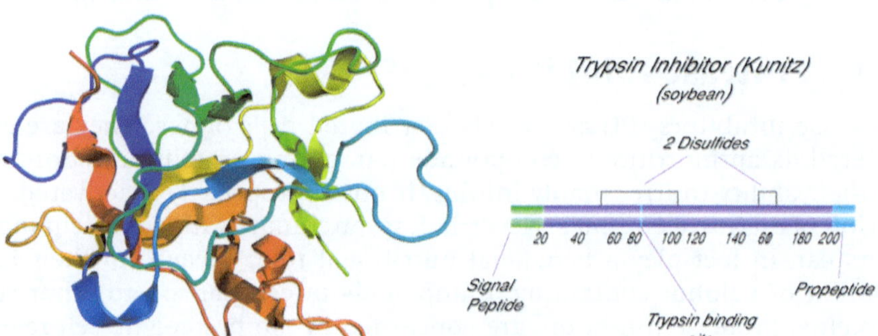

Figure 7.3 Soybean Kunitz trypsin inhibitor (KTI).

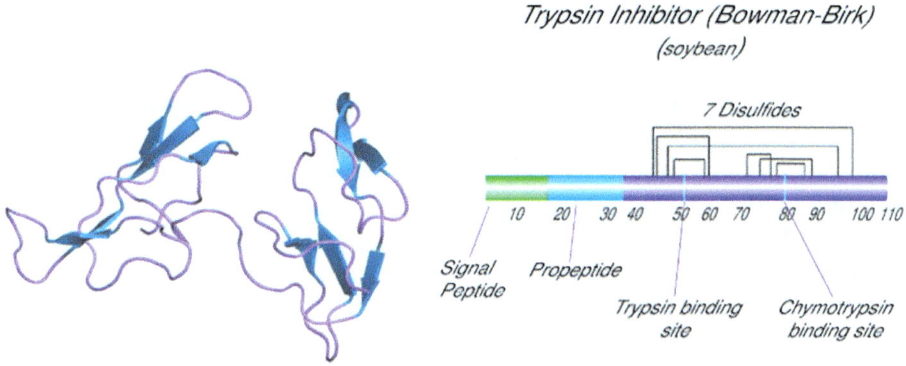

Figure 7.4 Soybean Bowman–Birk trypsin/chymotrypsin inhibitor (BBI).

Similarly, the activity of soybean BBIs are reduced 100-fold when trypsin is attached at the other site.[71]

The Bowman–Birk inhibitor from soybeans operates as an example of a family of inhibitors that are predominant in legume seeds; these proteins are the products of multi-gene families. Proteins of the Bowman–Birk class of inhibitors (isoforms with different properties depending on their chemical structure) have been characterised in the seeds of many leguminous species.[12] For instance, protease inhibitors from *P. sativum* (cv Frilene) include up to six isoforms.[72] The amino acid sequence of the main soybean BBI-type isoinhibitors shows high sequence homology. The expression of different genes and the post-translation of primary gene products, which mainly arises during desiccation of the seeds, is responsible for the extensive range of isoinhibitors. Substantial differences in the activity of protease inhibitors can be also due to legume species and varieties, cultivation area and conditions, harvest and treatment.[73,74] Isoinhibitors can show differences in their action against human and bovine proteases. For example, in case of the four isolated lentil BBI inhibitors, human trypsin was inhibited less (about 75%) than bovine trypsin, and human chymotrypsin was inhibited much more (about 250%) than bovine chymotrypsin.[11] Generally, the highest TIU (tripsin inhibitor units) mg^{-1} sample values were found in soybean (43–84) and common bean (21–25). The inhibitor levels of *Lathyrus* cultivars, varied from 19–30 TIU mg^{-1} sample. In chickpea (15–19 TIU mg^{-1} sample) and pea (6–15 TIU mg^{-1} sample) the values were lower and the lowest ones were found in cultivars of lentil and faba bean (3–8, and 5–10 TIU mg^{-1} sample, respectively).[71]

7.4.1 Inactivation of Protease Inhibitors

The activity of enzyme inhibitors were reduced, but were not totally eliminated by different food processes including dehulling, soaking, germination and heat treatments: boiling, autoclaving, toasting (wet heat treatment, 10–20 min, 110 °C), micronization, microwave cooking, dry roasting and more recently extrusion cooking (145 °C, 16 seconds).[66] These processing methods are not appropriate for industrial production of legume seeds, the only exception being the large-scale industrial production of heat-treated soybeans for the oil extraction process. However, heat treatment should be carried out under controlled conditions to avoid damage to the amino acids, especially lysine. For this purpose, instant controlled pressure drop (DIC) treatment proved to be effective for PI reduction and the optimal condition for all the seeds was 6 bar, 1 min. The main benefits of DIC are its short processing time and the ability to process whole seeds for industrial uses.[75] Trypsin inhibitor content has also been reduced by fermentation.[76] It is important to mention that cultivation of soybean species with KTI-free or reduced trypsin inhibitor content has been achieved.[77]

7.4.2 Nutritional and Physiological Effects of Protease Inhibitors

In terms of animal and human nutrition, most of the research regarding the nutritional effects of enzyme inhibitors has resulted from experiments with animals. The action of trypsin inhibitors on the human body is not well comprehended. Human trypsin has two forms: cationic, the main component of pancreatic juice, which is weakly inhibited; and anionic, about 10 to 20% of the total trypsin, which is totally inhibited.[78] The plant protease inhibitors restrain the action of the digestive enzymes (serine proteases such as trypsin and chymotrypsin) by competitive binding, so they diminish the action of the digestive enzyme in the lumen. When the duodenal content of trypsin/chymotrypsin is decreased below a certain level, the endocrine cells release the hormone cholecystokinin (CCK) in the duodenum, stimulating the pancreas to synthesize more serine-proteases (it means that KTI and BBI act by suppressing the negative feedback regulation mechanism of pancreatic secretions).[79] The KTI, BBI and the lectins interfere with neuroendocrine regulation through the release of CCK and/or other peptide hormones of the gut thus prompting the growth of the pancreas (hyperplasia). This can lead to carcinogenic processes.[80,81] The chronic hypersecretion of pancreatic enzymes (sulphur-rich trypsin and chymotrypsin) leads to the loss of the sulphur-rich amino acids that are used to synthesise tissue proteins and in the synthesis of these enzymes. This, coupled with changes in systemic metabolism, leads to a poor/negative overall nitrogen balance[82] and to a lower amount of essential amino acids, which decreases animal growth and impairs a critical state in relation with the protein of legumes, which are generally deficient in sulphur amino.[83,84] Trypsin inhibitors produced irrelevant growth depression in poultry and rats, but provoked pancreas enlargement (hypertrophy and hyperplasia) in poultry, rats and mice and hypersecretion of digestive enzymes.[85] Considerable amounts of BBI inhibitors are resistant to gastrointestinal hydrolysis in the swine, which is generally alleged to be an appropriate model for human digestive physiology.[86] PIs were identified in the stomach in functionally active form. However, they were partially degraded in the small intestine and formed enzyme complexes with the gut proteases, which were immunologically detectable. Several data indicated that BBI inhibitors survived the digestion in a lower rate, while the KTI was stable in a higher proportion. Both inhibitors were localised in the blood after 15 min of exposure.[77,87–89] The heat-treated (*e.g.* extrusion, dielectric treatment, microwave treatment) animal soy meal feed with effectively reduced functional activity (trypsin inhibitor activity, haemagglutination activity, antigenicity) improved the protein efficiency ratio in animal feeding trials.

On the other hand, the occurrence of trypsin inhibitors in the diet has been associated with health-beneficial properties. BBIs present in different legumes can resist both acidic conditions and the action of proteolytic enzymes, and transit through the stomach and small intestine without major degradation, permitting significant amounts to reach the large intestine in active form to exert their reported anti-carcinogenic and anti-inflammatory properties.[90] Several patents on the use of BBIs to treat obesity and autoimmune diseases and, in general, skeletal muscle atrophy have been released.

7.5 Allergens

Legumes are one of the conventional foodstuff with capacity to provoke allergenic response. Peanut, soybean, pea, lentil, lupin, chickpea, blackgram, *etc.*, are the most common allergenic legumes. Various legume allergenic proteins are storage proteins and have been characterised as belonging to the cupin superfamily.[91] The major allergens of peanut and soybean have been widely investigated and one of the best characterised is the vicilin Ara h 1 (65 kDa glycoprotein).[92] Allergens from lentils and chickpeas are less studied. Until now, only three lentil and two chickpea allergens have been identified and characterised. Researching is on-going by various groups for identification of lupin allergens.[93,94] Several legume allergens have shown high resistance to food processing and digestion, hence intensifying their allergenic capacity.[95,96]

The genetic status of individuals, their consumption habits, and possibly other factors, may be involved in the variation of the legume sensitization profile in different parts of the world. The prevalence of peanut allergy is higher in the North America, United Kingdom and France, whereas soybean allergy is mostly reported in Asia. In the Mediterranean area, lentil, chickpea and lupin allergic reactions are more important.[97] In the European Union, the regulations indicate that it is mandatory to label 14 groups of allergenic foods, among which, peanut and soybean were the first to be included, and lately, lupin as well as its derived ingredients have been added.[98]

7.5.1 Effect of Thermal Processing on Legume Allergens

After processing, proteins can be aggregated, denatured, fragmented and re-assembled and these changes most often reduce solubility.[99] The global IgE binding profiles of a specific extract can become more or less allergenic or even result in new allergens (neoallergens) after processing.[100] Thus, it is necessary to study the influence of processing when assessing the allergenicity of current and newly introduced foods.[101] The thermal processing effect mostly depends on temperature and time used. Furthermore, interactions with other food matrix constituents can also alter the protein structure. On the whole, temperatures around 70–80 °C produce loss of secondary structure, whereas at 80–90 °C, rearrangements of disulfide bonds and formation of new ones takes place. Higher temperatures (90–100 °C) result in aggregates formation.[102] Protein digestibility is probably increased by heat processing, so the absorption of protein in the gastrointestinal tract can also increase, and because of this, the ability of allergenic protein to provoke an allergenic reaction decreases. Nevertheless, thermal treatment may reduce the digestibility of a particular allergen in some circumstances or even some neoantigens may be formed. These neoantigens may be an additional problem and, in sensitized patients, this global phenomenon can increase the allergenic problem. The Maillard reaction *i.e.*, the interaction of protein component with sugar residues upon heating, generating sugar conjugated protein derivatives, is considered one of the major factors responsible for the formation of neoantigens, and therefore a potential enhancer of the protein allergenicity.[103]

The epitopes present on allergenic proteins are recognized by the IgE antibodies. Two types of IgE binding epitopes are possible, linear or conformational. The amino acids are linearly ordered in the polypeptide in the case of linear epitopes whereas in the conformational ones, distant amino acids in the primary sequence may approach during polypeptide chain folding. Linear epitopes are more resistant to heat treatment and so they can be more problematic as compared to the conformational ones. Heat treatment mostly affects conformational epitopes as the bonds may be broken down by heating. Alterations in protein folding have the potential to affect the stability of digestion, and hence the form in which allergens exist.[104] Consequently, thermal treatment as well as other processing procedures can considerably modify the structure, function and allergenicity of foods. Incomplete awareness of the allergenic proteins from processed foods gives rise to difficulties on the diagnosis of food allergies.

Peanut allergy is one of the most serious IgE-mediated reactions to food due to its severity and lifetime persistence.[105] Nowadays, eleven peanut allergenic proteins have been identified (Ara h 1-Ara h 11). Thermal treatment has a significant effect on peanut immunoreactivity. Studies with peanuts[103,106] established that roasted extracts bind IgE approximately 90-fold higher than raw peanuts. These findings demonstrated that the major allergenic proteins, Ara h 1 and Ara h 2, from roasted peanuts have undergone structural changes that increase their allergenic characteristics. In contrast, boiling decreases peanut allergenicity.[107] In accordance with Cabanillas *et al.*[108] autoclaving under extreme conditions (2.6 bar, 30 min) significantly decreased the IgE immunoreactivity of roasted peanuts. DIC treatment at 6 bars for 3 min caused a remarkable decrease of putative Ara h 1 (65 kDa) and no immunoreactive bands lower than 20 kDa were detected. Nevertheless, other bands, such as, at 37 kDa, possibly the Ara h 3 acid subunit, behave in a different way in raw and roasted peanuts. The DIC technique seems to be more effective with roasted peanut proteins immunoreactivity than with raw peanuts.

White lupin is a rich source of protein with a notable content of lysine and is being progressively used due to its functional properties and its likely hypocholesterolemic and hypoglycaemic actions.[109] On the other hand, lupin seed flour has been described as a relevant cause of allergenic reactions, mainly in patients with allergy to peanut.[110] Álvarez-Álvarez *et al.*[111] concluded that autoclaving at 2.6 bar (138 °C) for 20 min produces a relevant reduction in lupin allergenicity. Only two proteins maintain IgE-binding capacity after autoclaving for 20 min, at 23 and 29 kDa, while autoclaving for 30 min totally abolished IgE binding of these proteins, which suggest that lupin allergens are moderately heat stable. Later on, two main allergens were characterised as Lup 1 (conglutin β, 7S protein, 34.5 kDa) and Lup 2 (a basic subunit of conglutin α, 11S protein, 20 kDa).[93] Both were partly sequenced and the high homology level shown with major allergens from lentil, peanut, soybean and pea could explain their IgE cross-reactivity with these legumes. Guillamón *et al.*[112] analysed the influence of DIC on lupin allergenicity and confirmed the thermal resistance of the major lupin allergens previously reported from autoclave treatments.[111] Both findings suggest that processes combining heat and steam pressure, such as DIC and autoclave, on lupin seeds could nearly remove *in vitro* immunoreactivity of lupin. Moreover, the breadmaking properties of this hypoallergenic lupin flour has been studied[113] and the results showed that thermally-treated lupin flour had comparable breadmaking and sensorial properties to untreated lupin flour.

Lentil has also been described as a cause of IgE-mediated inmunoreactivity. Several IgE binding allergenic proteins have been detected in raw as well as boiled lentils although investigations on lentil allergenicity with well documented clinical sera are limited. Len c 1 (48 kDa vicilin)[114] and Len c 2 (66 kDa) are the two major lentil allergens that were first isolated from boiled lentils.[115] More recently, Len c 3, has also been characterised.[116] The effect of boiling on lentil does not appear to be effective in decreasing its allergenic capacity. Although autoclaving using high pressure (2.6 bar) for 30 min has a prominent effect on minimizing the global IgE-binding ability, particularly resistant immunoreactive proteins still persisted in this legume even after this severe process.[117] When the DIC-treatment effect on lentils has been studied,[118] authors observed that DIC processing under extreme conditions (6 bar, 3 min) produced a reduction in IgE-binding proteins comparable to that observed with autoclaving in harsh conditions. After such DIC processing conditions only the heat-stable allergenic proteins were still present (MW 29, 48, 57 ad 68 kDa). These findings corroborated the thermostability of some lentil allergens similarly to that detected with autoclave processing.

In several parts of the world, chickpeas are an essential protein source, mainly in Asian regions where they are widely consumed. In the Mediterranean area, the high intake rate of chickpeas has resulted in immunoreactivity problem for sensitised subjects.[97] In Spanish children, chickpeas and lentils are the most frequent cause of allergic reactions to legumes and -reactivity between both occur.[119] Symptoms have been reported when some individuals allergic to chickpeas inhale vapors from cooking this legume.[120] Multiple IgE-reactive bands in chickpea extracts (10–106 kDa) were previously detected, mostly heat stable.[121] So far, allergens of chickpea have not been immunologically typified, and only two immunoreactive polypeptides (2S albumin and 11S globulin) from chickpea have been characterised.[122] Cuadrado *et al.*[117] detected several IgE-binding proteins in chickpea boiled for 30 min. A reduction in the band number and intensity was found after autoclaving (1.2 bar, 12 min) and the immunoreactivity was decreased as pressure and time increased. Only two polypeptides (19 and 16 kDa) were still antigenic at 2.6 bar (30 min). The influence of DIC treatment on chickpeas and lentil allergenic proteins was similar. After DIC treatment at 6 bar for 3 min, the immunoreactive polypeptide profile also showed a marked decrease in the number and intensity of IgE-binging proteins. Nevertheless, when chickpea is analysed under different applied pressure and time conditions, no apparent distinctions were detected and more heat-stable IgE-binding proteins were still detected at the same extreme DIC conditions (6 bar, 3 min).[118]

Soybean is extensively consumed throughout the world, being mainly used as an ingredient in formulated foods, meat, bakery and pastry products, apart from its pharmaceutical and industrial applications.[123] The consumption of soybean has also been associated with high risk of allergy for consumers and the prevalence rate of allergy in the general population is around 0.3–0.4%. Around 21 known allergenic proteins have been described in soybean and various major allergens have been characterised, *i.e.*, P34 (Gly m Bd 30 K), Gly m 1, Gly m 2, Gly m 3, Gly m 4 and Gly m Bd 28 K.[94] Because of the commercial use of soybean protein based foodstuffs, which frequently comprise thermally processed proteins, it is of high importance to look into the effect of thermal treatment on the major allergenic proteins of soybean. Burks *et al.*[124] did not show any relevant reduction in IgE-binding after heating soy proteins at various temperatures and time intervals. Wilson *et al.*[125] concluded that various techniques are needed to eliminate soybean allergenicity, particularly P34, the major allergen. While DIC treatment at 3 bar for 1 and 3 min caused fewer soybean immunoreactive bands, the increase of pressure up to 6 bar for 3 min did not produce any immunoreactive proteins.[118] Comparable findings were obtained when autoclaving was applied (2.6 bar, 30 min) on this legume.[126] The DIC treatment applied has the strongest influence on the soybean antigenic proteins compared to the other legumes analysed. The short time of processing (3 min) is beneficial for future possible uses in the food industry. Removal of allergens through processing could eventually improve the safe use of soybean foodstuffs, making them suitable for soybean sensitised subjects.

7.5.2 Effect of Enzymatic Digestion on Legume Allergens

The effects of enzymatic hydrolysis on the allergic reactivity and digestibility of food proteins have been evaluated by a number of different studies. Hydrolysate of cow's milk formulae were developed to reduce its allergenicity.[127,128] Modification of IgE-binding of enzymatically hydrolysed legumes has been tested mainly in soybean[129,130] and peanut but other legumes are less studied.[130] The reduction in immunoreactivity to chickpea allergenic proteins after digestion applying the sequential action of Alcalase (endoprotease) and Flavourzyme (exoprotease) was evaluated by Clemente *et al.*[131] The IgE-binding capacity decreased by more than 90%. Additionally, the high protein quality of this sequential hydrolysate is an interesting food product.[132] The influence of trypsin and pepsin on the *in vitro* reactivity of 11S globulin of lentil, chickpea, and lupine, using rabbit antiserum for these proteins was investigated by Sormus de Castro Pinto *et al.*[133] A large amount of short peptides and free amino acids were obtained after enzymatic hydrolysis. The evaluation of enzyme activities in order to reduce allergenicity is relevant and such assessment should be carried out using sera from individuals with well documented clinical allergy to the legume source.[134] Enzymatic protein hydrolysates represent an alternative option to intact proteins in the production of special formulations conceived to give nutritional support to population groups with specific needs, such as food-allergic individuals, infants and the elderly. Moreover, the extensive enzymatic digestion combined with post hydrolysis food processing treatments, such as thermal procedures, ultrafiltration, *etc.*, is contemplated as the most effective method to achieve protein foodstuffs with reduced antigenicity and an added high value for human nutrition.[135]

Several studies have evaluated the effects of enzymatic hydrolysis on the allergenicity and digestibility of peanut proteins in order to explore strategies designed to decrease the ability of peanuts to elicit dangerous allergic responses. Cabanillas *et al.*[136] used endoprotease (Alcalase) and exoprotease (Flavourzyme), both individually and sequentially, to evaluate their effects in the IgE antibody reactivity to roasted peanut protein extract (RP). None of the sera recognised any RP proteins after sequential endo- and exo-protease digestion. These enzymatically treated protein hydrolysates could constitute an alternative to intact proteins in the development of different products. Studies evaluating the effect of the food-hydrolysis products on IgE reactivity with sera from patients with well-documented allergy to these food sources is scarce. In the study reported by Cabanillas *et al.*[137] lentil protein extract was hydrolysed by sequential action of Alcalase (endoprotease) and Flavourzyme (exoprotease). Clemente *et al.*[131] reported an important reduction of chickpea antigenicity by sequential treatment with Alcalase and Flavourzyme compared to individual treatments. The data with sequential lentil hydrolysis indicated a 24% DH (degree of hydrolysis) produced by Alcalase at 180 min; after addition of Flavourzyme, the DH increased to 56% in the final reaction time (480 min). These findings suggest that endoprotease Alcalase yielded an initial hydrolysis that raised the number of target sites for the action of the exoprotease Flavourzyme. Similar data are reported for pea[138] and soybean[139] protein hydrolysates. The allergenicity to raw and hydrolised lentil extract was measured by means of ELISA and IgE immunoblotting using sera from five subjects with clinical allergy to lentil. According to these *in vitro* experiments, the sequential hydrolysis of lentil produces an important proteolytic destruction of IgE-binding epitopes. Nevertheless, four out of five patients showed few immunoreactive proteins in the last step of sequential hydrolysis. Individual hydrolysis of chickpeas with Alcalase or Flavourzyme achieved 27% DH; Alcalase hydrolysates did not show visible protein bands, contrary to the Flavourzyme hydrolysates.

As *in vitro* IgE-reactivity is only an indicator of possible *in vivo* allergenicity, additional *in vitro* tests and *in vivo* clinical data are needed to settle that these treatments are useful to diminish the *in vivo* allergenicity of these foods. Only after such studies could these supposed hypoallergenic legumes be safely ingested and even recommended as desensitizing food.

Acknowledgements

The support from Plan Nacional I+D+I [grant numbers, AGL2017-83082 and RTA2015-00003] is acknowledged.

References

1. C. Dillard and J. B. German, *J. Sci. Food Agric.*, 2000, **80**, 1744.
2. WHO, *WHO Technical Report Series*, World Health Organization, Geneva, 2003, No. 916.
3. C. Nishida, R. Uauy, S. Kumanyika and P. Shetty, *Public Health Nutr.*, 2004, 7, 245.
4. S. Rochfort and J. Panozzo, *J. Agric. Food Chem.*, 2007, **55**, 7981.
5. M. Duranti, *Fitoterapia*, 2006, **77**, 67.
6. P. Leterme, *Br. J. Nutr.*, 2002, **88**, 239.
7. M. M.-J. Champ, *Br. J. Nutr.*, 2002, **88**, 307.
8. M. Muzquiz, *Jornada internacional sobre proteınas alimentarıas*, Sevilla, Spain, 2000.
9. C. Goyoaga, C. Burbano, C. Cuadrado, C. Romero, E. Guillamon, A. Varela, M. M. Pedrosa and M. Muzquiz, *J. Food Compos. Anal.*, 2011, **24**, 391.

10. C. Goyoaga, C. Burbano, C. Cuadrado, A. Varela, E. Guillamon, M. M. Pedrosa and M. Muzquiz, *Eur. Food Res. Technol.*, 2008, **227**, 1537.
11. M. M. Pedrosa, C. Cuadrado, C. Burbano, K. Allaf, J. Haddad, E. Gelencser, K. Takacs, E. Guillamon and M. Muzquiz, *Food Chem.*, 2012, **131**, 862.
12. M. Muzquiz, A. Varela, C. Burbano, C. Cuadrado, E. Guillamon and M. M. Pedrosa, *Phytochem. Rev.*, 2012, **11**, 227.
13. M. Muzquiz and J. Wood, *Chickpea Breeding and Management*, 2007, p. 143.
14. R. Campos-Vega, G. Loarca-Piña and B. D. Oomah, *Food Res. Int.*, 2010, **43**, 461.
15. *Recent Advances of Research in Antinutritional Factors in Legume Seeds and Oilseeds*, ed. M. Muzquiz, G. D. Hill, C. Cuadrado, M. M. Pedrosa and C. Burbano, Wageningen Academic Publishers, Wageningen, 2004.
16. G. Grant, *Recent Advances of Research in Antinutritional Factors in Legume Seeds and Rapeseed*, 1998.
17. S. Zucoloto, A. C. Scaramello, F. M. Lajolo and G. Muccillo, *Int. J. Exp. Pathol.*, 1991, **72**, 41.
18. M. Figueroa, F. J. Mancini and F. Lajolo, *Arch. Latinoam. Nutr.*, 1984, **34**, 488.
19. S. Bardocz, D. S. Brown, G. Grant, A. Pusztai, J. C. Stewart and R. M. Palmer, *Br. J. Pharmacol.*, 1992, **106**, 476.
20. S. Bardocz, G. Grant and A. Pusztai, *Br. J. Nutr.*, 1996, **76**, 613.
21. A. Pusztai, S. Bardocz and M. Martín-Cabrejas, *Publ. - Eur. Assoc. Anim. Prod.*, 2004, **110**, 87.
22. I. F. Pryme, A. Pusztai, S. Bardocz and S. W. Ewen, *Histol. Histopathol.*, 1998, **13**, 575.
23. A. Pusztai, G. Grant, W. C. Buchan, S. Bardocz, A. F. de Carvalho and S. W. Ewen, *Br. J. Nutr.*, 1998, **79**, 213.
24. G. Hajós, E. Gelencsér, A. Pusztai and S. Bardocz, *COST98-Effects of Antinutrients on the Nutritional Value of Legume Diets*, Office for Official Publications of the European Communities, Luxembourg, 1996, vol. 1, p. 130.
25. M. Muzquiz, C. Burbano, C. Cuadrado and M. Martin, *Handbook on Common Bean Related Laboratory Methods*, Galicia, Spain, 2001, p. 11.
26. L. C. Trugo, M. Muzquiz, M. M. Pedrosa, G. Ayet, C. Burbano, C. Cuadrado and E. Cavieres, *Food Chem.*, 1999, **65**, 85.
27. C. Cuadrado, C. Burbano, E. Gelencser, M. M. Pedrosa, G. Ayet, M. Muzquiz, A. Pusztai and G. Hajos, in *Cost 98-Effects of Antinutrients on the Nutritional Value of Legume Diets*, ed. G. J. Pierzynowski, J. M. Gee and J. Svendsen, Office for Official Publications of the European Communities, Luxembourg, 2000, vol. 9, p. 92.
28. C. Cuadrado, G. Hajos, C. Burbano, M. M. Pedrosa, G. Ayet, M. Muzquiz, A. Pusztai and E. Gelencser, *Food Agric. Immunol.*, 2002, **14**, 41.
29. G. Urbano, M. López-Jurado, P. Aranda, C. Vidal-Valverde, E. Tenorio and J. Porres, *J. Physiol. Biochem.*, 2000, **56**, 283.

30. F. Loewus, in *Food Phytates*, ed. N. R. Reddy and S. K. Sathe, CRC Press, Boca Raton, 2002, p. 53.
31. L. Bohn, A. S. Meyer and S. K. Rasmussen, *J. Zhejiang Univ., Sci., B*, 2008, **9**, 165.
32. M. M. J. Champ, *Br. J. Nutr.*, 2007, **88**, 307.
33. C. Burbano, M. Muzquiz, A. Osagie, G. Ayet and C. Cuadrado, *Food Chem.*, 1995, **52**, 321.
34. M. Muzquiz, C. Burbano, G. Ayet, M. M. Pedrosa and C. Cuadrado, *Biotechnol., Agron., Soc. Environ.*, 1999, **3**, 201.
35. H. A. Oboh, M. Muzquiz, C. Burbano, C. Cuadrado, M. M. Pedrosa, G. Ayet and A. U. Osagie, *J. Chromatogr. A*, 1998, **823**, 307.
36. P. M. Garcia-Lopez, M. Muzquiz, M. A. Ruiz-Lopez, J. F. Zamora-Natera, C. Burbano, M. M. Pedrosa, C. Cuadrado and P. Garzon-De la Mora, *J. Food Compos. Anal.*, 2001, **14**, 645.
37. K. Raes, D. Knockaert, K. Struijs and J. Van Camp, *Trends Food Sci. Technol.*, 2014, **37**, 32.
38. R. Greiner and U. Konietzny, *Food Technol. Biotechnol.*, 2006, **44**, 125.
39. A. S. Sandberg, M. Brune, N. G. Carlsson, L. Hallberg, E. Skoglund and L. Rossander-Hulthen, *Am. J. Clin. Nutr.*, 1999, **70**, 240.
40. I. Vucenik and A. M. Shamsuddin, *Nutr. Cancer*, 2006, **55**, 109.
41. M. Fredrikson, M. L. Alminger, N.-G. Carlsson and A.-S. Sandberg, *J. Sci. Food Agric.*, 2001, **81**, 1139.
42. H. A. Oboh, M. Muzquiz, C. Burbano, C. Cuadrado, M. M. Pedrosa, G. Ayet and A. U. Osagie, *Ecol. Food Nutr.*, 2002, **41**, 229.
43. U. Konietzny and R. Greiner, *Int. J. Food Sci. Technol.*, 2002, **37**, 791.
44. M. A. Mohammed, E. A. Mohamed, A. E. A. Yagoub, A. R. Mohamed and E. E. Babiker, *J. Food Process. Preserv.*, 2017, **41**, e12960.
45. H. A. Haileslassie, C. J. Henry and R. T. Tyler, *Int. J. Food Sci. Technol.*, 2016, **51**, 1947.
46. G. Ayet, C. Burbano, C. Cuadrado, M. M. Pedrosa, L. M. Robredo, M. Muzquiz, C. delaCuadra, A. Castano and A. Osagie, *J. Sci. Food Agric.*, 1997, **74**, 273.
47. R. Greiner, M. Muzquiz, C. Burbano, C. Cuadrado, M. M. Pedrosa and C. Goyoga, *J. Agric. Food Chem.*, 2001, **49**, 2234.
48. Y. Aguilera, M. Felicia Diaz, T. Jimenez, V. Benitez, T. Herrera, C. Cuadrado, M. Martin-Pedrosa and M. A. Martin-Cabrejas, *J. Agric. Food Chem.*, 2013, **61**, 8120.
49. X. Wang, R. Yang, X. Jin, Z. Chen, Y. Zhou and Z. Gu, *Food Sci. Biotechnol.*, 2015, **24**, 1829.
50. C. Cuadrado, G. Ayet, L. M. Robredo, J. Tabera, R. Villa, M. M. Pedrosa, C. Burbano and M. Muzquiz, *Z. Lebensm.-Unters. Forsch.*, 1996, **203**, 268.
51. R. Y. Khattab and S. D. Arntfield, *LWT–Food Sci. Technol.*, 2009, **42**, 1113.
52. Q. Chen, *J. Agric. Food Chem.*, 2004, **52**, 4604.
53. M. M. Pedrosa, C. Cuadrado, C. Burbano, M. Muzquiz, B. Cabellos, B. Olmedilla-Alonso and C. Asensio-Vegas, *Food Chem.*, 2015, **166**, 68.

54. M. A. Martin-Cabrejas, Y. Aguilera, M. M. Pedrosa, C. Cuadrado, T. Hernandez, S. Diaz and R. M. Esteban, *Food Chem.*, 2009, **114**, 1063.
55. N. Wang, D. W. Hatcher and E. J. Gawalko, *Food Chem.*, 2008, **111**, 132.
56. J. Haddad, R. Greiner and K. Allaf, *LWT–Food Sci. Technol.*, 2007, **40**, 448.
57. N. Nikmaram, S. Y. Leong, M. Koubaa, Z. Zhu, F. J. Barba, R. Greiner, I. Oey and S. Roohinejad, *Food Control*, 2017, **79**, 62.
58. R. Alonso, L. A. Rubio, M. Muzquiz and F. Marzo, *Anim. Feed Sci. Technol.*, 2001, **94**, 1.
59. A. Varela, E. Guillamón, C. Cuadrado, F. Marzo, C. Burbano, M. Muzquiz and M. M. Pedrosa, *Changes in nutritional active factors and protein in different legumes by extrusion/cooking, 6th European Conference on Grain Legumes*, Lisboa (Portugal), 2007.
60. P. Morales, J. D. J. Berrios, A. Varela, C. Burbano, C. Cuadrado, M. Muzquiz and M. M. Pedrosa, *Food Funct.*, 2015, **6**, 3135.
61. G. Lombardi-Boccia, G. D. Lullo and E. Carnovale, *J. Sci. Food Agric.*, 1991, **55**, 599.
62. Y. Birk, *Ann. N. Y. Acad. Sci.*, 1968, **146**, 388.
63. R. Kansal, M. Kumar, K. Kuhar, R. N. Gupta, B. Subrahmanyam, K. R. Koundal and V. K. Gupta, *Braz. Plant Physiol.*, 2009, **20**, 313.
64. L. Shi, K. Mu, S. D. Arntfield and M. T. Nickerson, *J. Food Sci.Technol.*, 2017, **54**, 1014.
65. E. A. E. Morsi, *Egypt. J. Biol.*, 2001, **3**, 164.
66. R. Roychaudhuri, G. Sarath, M. Zeece and J. Markwell, *Biochim. Biophys. Acta, Proteins Proteomics*, 2004, **1699**, 207.
67. E. B.-F. S. Rustgi, C. Reinbothe, D. von Wettstein and S. Reinbothe, *Commun. Integr. Biol.*, 2017, **9**(4), 9.
68. M. Carbonaro, P. Maselli and A. Nucara, *Food Res. Int.*, 2015, **76**, 19.
69. A. Tur Sinal, Y. Birk, A. Gertler and M. Rigbi, *Biochim. Biophys. Acta*, 1972, **263**, 666.
70. I. P. Gladysheva, T. S. Zamolodchikova, E. A. Sokolova and N. I. Larionova, *Biochemistry*, 1999, **64**, 1244.
71. E. Guillamón, M. M. Pedrosa, C. Burbano, C. Cuadrado, M. D. C. Sanchez and M. Muzquiz, *Food Chem.*, 2008, **107**, 68.
72. L. Quillien, E. Ferrasson, D. Molle and J. Gueguen, *J. Protein Chem.*, 1997, **16**, 195.
73. A. Clemente and C. Domoney, *Curr. Protein Pept. Sci.*, 2006, **7**, 201.
74. C. Domoney, T. Welham, N. Ellis, P. Mozzanega and L. Turner, *Plant Mol. Biol.*, 2002, **48**, 319.
75. K. Takacs, E. Guillamon, M. M. Pedrosa, C. Cuadrado, C. Burbano, M. Muzquiz, J. Haddad, K. Allaf, A. Maczo, M. Polgar and E. Gelencser, *Food Agric. Immunol.*, 2014, **25**, 173.
76. V. Kumar, A. Rani, R. Rawal and V. Mourya, *Breed. Sci.*, 2015, **65**, 447.
77. É. Gelencsér, *Acta Aliment.*, 2009, **38**, 61.
78. I. E. Liener and M. L. Kakade, in *Toxic Constituents of Plant Foodstuffs*, ed. I. E. Liener, Academic Press, New York, 1980, p. 7.
79. I. E. Liener, *Crit. Rev. Food Sci. Nutr.*, 1994, **34**, 31.

80. G. M. Green, V. H. Levan and R. A. Liddle, in *Nutritional and Toxicological Significance of Enzyme Inhibitors in Foods*, ed. M. Friedman, Plenum Press, New York, 1986, p. 123.
81. A. Pusztai, G. Grant, D. J. Brown, J. C. Stewart, S. Bardocz, S. W. Ewen, A. M. Gatehouse and V. Hilder, *Br. J. Nutr.*, 1992, **68**, 783.
82. G. Grant, N. H. Mckenzie, W. B. Watt, J. C. Stewart, P. M. Dorward and A. Pusztai, *J. Sci. Food Agric.*, 1986, **37**, 1001.
83. M. Friedman and D. L. Brandon, *J. Agric. Food Chem.*, 2001, **49**, 1069.
84. M. Carbonaro, *Pulse Foods: Processing, Quality and Nutraceutical Applications*, ed. B. K. Tiwari, A. Gowen and B. McKenna, Academic Press, Oxford, 2011, p. 385.
85. Y. Birk, *Arch. Latinoam. Nutr.*, 1996, **44**, 26S.
86. A. Clemente, E. Jimenez, M. C. Marin-Manzano and L. A. Rubio, *J. Sci. Food Agric.*, 2008, **88**, 513.
87. L. A. Rubio, M. M. Pedrosa, C. Cuadrado, E. Gelencser, A. Clemente, C. Burbano and M. Muzquiz, *J. Sci. Food Agric.*, 2006, **86**, 979.
88. G. Hajós and E. Gelencser, *J. Agric. Food Chem.*, 1995, **43**, 165.
89. A. Pusztai, G. Grant, S. Bardocz, K. Baintner, É. Gelencsér and S. W. B. Ewen, *Am. J. Physiol.*, 1997, **35**, 340.
90. A. Clemente and C. Arques Mdel, *World J. Gastroenterol.*, 2014, **20**, 10305.
91. E. Mills, J. Jenkins, N. Marigheto, P. Belton, A. Gunning and V. Morris, *Biochem. Soc. Trans.*, 2002, **30**, 925.
92. A. Burks, H. Sampson and G. Bannon, *Allergy*, 1998, **53**, 725.
93. E. Guillamón, J. Rodríguez, C. Burbano, M. Muzquiz, M. Pedrosa, B. Cabanillas, J. Crespo, A. Sancho, E. Mills and C. Cuadrado, *Mol. Nutr. Food Res.*, 2010, **54**, 1168.
94. A. K. Verma, S. Kumar, M. Das and P. D. Dwivedi, *Clin. Rev. Allergy Immunol.*, 2013, **45**, 36.
95. E. Mills, J. A. Jenkins, M. Alcocer and P. Shewry, *Crit. Rev. Food Sci. Nutr.*, 2004, **44**, 379.
96. C. Cuadrado and M. Pedrosa, in *Legumes for Global Food Security*, ed. A. Clemente and J. C. Jimenez-Lopez, Nova Science Publishers, Inc., New York, 2017, p. 223.
97. J. Crespo and J. Rodriguez, *Allergy*, 2003, **58**, 98.
98. Regulation (EU) No 1169/2011 of 25 October 2011 on the provision of food information to consumers.
99. S. J. Maleki, *Curr. Opin. Allergy Clin. Immunol.*, 2004, **4**, 241.
100. D. A. Schmitt, J. B. Nesbit, B. K. Hurlburt, H. Cheng and S. J. Maleki, *J. Agric. Food Chem.*, 2010, **58**, 1138.
101. J. M. Wal, *Allergy*, 2003, **58**, 727.
102. P. Davis and S. Williams, *Allergy*, 1998, **53**, 102.
103. S. J. Maleki, S. Y. Chung, E. T. Champagne and J. P. Raufman, *J. Allergy Clin. Immunol.*, 2000, **106**, 763.
104. S. K. Sathe and G. M. Sharma, *Mol. Nutr. Food Res.*, 2009, **53**, 970.
105. S. H. Sicherer and H. A. Sampson, *J. Allergy Clin. Immunol.*, 2007, **120**, 491.
106. S. Chung, C. Butts, S. Maleki and E. Champagne, *J. Agric. Food Chem.*, 2003, **51**, 4273.

107. K. B. Beyer, E. Morrow, X. M. Li and L. Bardina, *J. Allergy Clin. Immunol.*, 2001, **107**, 1077.
108. B. Cabanillas, S. J. Maleki, J. Rodriguez, C. Burbano, M. Muzquiz, M. Aranzazu Jimenez, M. M. Pedrosa, C. Cuadrado and J. F. Crespo, *Food Chem.*, 2012, **132**, 360.
109. M. Duranti, *Fitoterapia*, 2006, **77**, 67.
110. D. Moneret-Vautrin, L. Guérin, G. Kanny, J. Flabbee, S. Frémont and M. Morisset, *J. Allergy Clin. Immunol.*, 1999, **04**(4 Pt 1), 883.
111. J. Álvarez-Álvarez, E. Guillamón, J. F. Crespo, C. Cuadrado, C. Burbano, J. Rodríguez, C. Fernández and M. Muzquiz, *J. Agric. Food Chem.*, 2005, **53**, 1294.
112. E. Guillamón, C. Burbano, C. Cuadrado, M. Muzquiz, M. M. Pedrosa, M. Sanchez, B. Cabanillas, J. F. Crespo, J. Rodriguez, J. Haddad and K. Allaf, *Int. Arch. Allergy Immunol.*, 2008, **145**, 9.
113. E. Guillamón, C. Cuadrado, M. M. Pedrosa, A. Varela, B. Cabellos, M. Muzquiz and C. Burbano, *Spanish Journal of Agricultural Research*, 2010, **8**, 100.
114. G. Lopez-Torrejon, G. Salcedo, M. Martin-Esteban, A. Diaz-Perales, C. Y. Pascual and R. Sanchez-Monge, *J. Allergy Clin. Immunol.*, 2003, **112**, 1208.
115. R. Sanchez-Monge, C. Y. Pascual, A. Diaz-Perales, J. Fernandez-Crespo, M. Martin-Esteban and G. Salcedo, *J. Allergy Clin. Immunol.*, 2000, **106**, 955.
116. J. H. Akkerdaas, F. Schocker, S. Vieths, S. Versteeg, L. Zuidmeer, S. L. Hefle, R. C. Aalberse, K. Richter, F. Ferreira and R. van Ree, *Mol. Nutr. Food Res.*, 2006, **50**, 18.
117. C. Cuadrado, B. Cabanillas, M. M. Pedrosa, A. Varela, E. Guillamon, M. Muzquiz, J. F. Crespo, J. Rodriguez and C. Burbano, *Mol. Nutr. Food Res.*, 2009, **53**, 1462.
118. C. Cuadrado, B. Cabanillas, M. Pedrosa, M. Muzquiz, J. Haddad, K. Allaf, J. Rodriguez, J. Crespo and C. Burbano, *Int. Arch. Allergy Immunol.*, 2011, **156**, 397.
119. J. Crespo, C. Pascual, A. Burks, R. Helm and M. Martin-Esteban, *Pediatr. Allergy Immunol.*, 1995, **6**, 39.
120. P. V. Niphadkar, S. P. Patil and M. M. Bapat, *Allergy*, 1997, **52**, 115.
121. S. P. Patil, P. V. Niphadkar and M. M. Bapat, *Ann. Allergy, Asthma, Immunol.*, 2001, **87**, 140.
122. J. Vioque, R. Sanchez-Vioque, A. Clemente, J. Pedroche, J. Bautista and F. Millan, *J. Agric. Food Chem.*, 1999, **47**, 1405.
123. *Soy Protein Products-characteristics, Nutritional Aspects, and Utilization*, ed. J. Endres, AOCS Publication, Champaign, 2001.
124. A. W. Burks, L. W. Williams, R. M. Helm, C. Connaughton, G. Cockrell and T. O'Brien, *J. Allergy Clin. Immunol.*, 1991, **88**, 172.
125. S. Wilson, K. Blaschek and E. de Mejia, *Nutr. Rev.*, 2005, **63**, 47.
126. C. Cuadrado, B. Cabanillas, M. Pedrosa, A. Varela, E. Guillamón, M. Muzquiz, J. Crespo and J. Rodriguez, in *Integrating Legume Biology for Sustainable Agriculture. Proceedings of the 6th European Conference on Grain Legumes*, Lisbon, Portugal, 2007, p. 186.

127. L. Terracciano, P. Isoardi, S. Arrigoni, A. Zoja and A. Martelli, *Ann. Allergy, Asthma, Immunol.*, 2002, **89**, 86.
128. R. J. Merritt, M. Carter, M. Haight and L. D. Eisenberg, *J. Pediatr. Gastroenterol. Nutr.*, 1990, **11**, 78.
129. H. W. Lee, E. H. Keum, S. J. Lee, D. E. Sung, D. H. Chung, S. I. Lee and S. Oh, *J. Food Sci.*, 2007, **72**, C168.
130. A. W. Burks, L. W. Williams, W. Thresher, C. Connaughton, G. Cockrell and R. M. Helm, *J. Allergy Clin. Immunol.*, 1992, **90**, 889.
131. A. Clemente, J. Vioque, R. Sanchez-Vioque, J. Pedroche and F. Millan, *J. Agric. Food Chem.*, 1999, **47**, 3776.
132. A. Clemente, J. Vioque, R. Sanchez-Vioque, J. Pedroche, J. Bautista and F. Millan, *Food Chem.*, 1999, **67**, 269.
133. S. E. Sormus de Castro Pinto, V. A. Neves and B. M. Machado de Medeiros, *J. Agric. Food Chem.*, 2009, **57**, 1070.
134. J. S. Lucas and R. G. Atkinson, *Clin. Exp. Allergy*, 2008, **38**, 1095.
135. A. Clemente, J. Vioque, R. Sánchez-Vioque, J. Pedroche, J. Bautista and F. Millán, *J. Sci. Food Agric.*, 2000, **80**, 79.
136. B. Cabanillas, M. M. Pedrosa, J. Rodriguez, M. Muzquiz, S. J. Maleki, C. Cuadrado, C. Burbano and J. F. Crespo, *Int. Arch. Allergy Immunol.*, 2012, **157**, 41.
137. B. Cabanillas, M. M. Pedrosa, J. Rodriguez, A. Gonzalez, M. Muzquiz, C. Cuadrado, J. F. Crespo and C. Burbano, *Mol. Nutr. Food Res.*, 2010, **54**, 1266.
138. L. M. Humiski and R. E. Aluko, *J. Food Sci.*, 2007, **72**, S605.
139. W. H. Seo, H. G. Lee and H. H. Baek, *J. Food Sci.*, 2008, **73**, S41.

Part III

Processing of Legumes: Changes of Bioactive Compounds

CHAPTER 8

Obtaining, Chemically Characterizing and Nutritionally Evaluating Seasonal Legume Sprouts as a Feed Alternative

M. F. DÍAZ*[a], M. MARTÍNEZ[a], L. L. SAVÓN[a], V. TORRES[a] AND G. COTO[b]

[a]Instituto de Ciencia Animal (ICA), Cuba; [b]Histoterapia Placentaria de Cuba, Escuela Latinoamericana de Medicina, Cuba
*E-mail: mdiaz@ica.co.cu

8.1 Legumes: A Nutritional Alternative for Sustainable Development

Overcoming hunger and malnutrition in the 21st century implies increasing both feed amount and quality, and, at the same time, producing feed in a sustainable manner, with effectiveness and security. In September 2015, the Agenda 2030 for Sustainable Development was passed, an action plan in favor of people, the planet and prosperity. The document defines a list of Objectives of Sustainable Development addressed to eradicate hunger and

Food Chemistry, Function and Analysis No. 8
Legumes: Nutritional Quality, Processing and Potential Health Benefits
Edited by María Ángeles Martín-Cabrejas

Published by the Royal Society of Chemistry, www.rsc.org

malnutrition, as well as extreme poverty, while addressing climatic change, among other objectives.[1]

Legumes have been an essential part of human nutrition for centuries. Nevertheless, their nutritional value, in general, is not recognized and frequently their intake is not sufficiently appreciated. This lack of recognition is undeserved, since they play a crucial role in healthy feeding, sustainable feed production and, above all, food security. With the goal of turning legumes into essential feed for people and countries, 2016 was considered by the Food and Agriculture Organization (FAO) of the United Nations, International Year of Legumes.[2]

A healthy nutrition protects us from certain illnesses, thus, in recent years, more attention has been paid to incorporating, in the diet, feed that, besides having greater nutritive value, provides some bioactive compounds such as fibers and vitamins, which protect us against certain illnesses. Biotechnological processes play a predominant role in this research field. Fermentation and germination have been carried out for centuries and are transformation processes that improve the conservation of certain feedstuffs. More recently, they have been used to produce products with added value in respect to their culinary and nutritive properties.[3,4]

In germination, there are sequential and synchronized processes, organized in such a way that anabolic and catabolic processes are simultaneous. In general, with seed germination, the outer seed skin is broken and the radicle emerges. Seeds subject to the germination process with greatest frequency are peas, mung beans, soybeans, and chickpeas.[5,6]

Germination has been proven to make a series of positive changes in the content of nutrients in legumes, such as increasing the content of essential amino acids and soluble proteins, and improving the *in vitro* digestibility of protein;[7–10] causing a significant decline in the levels of alpha-galactosides and inositol penta and hexa-phosphates;[11,12] and a reduction in inhibitors of proteases and lectins, as well as in the degree of polymerization of tannins.[13–15] Other studies have provided evidence that these processes, in legumes, result in significant increments in vitamins B_1, B_2 and C.[16,17]

During the germination process, there are important changes in the content of carbohydrates in the seeds due to the action of various enzymes. Therefore, oligosaccharides and α-galactosides are hydrolyzed by the action of galactosidases, and starch, the main carbohydrate in most legumes, is hydrolyzed by the action of α- and β-amylases.[18–20]

Several authors have stated that as a consequence of germination there is decrease in total starch content, and an increase in the usable starch : total starch ratio. With respect to dietary fiber components, the germination process in lentils caused a rise in lignin content and a fall in cellulose and hemicellulose.[5,21]

Even though germination scarcely affects the content of minerals, with an increase in phosphorus, potassium, zinc, and copper, this process causes a growth in the bioavailability of certain minerals. It is known that phytic acid and the rest of the inositol phosphates constitute the largest reserve of phosphorus in legumes and cereals (60–90%) and form chelating compounds with the multivalent cations calcium, magnesium, zinc, and iron. The phosphorus from inositol phosphates and cations that have been chelated results in their bioavailability being quite low. As a consequence of the germination process, the phytases in the seeds are activated and hydrolysis of inositol phosphates occurs, incrementing both phosphorus digestibility and that of cations trapped by inositol phosphate molecules.[19,22,23]

This chapter is intended to explain the improved nutritive value of the grains of *Vigna unguiculata*, *Canavalia ensiformis*, *Mucuna pruriens*, *Lablab purpureus* and *Glycine max* achieved through germination and providing new products as alternatives to conventional protein sources.

8.2 Legumes as an Alternative for Production of Protein of Animal Origin

In developing countries such as Cuba, in the last decades, much work has been done in the search for different alternative feed sources in the production of protein of animal origin that do not affect human feeding and can substitute imported foodstuffs.[24] Previous studies indicated the important agronomical and nutritional potential of seasonal legumes as a feed source, and technologies of grain and forage production were developed.[25] In case of *Vigna unguiculata*, it was demonstrated that it is possible to replace 16 and 60% of wheat and soybean, respectively, in isoproteic diets, by feedstuff containing raw grains of this legume, which is equivalent to inclusion levels up to 20% of the diet. However, the presence in these plants of antinutritional compounds (ANFs) affects the availability and utilization of the nutrients for animal nutrition, primarily in monogastric species.[26,27]

In this context, the use of biological processes such as germination, which manages to reduce the content of ANFs and enhance the nutritive value of nationally produced legume grains, could be an alternative for small farmers, families, and backyard agriculture. The outcomes presented here are related to obtaining, chemically characterizing, and nutritionally evaluating sprouts of seasonal legumes as alternative feed, which will be described in three research stages:

- Obtaining and physico-chemically characterizing the legume sprouts
- Changes in the content of antinutritional factors and antioxidant activity during the germination process
- Biological assessment of germinated and non-germinated grains in rats as an experimental model and in broilers as species of economic interest

8.3 Obtaining and Physico-chemically Characterizing Legume Sprouts

Studies dealing with the optimization of the germination process were conducted through 21 trials in five legume species: *Vigna unguiculata* (cowpea) variety INIFAT 94, *Canavalia ensiformis* (jack bean) variety Blanco, *Lablab purpureus* (dolicho) variety Rongai, *Glycine max* (soybean) variety INCASOY 24 and *Mucuna pruriens* (mucuna) variety Cenizo. The legume grains were germinated for a period of 96 up to 120 hours, in different conditions (light intervals of twelve hours, total light or total darkness). A grain germination methodology was attained producing favorable changes in the physico-chemical properties, improving the quality of the final products, and enhancing their functionality; which turns it into an accessible and affordable choice for small and medium farmers.[28,29]

It was demonstrated that physical and chemical transformations during germination depend on both legume species and germination process conditions. In all the species under study, the weight of the germinated grains was incremented as the process went forward, as well as radicle length and germination percentage, which provoked changes in the resulting chemical composition of the germinated grains.[27–31]

Chemical composition: In cowpea, in the three germination variants the contents of protein and fiber increased as the germination process advanced. The most significant results were found in total darkness conditions, in which protein (CP) rose to 25%, compared to non-germinated cowpea.[29] In dolicho, the total darkness variant also showed the highest concentrations of CP (13%), compared with the non-germinated grain.[30] In jack bean, in the three variants CP rose to 11%, compared to the non-germinated control. In mucuna, in the light and total light intervals, there was an increase in protein and fiber indicators, the last variant being outstanding, with values superior to the non-germinated control. In soybean, there were variations in CP, with increases of up to 16%, compared with the non-germinated control, and a decrease of up to 48% in cellulose, compared to the control.[31–33]

Amino acids: Good amino acid balance (Table 8.1) was found, with acceptable concentrations of essential amino acids, according to feed requirements, which varied depending on species and germination variant.[34–38]

The germination process in soybean provokes changes in the pattern of amino acids depending on the light conditions and its impact was greater when performed in conditions of total darkness for cysteine and lysine and in total light for cysteine and arginine. The cysteine increment is of relevance for nutrition since sulfurous amino acids are limited in most legume seeds.[39,40] That of arginine is also of great importance due to the effect of this amino acid on cancerous cells. The methionine and arginine ratio has been analyzed for years due to its possible activity against neoplasia. Soybean has a high content of arginine and low of sulfurous amino acids, which could cause anti-mutagenic activity in the cells.[41–43] Cowpea

Table 8.1 Amino acid composition in unprocessed legumes grain and in grains germinated for 96 hours.[a]

Legume...	*Glycine max*			*Vigna unguiculata*			*Canavalia ensiformis*			*Lablab purpureus*			*Mucuna pruriens*		
g/100 g of aa	Control	Total light	Total dark	Control	Total light	Total dark	Control	Total light	Total dark	Control	Total light	Total dark	Control	Total light	Total dark
Aspartic acid	18.88	19.26	11.85	12.80	20.26	20.89	14.05	17.42	19.68	14.15	12.18	17.90	12.92	15.49	7.34
Threonine	11.96	8.72	11.12	10.18	15.11	12.01	3.75	4.47	8.39	8.51	12.53	9.34	4.55	1.86	1.48
Serine	5.63	4.76	5.69	10.15	6.67	6.48	5.12	7.19	2.45	8.17	3.84	3.53	6.00	6.17	8.22
Glutamic acid	11.62	6.42	13.52	8.67	8.02	9.90	11.20	12.18	10.26	6.26	1.15	9.48	5.97	2.99	1.06
Proline	3.01	1.00	7.08	5.50	6.64	6.26	11.22	11.53	8.21	4.90	2.90	6.77	2.23	1.47	5.71
Glycine	1.64	1.15	3.26	2.36	1.00	0.63	8.98	4.05	4.70	0.61	ND	1.05	5.57	9.17	13.11
Alanine	1.10	1.75	0.87	1.35	2.08	2.79	4.17	7.54	1.85	0.43	ND	ND	3.31	6.74	8.60
Cysteine	4.12	6.28	7.01	1.55	2.61	ND	0.44	2.32	0.42	1.88	7.20	2.58	2.27	2.89	2.00
Valine	5.00	5.88	5.17	6.83	4.08	5.91	1.50	2.14	0.95	6.51	9.81	3.28	7.35	3.20	5.60
Methionine	3.75	2.68	2.50	2.25	ND	ND	4.37	0.62	2.72	3.20	7.09	4.01	ND	2.87	1.03
Isoleucine	3.43	2.93	2.53	2.77	1.52	2.42	1.96	1.54	1.82	2.09	8.22	3.89	6.21	5.11	4.28
Leucine	5.09	7.80	3.79	5.81	2.00	5.95	3.49	2.21	1.48	4.66	10.32	3.74	7.19	7.02	6.45
Tyrosine	2.05	2.36	2.01	0.63	0.47	0.44	1.10	0.86	0.93	1.47	4.81	3.14	7.35	5.16	4.41
Phenylalanine	5.69	6.68	3.57	5.00	5.45	1.93	1.28	0.80	1.00	3.91	1.08	8.51	10.37	9.70	8.57
Histidine	4.16	5.63	4.79	5.21	5.58	6.21	6.68	7.83	8.72	4.34	7.13	6.73	4.75	2.34	3.83
Lysine	6.50	6.54	8.87	7.24	8.44	7.70	11.47	11.24	12.33	4.60	5.14	8.10	7.59	7.66	9.34
Arginine	6.38	10.17	6.31	11.69	10.05	10.45	9.19	6.03	14.07	5.41	6.56	8.00	6.36	10.14	8.96

[a]These values are the mean of three analyses. Standard deviation was lower than 5% of the mean for all, ND = non-detected.

showed an amino acid pattern similar to previous reports with respect to sulfurous and basic amino acids; however, the values for amino acids of an acidic nature such as threonine and serine were above reported figures for this species.[44–49] The germination process in cowpea brought about greater variation in methionine and in aspartic acid, because the first was not identified in the germinated variants and the values of the latter were high. It was demonstrated that soluble amino acids change during germination and there is also mobilization of stored proteins in legumes.[50] Specifically, in *Vigna unguiculata*, germination improves the bioavailability and digestibility of nutrients.[36,40,51] The germination process provoked changes in the pattern of basic amino acids, both in soybean and cowpea. They showed lower variations, except arginine, which presented important increments when the process occured in total darkness.[52,53] The high arginine values, in jack bean, are noteworthy since this species has canavanine, a toxic amino acid that is a structural analogue of arginine. The fact that arginine is found in high concentrations permits it to counteract the adverse effects provoked by this anti-nutritive compound because arginine supplementation in animals fed meals of jack bean grains is used to improve the physiological and productive performance of the animals.[34,54] In case of dolicho, germination in total light provoked important increments of cysteine, valine, methionine, leucine, tyrosine, lysine and histidine. In total darkness, the increases of phenylalanine, lysine, arginine and histidine are notable. Mucuna presented an amino acid pattern, with the exception of methionine, that fulfils the requirements for essential amino acids in feed.[47,53,55] Although the germination process of this legume changed the amino acid pattern, depending on the light conditions, changes were not as sharp as in the rest of the studied species. The most relevant variation was shown in methionine, in which the process permitted values to be obtained that could fulfill the requirements in the variation of total light.

Therefore, in soybean, there were important increase of cysteine and arginine; in cowpea, of aspartic acid; in jack bean, of arginine; in dolicho, of cysteine, valine, methionine, leucine, tyrosine, lysine, histidine, phenylalanine and arginine; and in mucuna, of methionine. The results showed the functional characteristics of the new produces obtained and the effectiveness of the procedure.

Dietary fiber: The main goal of the fiber content analysis lies in its beneficial physiological properties. Table 8.2 shows the results for total dietary fiber (TDF) and its fractions, soluble (SDF) and insoluble (IDF) in germinated and non-germinated legumes.

The TDF levels, in the non-germinated grains, were higher than those previously reported in cereal seeds and other legumes. The IDF content represents 93–97% of the TDF of the genotypes under assessment.[56] Germination increased TDF from 6% (mucuna) to 14% (cowpea), except for soybean in which TDF decreased by 10% because of the fall in IDF. For all the germination variants, the highest levels of TDF and IDF were found in total

Table 8.2 Content of soluble and insoluble fiber in unprocessed legumes grains and in grains germinated for 96 hours (g kg^{-1} DM).[a]

	Insoluble dietary fiber	Soluble dietary fiber	Total dietary fiber	IDF : SDF
Vigna unguiculata				
Control	303.2[a]	9.0[b]	312.2[a]	34.09
Light interval	324.4[ab]	11.9[b]	339.3[ab]	27.95
Total light	332.1[b]	6.7[a]	338.8[ab]	54.58
Total darkness	338.8[b]	17.1[c]	355.9[b]	19.81
SE ±	7.42*	1.00***	8.23*	
Canavalia ensiformis				
Control	317.4[b]	15.1[b]	332.5[ab]	21.41
Light interval	317.1[b]	9.5[ab]	326.5[a]	41.69
Total light	310.1[a]	29.8[c]	339.9[b]	10.43
Total darkness	355.5[c]	7.0[a]	362.5[c]	55.69
SE ±	1.85***	1.8***	3.26***	
Glycine max				
Control	520.6[d]	27.1[b]	547.6[d]	19.24
Light interval	441.9[b]	33.0[d]	473.8[b]	13.41
Total light	407.2[a]	28.9[c]	436.1[a]	14.06
Total darkness	473.2[c]	22.3[a]	495.5[c]	22.55
SE ±	4.93***	0.29***	5.20***	
Lablab purpureus				
Control	399.0[b]	21.2[c]	420.2[c]	19.26
Light interval	387.4[b]	9.1[a]	396.6[b]	42.49
Total light	364.7[a]	14.3[b]	379.0[a]	25.66
Total darkness	422.9[c]	38.1[d]	461.0[d]	11.14
SE ±	4.33***	1.40***	3.57**	
Mucuna pruriens				
Control	397.6[a]	29.0	426.6[a]	15.86
Light interval	417.9[b]	31.4	449.3[b]	13.41
Total light	388.8[a]	28.4	417.2[a]	14.88
Total darkness	415.2[b]	37.2	452.4[b]	11.23
SE ±	3.91**	3.37	5.35**	

[a]Means with different letters within each column differ at $P < 0.0$; 5 *$P < 0.05$;**$P < 0.01$; ***$P < 0.00$.

darkness. It was shown that there is a significant percentage of protein that still remains bound to total fiber, which represents between 23 and 43% of the total protein, and approximately 90% is associated with the insoluble fraction.[56]

Further studies confirmed that germination with light intervals has a significant impact on both fractions of dietary fiber and improves the IDF : SDF ratio. The content of total sugars rose, primarily in the IDF fraction, due to the release of cellulosic sugars, and these TDF changes varied the physical and chemical properties of the legumes. After germination, oil retention capacity, water retention and absorption capacity, and gelation capacity were enhanced, resulting from the growth of the dietary fiber fraction and the percentage of available starch. However, emulsifying and foaming capacities were reduced due to the hydrolysis of the proteins during germination.[57,58]

The profile of sugars composition in TDF and in IDF and SDF fractions in these germinated and non-germinated legumes showed the IDF was constituted, mainly, of glucose, uronic acids and arabinose, and the SDF, of glucose and arabinose, followed by uronic acids. The changes in the fiber fractions were mainly increased in polysaccharides, cellulose and peptic substances.[58] The germinated legumes kept the same pattern of sugars in the IDF; however, germination incremented the total sugars in all the legumes except in jack bean. Nevertheless, the lignin amount was maintained constant, not showing significant differences in lignin portion, between the germinated and the non-germinated legumes.[58] Germination modified the SDF composition, although these changes were not observed in all the legumes in the same form. Sugars were incremented only in cowpea (17%) and mucuna (15%), which reveals that after germination, a series of metabolic processes occur acting differently on legumes and their compounds.[21,59]

With respect to the composition of carbohydrates in the TDF, the findings were quite similar to that of the IDF (most abundant fraction), the principal sugars detected being glucose, arabinose and uronic acids. After germination, increments of the total sugars between 30% and 15% were observed in the germinated legumes, except in jack bean. This is, mainly, due to the increase in the level of cellulosic sugars and, to a lesser extent, of the rest of the sugars (xylose, galactose and arabinose). The significant increments observed after germination in total fiber in the legumes studied could be useful to achieve improvement in their beneficial functional traits for human and animal health.

Soluble Carbohydrates: The germination process increased the total of soluble sugars (20–61%) and reduced the content of the raffinose family of oligosaccharides (RFOs) (98–63%). The effect of germination on soluble sugars was greater in dolicho, jack bean and mucuna, whereas the RFOs reduction was superior in jack bean (98%) and cowpea (94%).[56] These figures are better than those of RFOs reductions obtained by soaking or cooking, and comparable to fermentative processes or addition of exogenous enzymes (Figure 8.1).[59]

Comparing the α-galactosides individually, stachyose was the main oligosaccharide that had the greatest reduction, followed by raffinose and ciceritol. There was a large increase of sucrose, maltose and manotriose. The sucrose levels were significantly higher in jack bean sprouts compared to the non-processed seeds, followed by dolicho, cowpea, mucuna and soybean. Glucose, galactose and fructose, which were not present in many of the legumes under study, had significant increase as a consequence of the hydrolysis of galactosides and starch.[56]

Starch: The study of the starch content in these legumes is important because it may constitute up to 40% of their nutritional composition. It is a direct energy source and provides health benefits, as well as favorable

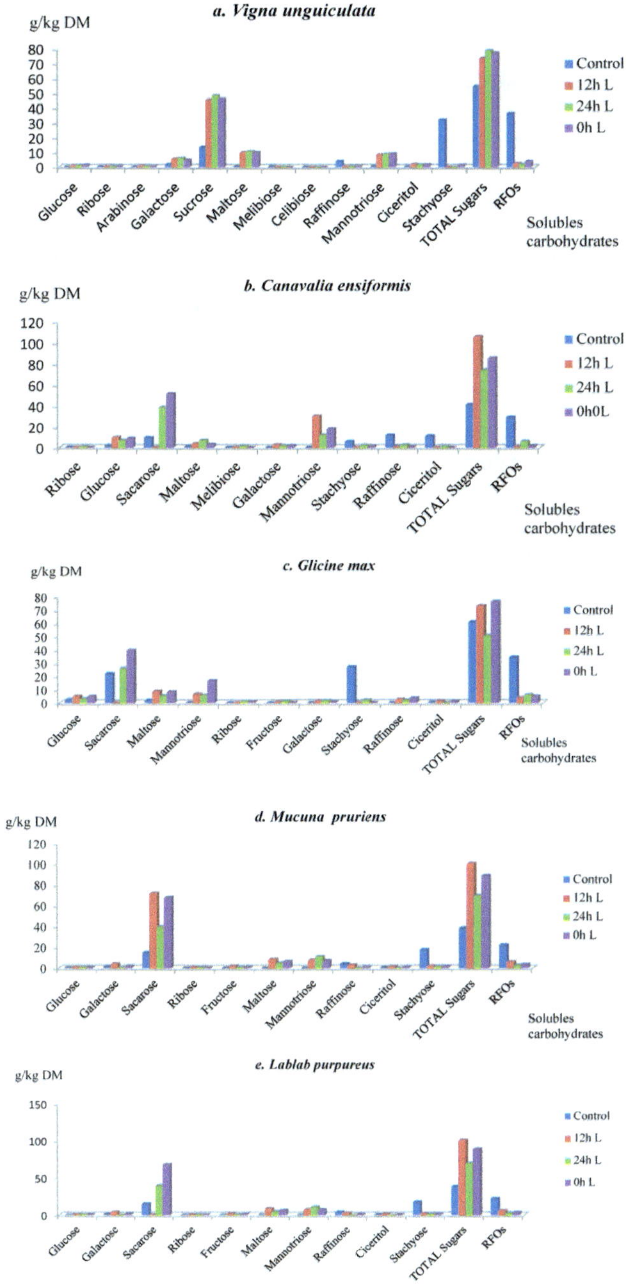

Figure 8.1 Content of soluble carbohydrate in germinated and non-germinated legume grains (g kg^{-1} DM).

organoleptic and functional properties for the feed industry.[60–63] The available starch, *i.e.*, starch that is able to be assimilated, represented between 81 and 90% of the total starch, varying in the different legumes. The rest of the starch of the total fraction was resistant starch (28.5–79 mg g^{-1} dry weight). These data agree with studies on some of these legumes[59] and on legumes in customary use.[64]

8.4 Changes in the Content of Antinutritional Factors and Antioxidant Activity During the Germination Process

The five non-conventional legumes being studied show large contents of enzymatic inhibitors that vary according to the species (Table 8.3).[65,66] Germination reduced the content of trypsin inhibitors in dolicho (53%) and jack bean (76%), whereas in cowpea it was increased to 61%. For the chymotrypsin inhibitors, the reductions were significant in mucuna (58%) and dolicho (36%). Hemagglutinating activity was found, without prior treatment of blood cells with trypsin, in jack bean, and germination reduced the levels of lectin to 50%. The rest of the legumes did not show relevant hemagglutinating activity.[65,66]

The content of total inositol phosphates (TIP) did not show great differences between the species evaluated, except in jack bean. Germination decreased the content of TIP, being more obvious in cowpea, and the IP6 content diminished by 8–37%, with jack bean having the lowest content of phytic acid after the germination process (5.7 mg g^{-1} DM).[65,66] In addition,

Table 8.3 Antinutritional factors of a protein nature in non-germinated legume grains and in grains germinated for 96 hours (g kg^{-1} DM).[e]

	α-Amylase inhibitor	Trypsin inhibitor	Chymotrypsin inhibitor	Lectins[c]	Lectins[d]
Vigna unguiculata					
Non-germinated	ND	2.1 ± 0.13^a	1.6 ± 0.08^a	ND	ND
Germinated	ND	3.4 ± 0.12^b	1.5 ± 0.04^a	ND	ND
Canavalia ensiformis					
Non-germinated	ND	3.2 ± 0.18^a	0.7 ± 0.07^a	5.1 ± 0.06^a	—
Germinated	ND	0.7 ± 0.06^b	0.4 ± 0.08^a	2.5 ± 0.06^b	—
Lablab purpureus					
Non-germinated	ND	6.9 ± 0.14^a	1.6 ± 0.10^a	ND	10.2 ± 0.03^a
Germinated	ND	3.2 ± 0.09^b	1.0 ± 0.10^b	ND	2.5 ± 50.03^b
Mucuna pruriens					
Non-germinated	ND	1.4 ± 0.01^a	1.5 ± 0.15^a	ND	ND
Germinated	ND	1.3 ± 0.10^a	0.9 ± 0.08^b	ND	ND

[ab] Means with different letters differ at $P < 0.05$ (Duncan (1955)).
[c]Hemaglutination units or lectins equivalent (g)/kg in legume seeds in rats cells.
[d]Hemaglutination units or lectins equivalent (g)/kg in legume seeds in trypsinised rats cells.
[e]ND = non-detected.

the germination process modified the antioxidant activity depending on the species, as well as the content of phenolic compounds. An increase in total polyphenols and catechins was exhibited in all the legumes except in cowpea, and proantocianidins were increased only in jack bean and mucuna.[66] In general, the samples with the largest phenolic content had greater antioxidant capacity with a high correlation coefficient.

8.5 Biological Assessment of Germinated and Non-germinated Grains in Rats as an Experimental Model and in Broilers as a Species of Economic Interest

Several studies were performed to establish an experimental procedure to measure protein quality, up to the cell level, in germinated and non-germinated seeds of the non-conventional legumes, as well as to discover the impact of germination on the protein metabolism of growing rats as an animal model for monogastric species. Previous research developed an experimental procedure based on the biological evaluation of two germinated grain meals of seasonal legumes (*Vigna unguiculata* and *Lablab purpureus*) using Sprague Doyle rats of 21 days of age, as an experimental model for non-ruminant species.[67] The rats received five experimental diets formulated out of a basal diet (nitrogen free), with the addition (until 10% of protein content) of 1 control: casein (fat and vitamin free), and diets 2, 3, 4, and 5 corresponding to germinated or non-germinated cowpea and dolicho grain meals, respectively. Also, a trial with albumin, giving a value of 4% crude protein was used to control for endogenous secretions.[68] It was shown, that regardless of the germination process utilized (total light, partial light intervals, and total darkness), the germinated cowpea grain meal showed greater protein content than the non-germinated grain meal. In the case of the germinated dolicho grain meal, only the light intervals and the total darkness trials had greater protein content than the non-germinated grain meal, and without differences between them. However, the germination process did not show any effect on the energy content of this grain meal. On the other hand, determination of the protein quality through biological methods, such as biological value and true digestibility, indicated that the casein control diet had high true digestibility (93%) and biological value within the range 72–84%, similar to those previously obtained.[69] The cowpea trial presented lower results for casein,[69] but they were adequate quality indexes, thus 77% of the total nitrogen was absorbed (true digestibility) and 62% was retained for growing and maintenance (BV). The germination process did not increase the BV and true digestibility in rats fed the cowpea grain meal.[70–72]

The hematological indicators were maintained within the physiological values established for rats fed germinated or non-germinated cowpea grain meal.[73] The germination process allowed the survival of the rats fed dolicho grain meal, although indicators such as neutrophils and lymphocytes, some biochemical blood indicators (total serum protein, globulin, glutamic pyruvic transaminase) were out of the range for this species.[58] The consumption of meals of these legume seeds could produce potential damage to digestive organs because of the presence of secondary metabolites.[34,56,58,70] It was observed that the germination process did not have a positive effect on the relative weight (organ weight/post mortem weight) of liver and spleen in the rats fed cowpea or dolicho grain meals. From the results obtained, a more critical analysis may be required in order to include these feed sources in the production process.

Evaluation of the germinated and non-germinated grains was performed in broilers as they are a species with a short biological cycle demanding high nutritive value feed. This also means that results can be obtained in a short time. Different studies were conducted to determine the effect of the meals of germinated and non-germinated legumes (*Canavalia ensiformis*) on the nutritive value and the digestive physiology of the animals. Thus, isoproteic and isoenergetic diets were used to substitute for the traditional source of protein. *Canavalia ensiformis* is an important non-conventional protein source that, submitted to a simple and economical process, such as germination, could be a good feedstuff as an alternative to poultry. Therefore, studies[74,75] were conducted with male broilers that received control treatment (maize/soybean meal) and two experimental treatments (replacement of maize soybean meal diet (10%) with non-germinated jack bean grain meal or germinated jack bean grain meal submitted to a 96-hour germination process of light intervals).[74] Chemical composition of non-germinated and germinated grain meal showed differences in protein and ash values favorable to the latter, but it was observed that insoluble dietary fiber (IDF) was similar for both grain meals.[3] Consumption, excretion and dry matter apparent retention did not differ between treatments.

The germination process resulted in intermediate values of nitrogen consumption that differed in the control and jack bean non-germinated meal treatments, probably due to the diversity of secondary metabolites that decreased the nutritive value of proteins.[76] Among them, there are concanavaline A, an hemagglutinating lectin and canavanine, a non-protein amino acid of arginine.[34] Apparent fecal N retention for broilers that consumed non-germinated grain meal had a value of 44.7% *versus* 65.5% for germinated jack bean grain meal, thus the latter was similar to the control. In the other experiments,[75] the effect of germination on morphometric indexes of gastrointestinal digestive tract and accessory organs was studied, as well as in lymphoid organs related to the immune response. The experimental treatments did not affect morphometric empty relative weights, except for empty small intestine and ceca relative weight, which was higher in non-germinated jack bean grain meal. It should be noted that left cecum relative length, in broilers that consumed non-germinated jack bean grain meal, was also higher than in those that received the control and germinated grain meal treatments that exhibit similar morphometric indexes. This result may be due to the content of oligosaccharides (29% as raffinose) present in non-germinated jack bean grain meal[63] with values of 11.8 *versus* 0.6 g kg^{-1} DM in germinated jack bean grain meal. Germination is an efficient process because it increases the activity of the enzyme α-galactosidase that could not be hydrolyzed and absorbed in non-ruminant animals. This enzyme reduces the level of RFOs. Thus, cecum micro-organisms could not utilize these sugars, leading to the production of gases, with the corresponding increase of cecum length. Pancreas relative weight increased in animals fed non-germinated jack bean grain meal. In this respect, it has been reported that legumes contain antinutritional factors that could cause morphological changes in the digestive tract and accessory organs.[34,66,70] It was observed that pancreas hypertrophy was eliminated with the germination procedure proposed (pancreas relative weight was similar in the control and germinated jack bean grain meal treatments). This process improves nutrient absorption because it favors trypsin inhibition and enzyme utilization, as has been demonstrated.[56,77]

According to morphometric indexes of lymphoid organs of animals, those receiving non-germinated jack bean grain meal had a similar relative weight of Fabricius bursa to the control. Germination was unable to improve Fabricius bursa and spleen relative weight of birds that received this treatment, but it was effective in increasing thymus relative weight to a level comparable to the control values. With regard to morphometric indexes of lymphoid organs, there is good evidence that this macroscopic evaluation is directly related to the immunological response in birds.[78] Relative weight of bursa, spleen and thymus could decrease as a consequence of exposure to stress or toxic factors. In our case, systemic effects due to the presence of the lectins mentioned above must be considered, as has been stated.[70,79] The development of antilectins circulating antibodies of the type of IgG may occur, due to thymus atrophy in animals fed germinated jack bean grain meal. Thus, thymus weight is a comfort indicator due to the tisular atrophy response in the presence of glucocorticoids and stress factors.[80] It is suggested that jack bean grain meal germination does not alter the morphometric indexes of the digestive organs of broilers except for the small intestine and ceca, but it provokes beneficial changes in pancreas and thymus morphometric indexes, which in the case of the latter, could lead to a positive immune response.

In conclusion, the germination process used was an effective and promising method for the legumes studied by increasing nutritional quality, reducing antinutritional factors, and increasing bioactive phenolic compounds and antioxidant activity, which shows their potential for animal feeding and as a possible ingredient for the formulation of human foods of high value added. The nutritional studies showed increases in digestibility and the possibility of the use of grains in bird feeding. The procedure and experimental sequence used in this study provide a methodology of selection, collection, characterization and evaluation of sprouts that can be applied to other species and varieties of legumes with possibilities of use in feeding.

References

1. ECAL, *Economic Commission for Latin America and the Caribbean*, United Nations, 2016, p. 48.
2. http://www.fao.org/publications, last accessed June 2018.
3. M. A. Martín-Cabrejas, N. Ariza, R. M. Esteban, E. Mollá, K. W. Waldron and F. J. López-Andréu, *J. Agric. Food Chem.*, 2003, **51**, 1254.
4. R. Kumar and V. Kumar, *LWT--Food Sci. Technol.*, 2017, **75**, 51.
5. Blázquez, *Tesis de Licenciatura en Ciencias Químicas*, Universidad Autónoma de Madrid, 1999, p. 75.
6. Y. Aguilera, M. A. Martín-Cabrejas, V. Benítez, E. Mollá, F. López-Andréu and R. Esteban, *J. Food Compos. Anal.*, 2009, **22**, 678.
7. U. D. Chavan, J. K. Chavan and S. S. Kadam, *J. Food Sci.*, 1988, **53**, 1574.
8. J. L. Kiers, R. M. J. Nout and F. M. Rombouts, *J. Sci. Food Agric.*, 2000, **80**, 1325.
9. B. M. Warle, C. S. Riar, S. S. Gaikwad and V. A. Mane, *J. Environ. Sci., Toxicol. Food Technol.*, 2015, **9**, 13.
10. A. Quinhone Júnior and E. I. Ida, *Food Chem.*, 2015, **166**, 173.
11. J. Frías, C. Díaz- Pollán, C. L. Hedley and C. Vidal-Valverde, *J. Agric. Food Chem.*, 1995, **43**, 2231.
12. C. Uwaegbute, *Food Chem.*, 2000, **68**, 141.
13. Vidal-Valverde, J. Frías, I. Estrella, M. J. Gorospe, R. Ruis and J. Bacon, *European Conference on Grain Legumes*, Valladolid, Spain, 1988, p. 388.
14. N. M. Nnam, *Ecol. Food Nutr.*, 1999, **38**, 21.
15. X. Huang, W. Cai and B. Xu, *Food Chem.*, 2014, **143**, 268.
16. C. León, M. E. Torija and M. C. Matallana, *Bol. R. Soc. Esp. Hist. Nat., Secc. Biol.*, 2013, **107**, 47.
17. T. Mohsin and A. Abid, *J. Food Meas. Charact.*, 2017, **11**, 2076.
18. J. Frías, C. Díaz-Pollan, C. L. Hedley and C. Vidal-Valverde, *Z. Lebensm.-Unters. Forsch.*, 1996, **202**, 35.
19. V. Kalpanadevi and V. R. Mohan, *LWT--Food Sci. Technol.*, 2013, **51**, 455.
20. V. Uppal and K. Bains, *Nutr. Food Sci.*, 2012, **49**, 184.
21. R. M. R. Megat, A. Azrina and M. E. Norhaizan, *Int. Food Res. J.*, 2016, **23**, 257.
22. M. A. Azeke, S. J. Egielewa, M. U. Eigbogbo and I. G. Ihimire, *J. Food Sci. Technol.*, 2011, **48**, 724.
23. Y.-W. Luo, W.-H. Xie, X.-X. Jin, Q. Wang and Y.-J. He, *CyTA--J. Food*, 2014, **12**, 22.
24. Anuario Estadístico de Cuba (AEC), Oficina Nacional de Estadística e Información, 2017.
25. M. F. Díaz, C. Padilla, A. González and F. Curbelo, *Cuban J. Agric. Sci.*, 2001, **35**, 271.

26. M. F. Díaz, L. Savón, M. Martínez, V. Torres, G. Coto, M. A. Martín-Cabrejas, E. Lon Wo and M. Castro, in *Avances en Producción Sustentable de alimentos y Biotecnología Reproductiva*, Universidad Autónoma de Sinaloa, México, 2014, p. 147.
27. M. F. Díaz, M. Martínez, L. Savón and M. A. Martín-Cabrejas, in *Repensando la Ganadería*, Universidad Autónoma Metropolitana, Unidad Xochimilco, México, 2016, p. 51.
28. M. F. Díaz, V. Torres, A. González and A. Noda, *Cuban J. Agric. Sci.*, 2004, **38**, 87.
29. M. F. Díaz, M. A. Martín-Cabrejas, A. González, V. Torres and A. Noda, *Cuban J. Agric. Sci.*, 2007, **41**, 161.
30. M. F. Díaz, M. A. Martín-Cabrejas, A. González, V. Torres and A. Noda, *Cuban J. Agric. Sci.*, 2011, **45**, 433.
31. M. F. Díaz, V. H. Vega, M. A. Martín-Cabrejas and A. González, *Revista Computadorizada de Producción Porcina*, 2012, **19**, online version.
32. V. Torres, R. Cobo Cuña, L. Sánchez and N. Raez, *Livestock Research for Rural Development.*, 2013, **25**, online version.
33. M. F. Díaz Sánchez, M. Á. Martín-Cabrejas, M. Martínez Pérez, L. L. Savón Valdés, Y. Aguilera, V. Benítez, V. T. Cárdenas, G. Coto Valdés, A. González Conde, M. Sarmiento Menéndez and Y. Hernández Herrera, *Cuban J. Agric. Sci.*, 2017, **51**, 381.
34. J. P. F. D'Mello, in *Tropical Legumes in Animal Nutrition*, CAB International, Walingford, UK, 1995, p. 135.
35. M. F. Díaz, G. Coto, M. A. Martín-Cabrejas, A. González, M. Sarmiento, V. Torres, A. Noda, Y. Aguilera and V. Benítez, *INFOVET Especial*, 2013, **3**, 36.
36. Y. H. Kuo, P. Rozan, F. Lambein, J. Frias and C. Vidal-Valverde, *Food Chem.*, 2004, **86**, 537.
37. A. Iqbal, I. A. Khalil, N. Ateeq and M. S. Khan, *Food Chem.*, 2006, **97**, 331.
38. M. C. Carrao and J. M. Gontijo, in *El cultivo de la soja en los trópicos: mejoramiento y producción*, FAO, Argentina, 1995, p. 254.
39. A. C. Laurena, F. M. Rodríguez, N. G. Sabino, A. F. Zamora and E. M. T. Mendoza, *Plant Foods Hum. Nutr.*, 1991, **41**, 59.
40. M. Vasconcelos, F. M. Machado Maia, D. F. Farias, C. C. Campello, A. F. Urano Carvalho, R. de Azevedo Moreira and J. T. Abreu de Oliveira, *J. Food Compos. Anal.*, 2010, **23**, 54.
41. R. Campos-Vega, G. Loarca-Piña and B. D. Oomah, *Food Res. Int.*, 2010, **43**, 461.
42. S. Wang, P. John, R. Melnyk, M. Tsao and F. Marcone, *Food Res. Int.*, 2011, **44**, 14.
43. C. N. Kunyanga, K. I. Jasper, M. W. Okotha, K. B. Hans and V. Vellingiri, *Food Sci. Technol.*, 2012, **45**, 269.
44. N. M. Nnam, *Ecol. Food Nutr.*, 1999, **38**, 21.
45. S. K. Arora, in *Tropical Legumes in Animal Nutrition*, ed. J. P. F. D'Mello and C. Devendra, CAB International, 1995, 96p.
46. S. S. Nielsen, T. A. Ohler and C. A. Mitchell, in *Advances in Cowpea Research*, Jackai, Int. Res. Center Agric Sc., Japan, 1997, p. 326.

47. M. F. Díaz, *Doctoral Thesis*, Instituto de Ciencia Animal, La Habana, 2000, p. 100.
48. W. A. Plahar, N. T. Annan and C. A. Nti, *Plant Foods Hum. Nutr.*, 1997, **51**, 343.
49. V. A. Onwuliri and J. A. Obu, *Food Chem.*, 2002, **78**, 1.
50. C. Vidal-Valverde, J. Frías, I. Estrella, M. J. Gorospe, R. Ruis and J. Bacon, *J. Agric. Food Chem.*, 1981, **42**, 2291.
51. R. A. Ghavidel and J. Prakash, *LWT--Food Sci. Technol.*, 2007, **40**, 1292.
52. C. Michelangeli and R. E. Vargas, *J. Nutr.*, 1994, **124**, 1081.
53. J. O. Agbede and V. A. Aletor, *J. Food Compos. Anal.*, 2005, **18**, 89.
54. A. B. I. Udedibie and C. R. Carlini, *Anim. Feed Sci. Technol.*, 1998, **74**, 95.
55. V. Ravindran and G. Ravindran, *J. Sci. Food Agric.*, 1988, **50**, 71.
56. M. A. Martín-Cabrejas, M. F. Díaz, Y. Aguilera, V. Benítez, E. Mollá and R. M. Esteban, *Food Chem.*, 2007, **107**, 1045.
57. V. Benítez, S. Cantera, Y. Aguilera, T. Jiménez, R. T. Sarmento, M. F. Díaz and M. A. Martin-Cabrejas, *Nutr. Clin. Diet. Hosp.*, 2011, **31**, 75.
58. V. Benítez, S. Cantera, Y. Aguilera, E. Mollá, R. M. Esteban, M. F. Díaz and M. A. Martín-Cabrejas, *Food Res. Int.*, 2013, **50**, 64.
59. P. Kadlec, J. Dostalova, J. Bernaškova and M. Skulinova, *Czech J. Food Sci.*, 2008, **26**, 99.
60. S. Seena and K. R. Sridhar, *Food Res. Int.*, 2006, **38**, 803.
61. D. Jezierny, R. Mosenthin and E. Bauer, *Anim. Feed Sci. Technol.*, 2010, **157**, 111.
62. G. Ermetice de Almeida, K. Queiroz-Monici, R. Machado and A. Costa de Oliveira, *Food Chem.*, 2006, **94**, 327.
63. M. A. Martín-Cabrejas, Y. Aguilera, M. M. Pedrosa, C. Cuadrado, T. Hernández, S. Díaz and R. M. Esteban, *Food Chem.*, 2009, **114**, 1063.
64. Y. Aguilera, R. M. Esteban, V. Benítez, E. Mollá and M. A. Martín-Cabrejas, *J. Agric. Food Chem.*, 2009, **57**, 10682.
65. Y. Aguilera, V. Benítez, M. F. Díaz, T. Jiménez, S. Cantera and M. A. Martin-Cabrejas, *Nutr. Clin. Diet. Hosp.*, 2011, **61**, 75.
66. Y. Aguilera, M. F. Díaz, T. Jiménez, V. Benítez, T. Herrera, C. Cuadrado, M. Martín-Pedrosa and M. A. Martín-Cabrejas, *J. Agric. Food Chem.*, 2013, **61**, 8120.
67. L. Savón, B. González, M. E. Arteaga, C. Forte, M. Orta, V. Torres and M. F. Díaz, in *Memorias del Evento Internacional Porcinocultura*, La Habana, 2008.
68. I. R. Lloyd, B. E. Mc Donald and E. W. Crampton, *Fundamentos de Nutrición*, Acribia, Zaragoza, Spain, 1982.
69. L. A. Aguirre, L. Savón, A. Oramas, L. E. Dihigo and V. Rodríguez, *Cuban J. Agric. Sci.*, 1998, **32**, 75.
70. R. Belmar-Casso and R. Navas, in *VII Encuentro de Nutrición y Producción de Especies Monogástricas*, Universidad Experimental de los Llanos Ezequiel Zamora (UNELLEZ) Guanare, Portuguesa, Venezuela, 2015, p. 51.
71. D. León, A. Blanco, M. B. O. Peña, M. E. González, A. M. Arteaga, Y. Bada, González and A. Mancebo, *Revista electrónica de veterinaria. RED Vet*, 2011, **12**, http://www.veterinaia.org/revistas/redvet.

72. M. Martínez, M. F. Díaz, Y. Hernández, M. Sarmiento and F. Sierra, *Livestock Research for Rural Development*, 2013, **25**, 120.
73. A. Emiola, A. D. Ologhobo and R. M. Gous, *Poult. Sci.*, 2007, **86**, 1234.
74. L. Savón, M. F. Díaz, Y. Hernández, M. Sarmiento and F. Sierra, in *Memorias XXIV Congreso Asociación Latinoamericano de Producción Animal*, Palacio de Convenciones, La Habana, Cuba, 2015, ISBN 978-607-715-305-4.
75. L. Savón and M. F. Díaz, in *Convención Agrocentro, Ciego de Ávila, Cuba*, 2018, ISBN 978-959-250-973-3.
76. M. Múzquiz, A. Varela, C. Burbano, C. Cuadrado, E. Guillamón and M. Pedrosa, *Phytochem. Rev.*, 2012, **11**, 1.
77. C. Vidal- Valverde, J. Frías, I. Sierra, L. Blázquez, F. Lambein and Y. H. Kuo, *Food Res. Technol.*, 2002, **215**, 472.
78. F. Perozo-Marín, J. Nava, Y. Marváez, E. Arenas, P. Serje and M. Briceño, *Revista Científica FCV-Luz*, 2004, **14**, 217.
79. A. Puztai, in *Recent Advances of Research in Antinutritional Factors in Legume Seeds*, ed. J. Huismana, T. F. B. Van der Poel and I. E. Liener. PUDOC, Wageningen, 1989, p. 17.
80. M. Pulido, S. Barceló and C. Perera, *Revista Cubana de Ciencia Avícola*, 2000, **25**, 45.

CHAPTER 9

Impact of Fermentation on the Nutritional Quality, Bioactive Compounds and Potential Health Properties of Legumes

J. FRIAS*, C. MARTÍNEZ-VILLALUENGA AND E. PEÑAS

Institute of Food Science, Technology and Nutrition (ICTAN-CSIC), Department Characterization, Quality and Safety of Foods, Jose Antonio Novais 10, Madrid, 28040, Spain
*E-mail: frias@ictan.csic.es

9.1 Introduction

Fermented foods are one of the key components of the traditional cuisine of every community in the world with a preponderant position in contemporary food markets.[1] Although fermentation was developed by ancient populations to preserve perishable foods for consumption, the enhancement of the sensory attributes and the variety of nutritive benefits are globally acclaimed.[2] Wine, cheese, yogurt, bread, beers, miso, tempeh and idli are typical fermented foods that remain an important part of diets throughout the world. These foods are not only popular, but they are also considered healthy. Many of these claims revolve around the microbial and biochemical changes that occur during fermentation through positive modifications that

Food Chemistry, Function and Analysis No. 8
Legumes: Nutritional Quality, Processing and Potential Health Benefits
Edited by María Ángeles Martín-Cabrejas

Published by the Royal Society of Chemistry, www.rsc.org

give rise to nutritive and health benefits.[3] Among the attractive fermented foods that are consumed worldwide, legume-derived fermented food production has emerged as a strategy to achieve food security by providing rich nutrient-dense derivatives with several advantages over other conventional feasible methods of legume processing.[4]

Grain legumes, also called pulses, are considered staple food for a large part of the world population. They have been cultivated for centuries and consumed as part of the basic human diet. Pulses are nutrient-dense foods that provide not only valuable amounts of carbohydrates, proteins, low amounts of lipids (with the exception of soybean in which oil is the main product), fibre, B vitamins, tocopherols and minerals but also bioactive compounds with healthy attributes.[5,6]

The antiquity of fermented legumes has been documented precisely in ancient Chinese literature called *Shi-ji* written by Si-Ma-Qian in S. II.[7] Through the centuries, the transformation of legume dishes around the world has been mostly due to the migration of ethnic Chinese to Japan and other surrounding countries and the acceptance of legumes as a cultural food by non-Asiatic communities. Low-cost, high-value and culturally accepted legume fermented foods are consumed as staple foods in many countries of the globe. They form part of the diet as main course dishes, pastes and accompaniment, seasoning and, sometimes, as meat substitutes for vegetarian or vegan individuals worldwide. Although fermented soy products are widely consumed around the world, soybeans are not considered a legume crop,[8] therefore these products will not be included in this chapter.

A significant proportion of the world legume production is processed by fermentation prior to consumption. As a catabolic process, fermentation causes desirable biochemical changes in legume composition due to the metabolism of the microorganisms involved, and also the activity of hydrolytic endogenous legume enzymes that degrade macro- and micro-components. The microorganisms implicated in fermentation activate their enzymatic machinery to hydrolyse and metabolise legume constituents resulting in the production of valuable derived compounds of nutritive and functional interest. In addition, microorganisms have the ability to produce antimicrobial compounds to preserve fermented food from undesirable microflora and develop desirable organic acids that contribute to the improvement of legume flavour, appearance, texture, nutrient digestibility and nutritional quality as well as health properties.[9] As a consequence of the anabolic process, fermentation brings about a decrease in the non-nutritional components, *i.e.*, legume constituents considered as antinutritional factors in the past. These compounds show different bioactivities when they are consumed in certain doses, as antioxidants, hypolipidemic, hypoglycemic and chemopreventive agents, contributing to the desirable composition of fermented legume products and providing a potential health benefit for consumers.[10] All these features have been attributed to the occurrence of their nutrients and phytochemicals, as well as the presence of alive microorganisms involved in fermentation that can exert probiotic properties.[11] Therefore, fermented legume-derived products are gaining tremendous attention nowadays and they are considered natural, nutritive and healthy foods.[12]

In this context, the aim of this chapter is to collect the most recent and remarkable scientific literature describing the biochemical changes in nutritional and bioactive compounds of legumes that occur during fermentation and to highlight the beneficial effects to health of fermented foods using the available scientific evidence.

9.2 Traditional Fermented Legumes Around the World

Preservation and safeguarding of foods and beverages remain the principal objective of fermentation, with wholesomeness, acceptability and overall quality, having become increasingly valued features to consumers, especially in rural areas where old traditions and cultural particularities in food fermentations are generally well maintained. Fermented legume products have been traditional symbols of cuisines worldwide over the centuries. They have been the result of simple household procedures that generally require very few ingredients and minimal processing. Eventually, fermentation of foods has been developed into a skilled practice offering improved nutritional and sensory qualities with healthy enhanced/therapeutic properties.[3]

A large variety of pulses are consumed around the globe as staple foods and part of the world legume production is processed by fermentation prior to consumption. Fermentation makes raw grain legumes into edible attractive products full of protein and energy as staple foods especially for low-income populations. It is usually pursued as a family business that becomes an art handed on from one generation to another. Diversity of legume grains and strains have created a wide variety of ethnic foods with unique sensory and nutritious features. Different countries in Asia, India, Africa and Latin America have traditional names for fermented legume food products, based on preparation and type of microbial strain involved in fermentation, resulting in differences in consistency, flavour, and nutritional attributes, and removal of antinutritional factors. Persistent references to health properties and therapeutic effects of artisan and traditional fermented pulses are gaining significance day by day for their preventative/alleviating effects on various degenerative and chronic diseases.[5,12] The alleged hypotensive and antidiabetic effects of *ugba*[13] a Nigerian traditional fermented food condiment, have gained attention. Similarly, *Dawadawa* from locust bean, *dhokla* and *khaman* from bengal gram or *dosa* and *idli* from black gram are some legume-based fermented foods of different countries whose mode of preparation has been extrapolated to other autochthonous legumes that form an important part of traditional diets.[14] Initially, the traditional manufacture of these fermented products included soaking of the seeds and fermentation initiated by the indigenous microorganisms present in the seed or inoculated during fermentation. Fermented legumes harbour a diversity of microorganisms responsible for their chemical changes and organoleptic properties, mostly mycelial *Rhizopus* ssp moulds, yeasts and bacteria including lactic acid bacteria (LAB) and *Bacillus* ssp.[5] *Tempeh* is one of the fermented legume products that initially originated in Central Java (Indonesia) and spread worldwide over the years. It is a well-known traditional fermented soybean food consumed as a staple source of protein in Asia, however other grain substrates such as chickpea, ground bean, horse gram, mung bean, lima bean and pea are used as a substitute for soybean. Therefore, *tempeh* is the collective name for cooked and fermented beans penetrated and bound together by the mycelium of *Rhizopus* mold that has been traditional in the diets of vegetarians in developing countries and nowadays is part of the diet of large populations around the globe.

In general, microorganisms involved in legume fermentation are responsible in one way or another for (i) dietary enrichment through development of a diversity of flavours, aromas and textures of the legume matrix, (ii) preservation through lactic acid, alcohol, acetic acid and alkaline fermentations, (iii) biological fortification of fermentation derived foods with proteins, essential amino acids, essential fatty acids, vitamins, and minerals, (iv) detoxification by degrading undesirable compounds, (v) enhancement with antioxidant and antimicrobial properties, (vi) probiotic vehicles and (vii) decreasing cooking times and fuel requirements.[14] All of these features have made traditional fermentation-based pulses contributors to the nutrition and well-being status of large population groups around the world, mostly in developing countries, and are having large repercussions in developed ones.[13] Notwithstanding the transcendence in safety and nutrition of traditional fermented legume products, their importance in modern-day life is underlined by the wide spectrum of such foods marketed both in developing and industrialized countries. Fermented legume foods are treasured as major dietary constituents in numerous developing countries because they maintain their quality under ambient conditions – thereby contributing to food security – and because they add value, enhance nutrient-dense quality and digestibility, improve food safety, and are traditionally acceptable and accessible worldwide. Fermentation is a low-input enterprise and provides individuals with limited purchasing power access to safe, inexpensive and nutritious foods, contributing to food security in all corners of the planet.

9.3 Modifications in Legume Composition by Fermentation and Associated Health Benefits

Fermentation is a biological process in which the metabolic activity of the responsible microbiota, in addition to the endogenous enzymatic activity of pulses, can transform their nutritive and non-nutritive compounds resulting in positive benefits for human nutrition and health. Grain legumes are nutritionally attractive due to their high protein content (20% in chickpeas to 40% in lupins), starch (60% in cowpeas to 64% in lentils, and, exceptionally, low in lupins), dietary fiber (11% in cowpeas to 25% in dry peas) and low content of lipids (with the exception of lupins) providing around 300–400 kcal/100 g. They are important sources of minerals (Fe, Ca, Zn, Se), vitamins (mostly water soluble vitamins) and phytochemicals of physiological relevance (such as phenolic compounds or prebiotic oligosaccharides, among others). The whole composition of pulses contributes to their health benefits, such as prevention of chronic diseases including cardiovascular diseases, diabetes, obesity and metabolic syndrome and, even, as chemopreventive agents (see included chapters).

Nevertheless, in the complex composition of grain legumes, there are several compounds, called antinutritional factors in the past due to their detrimental effects, with the dual features of being non-nutritive compounds but conferring healthy properties when they are consumed in certain doses. The processing of legumes before consumption is generally recommended in order to remove those negative compounds that, although minor in content are large in chemical diversity and limit the full nutritive potential of pulses. This group is led by galacto-oligosaccharides (flatulence-causing compounds), protease inhibitors (reduce protein digestibility) and phytic acid (limit mineral bioavailability). Additionally, some of these constituents might alter the sensory qualities, contributing to bitterness and astringency (tannins) or texture (dietary fibre may be hard to cook), reducing the acceptance of legume food products.[15]

The positive features in protein quality and composition during lactic-acid fermentation have been well documented. In the production of *idli*, protein digestibility and limiting amino acid scores improved.[16] During the fermentation of black gram (chickpeas), the proteinase activity increased, producing an increase in total nitrogen and available soluble proteins, with an enhanced balance of essential amino acids.[17] Similarly, fermentation of mung bean in a *dosa*-like product resulted in a higher nitrogen, protein and free amino acid content,[18] while kidney bean *tempeh*-like derivative was enriched in protein (+21.7%) and provided more sulphur amino acids (Met + Cys) as well as six other essential amino acids (score = 0.91 *versus* 0.76 in unfermented beans), improving the *in vitro* protein digestibility and the calculated protein efficiency ratio of the fermented pulse product.[19]

Fermentation of lentils by *Lactobacillus plantarum* causes the release of antihypertensive peptides and the biosynthesis of γ-amino butyric acid (GABA),[20] a non-protein amino acid with key physiological functions as an inhibitory neurotransmitter of the central nervous system with parallel actions as a hypotensive, antidiabetic, diuretic and tranquilizer.[21] *Udga* products from Nigeria local kidney beans provide essential amino acids to meet the FAO requirements, improving the nutritional status of a young population.[17] In *tempeh*-like foods from common beans, chickpea and Bambara groundnut, higher *in vitro* protein digestibility, total amino acid content, available Lys and protein efficiency ratio (PER) were found. At the same time, the levels of the trypsin inhibitors included in the Bowman–Birk class of proteins were largely reduced during fermentation and, hence, the protein quality was enhanced.[22] Our group has led the initiative of reducing trypsin inhibitors during lactic acid fermentation of legumes.[23] LAB provide a complex and active proteolytic system that plays an important role in legume fermentation, degrading protein inhibitors leading to bioactive peptides in different fermented faba bean products.[24] Similarly, fermentation carried out by *Bacillus* ssp. in cowpea, chickpeas and ground beans led to a large decrease in trypsin inhibitors and, hence, an increase in protein digestibility.[25] On the other hand, fermentation-derived products from dry peas, cowpeas and kidney beans with *Saccharomyces cerevisiae* maintained the activity of trypsin inhibitors at more than 50%.[26] These results show the different effect of fermentation conditions on these kinds of proteins, which have to be considered as bioactive compounds with dual effects since recent research has shown the ability of Bowman-Birk proteins to inhibit the growth of different cancer cell lines *in vitro*, and hence increased the interest in this class of protein for use as nutraceutical products.[27,28]

Fermented legumes are a good source of carbohydrates.[29] During natural fermentation, microbial and endogenous amylases are activated to hydrolyze starch. Although hydrolysis of starch during fermentation led to a decrease in amylose and amylopectin, the remaining amount is still difficult to digest. Fermented legumes have shown low values of hydrolysis index and, hence low glycaemic index, an attractive property of legumes for diabetic patients.[30] Hydrolysis of starch during fermentation concomitantly increases reducing sugars that favour the fermentative process[31,32] promoting lactic acid growth and, hence, a drop in pH associated with lactic acid production.[1] An enhanced digestibility of starch has been reported in naturally fermented pulses[31] and in different *tempeh*-like products originating from *Bacillus* ssp.[33] while fermentation with *Rhizopus oligosporus* and *Aspergillus oryzae* led to lower sugar bioavailability.[34,35] In addition, fermented legumes provide larger resistant starch than the unprocessed seeds, contributing to the total content of dietary fiber whose consumption is recommended for intestinal health, and they also enhance satiety, reduce energy intake and control post-prandrial glycaemia.[29]

Dietary fiber is one of the attractions of fermented pulses. As well as resistant starch, insoluble dietary fiber increased significantly during natural and induced lactic acid fermentation of common kidney beans while minor changes have been reported for the soluble fraction.[36,37] In naturally fermented lentils, a decrease in soluble pectins and hemicelluloses was found while insoluble cellulose and lignin content remained almost constant.[38] Natural fermentation contributed also to drop in total dietary fibre in a bean cake, and this was mostly attributed to the decrease in the soluble fraction.[36,39] The content of dietary fiber is associated with benefits in lipid metabolism, the fiber acting as an efficient lipid lowering compound by trapping fatty acids and bile salts while passing through the small intestine without digestion, thus inhibiting fat absorption in the small intestine.[40] In this sense, *idli* products have been recognized as a dietary fiber source with concomitant cardiovascular disease prevention attributes in India.[41]

Pulses consumed by humans present low lipid content ranging from 2% in lentils and peas to 6% in chickpea. However, most of the 70% of lipids are constituted by polyunsaturated fatty acids (PUFA) of relevant physiological importance such as oleic (18:1), linoleic (18:2) and linolenic (18:3) acids.[42] Different results have been reported regarding lipid composition after legume fermentation. *Tempeh*-like chickpea presented 50% less lipid content than the original flour,[43] however in *tempeh*-like cowpea and canavalia products, lipid content increased slightly but PUFA were mostly represented by linoleic (Ω-3) and linolenic (Ω-6) acids.[44,45] It has been shown through a meta-analysis study that diets rich in these PUFA contribute to lower risk of cardiovascular diseases associated with lower levels of serum lipids and apolipoproteins[46] and thus fermented legumes could contribute to improved heart health status in the world population.

Legumes are recognized to be a source of low molecular weight carbohydrates (2–12%) so called α-galactosides, galacto-oligosaccharides as well as raffinose family oligosaccharides (RFOs) that are strongly associated with the flatulence, an issue that sometimes arises after legume consumption. Such carbohydrates are mostly represented by the trisaccharide raffinose, the tetrasaccharide stachyose and the pentasaccharide. RFOs are non-digestible sugars resistant to the digestive enzymes that pass entirely to the large intestine where they are fermented by colon microbiota producing gasses and, hence, abdominal discomfort. In contrast, α-galactosides exert prebiotic activity stimulating the growth and activity of beneficial intestinal bacteria potentially associated with gut health and wellbeing. Besides these positive effects, the presence of α-galactosides is one of the reasons why people limit their consumption of legumes. Fermentation is clearly one of the most studied processes to reduce the content of α-galactosides to relieve the associated abdominal discomfort and flatulence and, hence, directly convince people to increase their legume consumption.[47] Our group have made a significant contribution in studying the nutritional composition of naturally and controlled lactic acid fermented legumes and it was observed that these processes bring about a sharp reduction in the content of α-galactosides (70–100%) in lentils, kidney beans, cowpeas and pigeon pea.[48–55] As an example, lactic acid fermentation of kidney beans by naturally occurring bacteria and LAB-controlled fermentation by the addition of a commercial starter (a mixture of *Lactobacillus acidophilus*, *Bifidobacterium* and *Streptococcus thermophilus*) led to a significant reduction in galacto-oligosaccharides, but natural fermentation was more effective than induced fermentation in reducing the flatulence-producing factors (95% and 11%, respectively, after 96 h of fermentation).[56] Similar results have been published in *natto*-like cowpeas produced with *Bacillus subtilis*,[57] while the production of *tempeh*-like grass pea by fungal fermentation only reduced α-galactosides partially and it was less effective in the prevention of intestinal cramps.[58]

Fermentation can also reduce the content of phytic acid in legumes, the main phosphorus storage compound in seeds responsible for chelating minerals of physiological relevance such as Ca, Fe and Mg and, therefore, reducing their bioavailability.[55] Phytic acid content of kidney beans underwent a significant decrease of 58% after solid state fermentation.[59] Native phytase enzymes from pulse flour are activated during fermentation and, in addition to those provided by the microorganisms involved, are able to hydrolyse the reactive phosphate groups and thus enhancing the bioavailability of essential minerals. Lactic acid fermentation caused a large reduction of phytic acid in Indian *idli* and *dosa* breakfast and, consequently, the bioaccessibilities of Ca, Fe and Zn were improved.[60] With this premise, the traditional consumption of these staple foods is recommended for Indian children to prevent mineral malnutrition and anaemia among the young populations.[61]

Fermented legume products are a rich source of B group vitamins.[62] *Idli* and *dosa* products provide larger amounts of thiamine, riboflavin, niacin, folic acid and cobalamin than the unfermented black gram flours, and fortification with mung bean led to higher content of these water-soluble vitamins.[63] Similarly, *tempeh*-like Bambara groundnut products contain higher levels of riboflavin, folic acid, niacin and biotin than the raw seeds.[64] In addition, naturally fermented cowpea products are a source of vitamin E, the most active natural liposoluble antioxidant.[55]

Well-known bioactive components produced and made available through fermentation include phenolic compounds, which may act as natural antioxidants.[65,66] In legumes, polyphenols are present in free form and conjugated with other organic acids or sugars groups as soluble phenolics and covalently bound with cell wall macromolecules such as polysaccharides and proteins as bound phenolics.[66] The former can be easily hydrolysed and absorbed in the upper intestinal tract while the latter reach the colon and are fermented and released by the colonic microbiota, exerting versatile health effects.[67] Phenolic solubilization has been described during lactic acid fermentation of chickpeas, faba beans, grass pea, lentils and black bean[68] and an enhancement of phenolic bioavailability by the release of free soluble forms from the esterified bound compounds in pigeon peas, Bambara groundnut, African yam bean and kidney beans has been also reported.[69] Natural and LAB-mediated fermentation significantly enhanced the phenolic content in the soluble and bound fractions of black gram, mottled cowpea, speckled kidney bean, lentil and small runner bean compared with non-fermented seeds. Among them, mottled cowpeas showed the highest phenolic content and their fermentation-derived products contained increased amounts of catechin, quercetin, ferulic acid and *p*-coumaric acid, while protocatechuic acid was identified in both soluble and bound fractions.[70] These authors explained that phenolic compounds in mottled cowpea could be transformed or metabolised differently depending on the naturally present microorganisms and LAB-inoculated strains involved in the fermentation process since they produce different enzymes such as β-glucosidase, decarboxylases, reductases and esterases that can metabolise, transform and release free forms of polyphenols and other products that are more easily absorbed.[71] Similar results were found in naturally fermented lentils where the content of *p*-hydroxybenzoic acid, protocatechuic acid and catechin increased, while hydroxycinnamic acid and procyanidin dimers decreased.[72] Lentils contain large amounts of proanthocyanidins, many of which are catechin polymers and derivatives that throughout the fermentation process could result in an increase in catechin monomers. The release of free shikimic acid, chlorogenic acid, rutin, daidzein, genistein and biochanin A has been also reported in fermented chick peas.[73]

The improvement of legume antioxidant activity due to the fermentation process has been widely reported and it has been mostly attributed to an increase in phenolic compound content.[74] Naturally fermented cowpeas brought about an increased inhibition of lipid oxidation and Trolox equivalent antioxidant capacities that were correlated to the rise observed in total phenolic compounds and vitamin E.[75] Tyrosol, vanillic, *p*-coumaric and ferulic acids increased significantly in fermented cowpeas, highlighting the large amount of quercetin (11–22 $\mu g\ g^{-1}$), a flavonoid with antioxidant implications that was not identified in unprocessed seeds.[74] In this line, fermented chickpea products presented an increased antioxidant capacity which correlated positively ($r = 0.88$, $P \leq 0.05$) with the total phenolic content.[76] Similarly, kidney bean *tempeh*-like products exhibited more antiradical activity (+43%) and antioxidant capacity (+38%) than unfermented seeds, values that were correlated with the large phenolic content.[77] A significant higher free radical scavenging activity and inhibition of lipid oxidation power during fermentation of underutilized varieties of pigeon pea, Bambara groundnut, African yam bean and kidney beans was associated with the content of free soluble phenolics.[69] Similarly, an enhanced antioxidant activity in LAB-mediated fermentation of chickpeas, faba beans, grass pea, lentils and black bean was also correlated with an increase in phenolic solubilisation.[68]

The comprehensive scientific evidences of fermentation-derived legumes and how they may provide health benefits had led to targeted identification of certain vitamins, minerals, amino acids, essential fatty acids, dietary fiber, probiotics and phytochemicals that distinguish fermented legumes from their unfermented forms, as described above. However, only a few recent studies have explored the role of fermentation-derived foods and their ingredients in health promotion such as antihypertensive activity,[20] glucose management,[78] lipid-lowering properties and hepatic activity of antioxidant enzymes,[79,80] chemoprotective agents,[27,81] antiinflammatory and healthy aging[82] and probiotic properties.[83] These findings for fermented legumes substantially advance the existing evidence and identify the consumption of fermented legume-derived products as a dietary strategy to fight non-communicable diseases.

9.4 Fermentation of Legumes is an Ancient Technology for Modern Times

Fermentation as a diversification enterprise offers many opportunities as a result of the global popularity of legumes. Fermented products are part of many social, cultural and consumption patterns. Despite the nutritional benefits of legumes being well recognised, the downward trend in their consumption in developed countries is a matter of concern nowadays. This fact seems to be associated with long cooking time, modern life style, low consumer acceptance of the flavours and textures of legumes, flatulence and adherence of consumers to calorific and ready to eat food products, among others. Nevertheless, the potential health benefits of consuming more pulse foods is challenging and new pulse-based products are being extensively studied.

It is firmly recognized that fermentation can serve as a valuable technology to produce natural, traditional and healthier foods of nutritional interest. Furthermore, in recent years, fermented legumes have emerged with significant strength as functional ingredients to develop new and desirable foods attractive for consumers. Sourdough fermentation has a well-established role in improving texture, flavour and palatability of breads, one of the foods about which consumers around the world are most demanding, and, hence, there is a growing interest in improving its nutritional quality and bioactive compounds profile.[84] Wheat sourdough fermentation and inclusion of legumes carried out with LAB is becoming one of the most efficient tools to improve the quality of breads and their healthiness by the supplementation or replacement of wheat flour with legumes. The sourdough legume fermentation can provide higher protein amounts of better nutritional quality with improved amino acid profile (lysine, glutamic and aspartic acids, among others) compared to cereal grains, the products produced are fibre-rich and gluten-free, retard starch digestion (low glycaemic index products), improve mineral bioavailability and also provide aromatic and pleasant flavours.[85] Beyond the nutritional benefits, legume sourdough can serve as a source of probiotic and bioactive compounds with protective effects against hypertension, cardiovascular diseases, diabetes, overweight, obesity and other highly prevalent cardiometabolic diseases.[86]

The Italian group headed by Dr M. Gobbetti has performed extensive research work during the last decade exploring the suitability of different species of common and underutilized legumes such as kidney beans, chickpeas, *Lathyrus*, lentils, faba beans, and peas for application in sourdough fermentation with LAB. Besides the mentioned enhancement of the nutritional value, α-galactosides were degraded, pyrimidine glycosides from faba bean almost eliminated, trypsin inhibitor activity reduced, condensed tannins hydrolysed and phytic acid degraded by the action of pulse endogenous and microbial enzymes.[85,87–92] In addition to the reduction of non-nutritive compounds, large amounts of phenolic compounds and soluble fibre were reported.[89] Legume sourdough includes hydrolysed proteins with improved digestibility, bioactive peptides and free amino acids, especially essential amino acids and GABA.[91] As a result of sourdough fermentation of nineteen Italian legumes, Rizzello *et al.*[93] found an increase in the number and intensity of lunasin-like polypeptides bands as a consequence of the proteolysis of the native proteins by selected LAB. Nine of them exhibited similarity to lunasin, a 43-amino acid peptide with anticancer, anti-inflammatory, antioxidant and cholesterol lowering activities.[94,95]

Fermented legumes have promising potential as functional ingredients. This concept has been fueled by the concern of informed consumers who demand added-value food products providing components with potential health benefits. As health concern is increasing and the functional food industry is growing steadily worldwide, fermented legume ingredients can be tailored to maximize biofunctional activities. Water soluble fractions of fermented lupins, lentils and beans both naturally or by LAB exhibited larger proteolysis and were proven as particularly promising functional ingredients in preventing hypertension.[20,96] Our research group is one of the leading teams showing the effectiveness of the serine-protease enzyme *Savinase*® 16L, a proteolytic subtilisin from *B. subtilis* acting under pH-controlled fermentation conditions, that simultaneously releases multiactive peptides and phenolic compounds from legumes.[97,98] In pinto beans, identified bioactive peptides derived mainly from vicilin, convicilin and legumin showed high dual antioxidant and angiotensin converting enzyme (ACE) inhibitory activity.[97] In addition, the combined esterase and protease activities of *Savinase* released phenolic compounds bound to polysaccharides and proteins, increasing their bioaccessibility in pinto bean.[98]

In lentils, just recently, we have explored the potential of pH-controlled fermentation by LAB combined with *Savinase* as a tailored strategy for the production of multifunctional ingredients containing bioactive peptides and phenolic compounds, both of which are involved in the prevention of metabolic syndrome, and made efforts to elucidate the mechanisms of action by which changes in the biological activities occur. As mentioned previously, natural LAB fermentation reduces the pH to acidic conditions making proteins, peptides and phenolic compounds almost insoluble.[20] Moreover, the proteolytic system of *L. plantarum* does not encode the extracellular protease *prtP*, that is involved in the primary breakdown of proteins into oligopeptides, within the cell envelope proteinases (CEPs).[99] However, LAB are equipped with specific oligopeptide transport (*Opp* and *Dtp*) systems and intracellular proteases that easily degrade oligopeptides to peptides and amino acids that are transported further on into the bacterial cell allowing growth.[100] Therefore, our fermentation strategy was to provide neutral and soft-alkaline conditions to enhance protein solubility (enzymes included) so that, throughout the *Savinase* proteolytic system, an enrichment in peptides in the soluble fraction (ingredient) can be achieved.[101] In addition, the metabolic system of *L. plantarum* produces extracellular carbohydrolases, β-glucosidases and a wide range of esterases which are highly specific and effective in releasing bound phenolic compounds that are cross-linked to cell-wall components in legumes and are also soluble at pH 7–10[102,103] and, hence, is responsible for the bioaccesibility and enhanced biological activity of the phenolic compounds. The feasibility of these combined fermentation conditions with the addition of *Savianse* enzyme reinforces the proteolytic activity of LAB (with reduced proteolytic activity in comparison with *Bacillus*)[104] providing, at the same time, larger total phenolic content and composition.[105] Our outcomes demonstrated that combined fermentation with *L. plantarum* CECT 748 along with *Savinase*® 16 L in alkaline conditions (FLPS) increased peptide content and some flavonoids as well as phenolic acids like *p*-hydroxybenzoic acid, vanillic acid, isorhamnetin glucuronide and kaempferol glucosides,[98,101] possibly due more to microbial enzymatic activities than to exogenous enzyme activity from *Savinase*. These results match with those reported by Khaw *et al.*,[106] in fermented mulberry juice. Free *p*-coumaric acids may be decarboxylated or reduced by *L. plantarum* phenolic acid decarboxylases or reductases to the corresponding phenol or vinyl derivatives that are more active biologically.[107]

Both peptides and released phenolic compounds display antioxidant activities contributing to the delay of different pathologies, including MetS and numerous non-communicable diseases.[79] Lentil ingredients FLPS, obtained by the combined pH-controlled fermentative process with *Savinase*, exhibited antioxidant and chemopreventive effects as radical peroxyl scavengers through hydrogen donating mechanisms attributed to bioaccessible compounds generated during processing.[108] *Savinase* seems to contribute to a greater extent to antioxidant activity of FLPS due to the generation of small peptides able to donate hydrogen atoms[19,97] whilst pH-controlled alkaline fermentation with *L. plantarum* promotes the bioconversion of native phenolic compounds with antioxidant activity.[20,96] The lentil ingredient also exhibited cytoprotective effects in stressful conditions tested on RAW 264.7 macrophages, reducing ROS overproduction induced by *tert*-butyl hydroperoxide (*t*-BOOH), and (+) catechin, *trans*-*p*-cumaric acid, and flavonols and flavon glucosides were the main contributors to such activity.[109,110] Xuan *et al.*[111] demonstrated that lentil soluble fraction attenuated angiotensin II-induced cardiomyocytes hypertrophy by decreasing the ROS level, which was attributed to phenolic compounds that provide parallel protection by ROS quenching and activation of the cell antioxidant defense system. In addition to antioxidant activities, lentil FLPS ingredients exhibited antihypertensive effects through the inhibition of angiotensin converting enzyme, an efficient strategy to reduce blood pressure, as well as inhibition of α-glucosidases and pancreatic lipase, with multiple positive health implications related to control of the glycemic index and hyperlipidaemia, respectively.[112–115] These effects were also attributed to bioactive peptides released by *Savinase* hydrolysis, those with specificity for aromatic and hydrophobic residues at position P1 of ACE[99] with low IC_{50} (140 µg mL^{-1}), and *L. plantarum* also contributes to the ACE inhibitory activity through the generation of ACE inhibitory peptides as reported previously.[96,113,116] Peptides released during the FLPS production contribute also to the inhibition of intestinal α-galactosidase enzyme, as do phenolic compounds such and kaempferol and quercetin glycosides previously described are good inhibitors of related enzymes.[110] Simultaneously, it has been observed that individual phenolic compounds are selective inhibitors of mucosal α-galactosidase subunits,[114] as reported for (+) catechin, a phenolic compound identified in FLPS.[108]

The composition of the FLPS ingredients responsible for the bioactivity behind the prevention of cardiometabolic diseases and their identity and characterization justified further studies. For this purpose, fractionation was performed and the most active soluble fraction was screened for its peptide and phenolic profile as well as bioactivity involved in health promotion potential.[108] Fraction F1 was the elective fraction, remarkable for having the highest antioxidant potential, measured as inhibition of ROS generation in stressed RAW 264.7 macrophages and as radical scavengers, as well as the highest ACE, pancreatic lipase and intestinal α-galactosidase inhibitory activities. An exhaustive search to gain insight into the content and composition of this fraction showed a total of 30 peptides with molecular masses within the range of 1100 to 3200 Da derived mainly from the major lentil proteins convicilin, vicilin (allergen Len c 1.0101 and 1.0102) and lectin. Some of them exhibited strong antioxidant and ACE inhibitory activities, as previously observed in hydrolysed lentils and recorded in bioactive peptide databases.[117] Among them, three main peptides (DLAIPVNNPGQLESF, LVNEGKGNLELVGF and ITPEKNPQLQDLDIF) were recently reported to have dual antioxidant and ACE inhibitory activities.[117] Analysing the amino acid sequences, proline, histine and tyrosine were contained within most sequences, and N-terminal and/or C-terminal hydrophobic amino acids were determinants for antioxidant activity. Antioxidant features of histidine residues seem to be due to the hydrogen-donating and radical-trapping mechanisms, whereas tyrosine residues can donate hydrogen to reduce free radicals and hydrophobic amino acids may increase interactions between peptides and fatty acid radicals.[118] FLPS fraction also contained aromatic (phenylalanine), hydrophobic (valine, leucine, glycine, proline, phenylalanine) and positively charged (arginine and histidine) amino acids in the C_t-tripeptide that has been ascribed as the determinant residue for ACE inhibitory activity.[118] Additionally, lentil F1 fraction exhibited α-galactosidase inhibition activity attributed to the presence of peptides whose structure contained hydroxyl or basic side chain amino acids and alanine or methionine at the C_t position,[119] relevant positions found in peptide REQSPGQWRPSHGKEEDEEEKEQKEAQ derived from lentil convicilin.[108] In F1 fractioned from FLPS ingredient, five flavonoids were identified as prodelphinidin dimer, (+)-catechin-3-*O*-hexoside, kaempferol rutinoside-hexoside, isorhamnetin glucoronide and kaempferol rutinoside-rhamnoside (II). The kaempferol rutinoside-hexoside was the most abundant compound (1 μg g^{-1}) accounting for 58% of the total phenolics. Flavonoids are well known antioxidants that exert biological action through several mechanisms. Lentil flavonoids seem to neutralize radicals and chelate metal ions through hydrogen, electron donating and mixed mechanisms.[110] Kaempferol exhibits protective effects against oxidative stress through activation of nuclear factor erythroid 2 (Nrf2)-mediated defensive response in lung tissues.[120] Additionally, kaempferol glucosides and aglycone have proven to be good inhibitors of intestinal α-glucosidase and pancreatic lipase activities.[110,121] Moreover, prodelfinidin dimers (0.33 $\mu g\ g^{-1}$ in FLPS ingredient) exhibited binding and inhibition activity on pancreatic lipase[122] and (+) catechin (0.25 $\mu g\ g^{-1}$ in FLPS ingredient) inhibited intestinal sucrose-isomaltase more potently than maltose-glucoamylase subunits.[114] All this information retrieved from the literature supports the idea that fermented legume-derived products can be a source of nutritious novel ingredients in the design of healthy functional foods to benefit modern society, contributing to the reduction of highly prevalent diseases and improving quality of life.

Acknowledgements

This research was supported by the Spanish ministry of Economy, Industry and Competitiveness and FEDER funding through the project AGL2013-43247-R. E. Peñas is indebted to Spanish Ramon y Cajal Programme for financial support.

References

1. F. Leroy and L. De Vuyst, *Trends Food Sci. Technol.*, 2004, **15**, 67.
2. M. J. Nout, *Food Microbiol.*, 2009, **26**, 698.
3. M. L. Marco and B. L. Golomb, *Microbiome*, 2016, **11**, 349.
4. J. Mubaiwa, V. Fogliano, C. Chidewe and A. R. Linnemann, *Food Rev. Int.*, 2017, **33**, 167.
5. J. Frias, C. Martínez-Villaluenga and E. Peñas, in *Fermented Foods in Health and Disease Prevention*, ed. J. Frias, C. Martínez-Villaluenga and E. Peñas, Elsevier, San Diego, CA, 2017, p. 385.
6. L. Kan, S. Nie, J. Hu, S. Wang, Z. Bai, J. Wang, Y. Zhou, J. Zeng and K. Song, *Food Chem.*, 2018, **260**, 317.
7. H. L. Wang and S. F. Suang, in *Indigenous Fermented Food of Non-Western Origin*, ed. C. W. Hesseltine, H. L. Wang and J. Cramer, Berlin, Germany, 1986. p. 23.
8. FAO, *Pulses, Nutritious Seeds for a Sustainable World*, Food and Agriculture Organization of the United Nations, 2016, ISBN 978-92-109463-1.
9. M. L. Marco, D. Heeney, S. Binda, C. J. Cifelli, P. D. Cotter, B. Foligne, M. Gänzle, R. Kort, G. Pasin, A. Pihlanto, E. J. Smid and R. Hutkins, *Curr. Opin. Biotechnol.*, 2017, **44**, 94.
10. X. Sanchez-Chino, C. Jimenez-Martínez, G. Dávila-Ortiz, I. Alvarez-Gonzalez and E. Madrigal-Bujaidar, *Nutr. Cancer*, 2015, **67**(3), 401.
11. K. H. Steinkraus, *Food Res. Int.*, 1994, **27**, 259.
12. E. C. Borresen, A. J. Henderson, A. Kumar, T. L. Weir and E. P. Ryan, *Recent Pat. Food, Nutr. Agric.*, 2012, **4**, 134.

13. N. A. Olasupo, C. P. Okorie and F. A. Oguntoyinbo, *Frontiers Microbiol.*, 2016, 7, 1153.
14. FAO, *Fermented Grain Legumes, Seeds and Nuts: A Global Perspective*, ed. S. S. Deshpande, D. K. Salunkhe, O. B. Oyewole, S. Azam-Ali, M. Battcock and R. Bressani, Food and Agriculture Organization, Rome, Italy, 2000.
15. C. G. Rizzello, M. Verni, H. Koivula, M. Montemurro, L. Seppa, M. Kemell, K. Katina, R. Coda and M. Gobbetti, *Food Funct.*, 2017, **8**, 860.
16. P. Riat and B. Sadana, *J. Food Sci. Technol.*, 2009, **46**, 247.
17. S. Balasubramanian, M. G. Jincy, M. Ramanathan, P. Chandra and S. D. Deshpande, *Int. Food Res. J.*, 2015, **22**, 139.
18. S. K. Soni and D. K. Sandhu, *J. Ferment. Bioeng.*, 1989, **68**, 52.
19. M. Reyes-Bastidas, E. Z. Reyes-Fernandez, J. Lopez-Cervantes, J. Milan-Carrillo, G. F. Lorca-Piña and C. Reyes-Moreno, *Food Sci. Technol. Int.*, 2010, **16**, 427.
20. R. I. Limon, E. Peñas, M. I. Torino, C. Martínez-Villaluenga, M. Dueñas and J. Frias, *Food Chem.*, 2015, **172**, 343.
21. M. Diana, J. Quílez and M. Rafecas, *J. Funct. Food*, 2014, **10**, 407.
22. A. Bujang and N. A. Taib, *Sains Malays.*, 2014, **43**, 551.
23. J. Tabera, J. Frias, I. Estrella, R. Villa and C. Vidal-Valverde, *Z. Lebensm.-Unters. Forsch.*, 1995, **201**, 591.
24. R. Coda, L. Melama, C. G. Rizzello, J. A. Curiel, J. Sibakov, U. Holopainen, M. Pulkkinen and N. Sozer, *Int. J. Food Microbiol.*, 2015, **193**, 34.
25. S. Halamatha, K. Platel and K. Srinivasan, *Eur. J. Clin. Nutr.*, 2007, **61**, 342.
26. R. Y. Khattab and S. D. Artfield, *LWT--Food Sci. Technol.*, 2009, **42**, 1113.
27. A. Clemente and M. C. Arques, *World J. Gastroenterol.*, 2014, **20**, 10305.
28. A. Clemente and R. Olias, *Curr. Opin. Food Sci.*, 2017, **14**, 32.
29. M. Granito and G. Alvarez, *J. Sci. Food Agric.*, 2006, **86**, 1164.
30. F. S. Atkinson, K. Foster-Powell and J. C. Brand-Miller, *Diabetes Care*, 2008, **31**, 2281.
31. C. Sotomayor, J. Frias, J. Fornal, J. Sadowska, G. Urbano and C. Vidal-Valverde, *Starch/Staerke*, 1999, **51**, 152.
32. S. S. Audu and M. O. Aremu, *Pak. J. Nutr.*, 2011, **10**, 1069.
33. A. Bhandal, *Int. J. Food Sci. Technol.*, 2008, **43**, 2090.
34. F. M. Abu-Salem and E. A. Abou-Arab, *J. Am. Sci.*, 2011, **7**, 107.
35. A. Starzynska-Janiszewska, B. Stodolak, R. Duliński and B. Mickowska, *Food Sci.Technol. Int.*, 2012, **18**, 113.
36. M. Granito, J. Frias, R. Doblado, M. Guerra, M. Champ and C. Vidal-Valverde, *Eur. Food Res. Int.*, 2002, **214**, 226.
37. M. A. Martin-Cabrejas, B. Sanfiz, A. Vidal, E. Molla, E. Esteban and F. J. Lopez-Andreu, *J. Agric. Food Chem.*, 2004, **52**, 261.
38. J. Frias, R. Esteban and C. Vidal-Valverde, *J. Food Sci.*, 1993, **56**, 1350.
39. A. Torres, J. Frias, M. Granito and C. Vidal-Valverde, *J. Agric. Food Chem.*, 2006, **54**, 6685.

40. B. Drzikova, G. Dongowski, E. Gebhardt and A. Habel, *Food Chem.*, 2005, **90**, 181.
41. V. Kavita, S. Verghese, G. R. Chitra and J. Prakash, *J. Food Sci. Technol.*, 1998, **35**, 299.
42. N. Kalogeropoulos, A. Chiou, M. Ioannou, V. T. Karathanos, M. Hassapidou and N. K. Andrikopoulos, *Food Chem.*, 2010, **121**, 682.
43. C. Reyes-Moreno, E. O. Cuevas-Rodríguez, J. Milán-Carrillo, O. G. Cárdenas-Valenzuela and J. Barrón-Hoyos, *J. Sci. Food Agric.*, 2004, **84**, 271.
44. W. Prinyawiwatkul, R. L. Beuchat, K. H. McWatters and R. D. Phillips, *Food Chem.*, 1996, **57**, 405.
45. V. R. Niveditha, D. V. Krishna and K. R. Sridhar, *Int. Food Res. J.*, 2012, **19**, 1401.
46. R. P. Mensink, P. L. Zock, A. D. M. Kester and M. B. Katan, *Am. J. Clin. Nutr.*, 2003, **77**, 1146.
47. C. Martínez-Villaluenga, J. Frias and C. Vidal-Valverde, *Crit. Rev. Food Sci. Nutr.*, 2008, **48**, 301.
48. G. Urbano, M. Lopez-Jurado, J. Hernandez, M. Fernandez, M. C. Moreu, J. Frias, C. Diaz-Pollan, M. Prodanov and C. Vidal-Valverde, *J. Agric. Food Chem.*, 1995, **43**, 1871.
49. J. Frias, C. Vidal-Valverde, H. Kozlowska, J. Tabera, J. Honke and C. L. Hedley, *J. Agric. Food Chem.*, 1996, **44**, 579.
50. C. Vidal-Valverde, J. Frias, C. Sotomayor, C. Díaz-Pollán, M. Fernandez and G. Urbano, *Z. Lebensm.-Unters. Forsch.*, 1998, **207**, 140.
51. J. Sadowska, J. Fornal, C. Vidal-Valverde and J. Frias, *Die Narhung*, 1999, **43**, 396.
52. J. Frias, C. Vidal-Valverde, C. Sotomayor, C. Diaz-Pollan and G. Urbano, *Eur. Food Res. Int.*, 2000, **210**, 340.
53. R. Doblado, J. Frias, R. Muñoz and C. Vidal-Valverde, *Pol. J. Food Nutr. Sci.*, 2002, **11/52**(S1), 73.
54. A. Torres, J. Frias, S. Frejnagel, K. Gulewicz and C. Vidal-Valverde, *Pol. J. Food Nutr. Sci.*, 2002, **11/52**(S1), 104.
55. R. Doblado, J. Frias, R. Muñóz and C. Vidal-Valverde, *J. Food Prot.*, 2003, **65**, 2313.
56. M. Granito, M. Champ, M. Guerra and J. Frias, *J. Sci. Food Agric.*, 2003, **83**, 1004.
57. Y. E. Madodé, M. J. R. Nout, E. J. Bakker, A. R. Linnemann, D. J. Hounhouigan and M. A. J. S. van Boekel, *LWT--Food Sci. Technol.*, 2013, **54**, 186.
58. A. Starzyńska-Janiszewska, B. Stodolak and Mickowska, *J. Sci. Food Agric.*, 2014, **94**, 359.
59. S. Yadav and N. Khetarpaul, *Food Chem.*, 1994, **50**, 403.
60. S. Krishnamoorthy, S. Kunjithapatham and L. Manickam, *Nutr. Diet.*, 2013, **70**, 241.
61. P. K. Dahiya, M. J. R. Nout, M. A. van Boekel, N. Khetarpaul, R. B. Grewal and A. Linnemann, *Br. Food J.*, 2014, **116**, 1031.

62. M. Prodanov, I. Sierra and C. Vidal-Valverde, *Eur. Food Res. Technol.*, 1997, **205**, 48.
63. S. K. Soni and D. K. Sandhu, *J. Ferment. Bioeng.*, 1989, **68**, 52.
64. I. F. Fadahunsi, *Pak. J. Nutr.*, 2009, **8**, 835.
65. S. Martins, S. I. Mussatto, G. Martinez-Avila, J. Montanez-Saenz, C. N. Aguilar and J. A. Texeira, *Biotechnol. Adv.*, 2011, **29**, 365.
66. B. A. Acosta-Estrada, J. A. Gutierrez-Uribe and S. O. Serna-Saldívar, *Food Chem.*, 2014, **152**, 46.
67. I. Hayat, A. Ahmad, T. Masud, A. Ahmad and S. Bashir, *Crit. Rev. Food Sci. Nutr.*, 2014, **54**, 580.
68. V. Vadivel and H. K. Biesalski, *Int. Food Res. J.*, 2012, **19**, 593.
69. G. Oboh, A. O. Ademiluyi and A. A. Akindahunsi, *Food Sci. Technol. Int.*, 2009, **15**, 41.
70. R.-Y. Gan, N. P. Shah, M.-F. Wang, W.-Y. Lui and H. Corke, *Int. J. Food Technol.*, 2016, **51**, 875.
71. N. T. Huynh, J. Van Camp, G. Smagghe and K. Raes, *Int. J. Mol. Sci.*, 2014, **15**, 19369.
72. B. Bartolome, I. Estrella and T. Hernandez, *Z. Lebensm.-Unters. Forsch.*, 1997, **205**, 290.
73. Y. Xiao, G. Xing, X. Rui, W. Li, X. Chen, M. Jiang and M. Dong, *J. Funct. Foods*, 2014, **10**, 210.
74. M. Dueñas, D. Fernandez, T. Hernandez, I. Estrella and R. Muñoz, *J. Sci. Food Agric.*, 2005, **85**, 297.
75. R. Doblado, H. Zielinski, M. Piskula, H. Kozslowska, R. Muñoz, J. Frias and C. Vidal-Valverde, *J. Agric. Food Chem.*, 2005, **53**, 1215.
76. R. Fernandez-Orozco, J. Frias, H. Zielinski, M. K. Piskula, H. Kozlowska and C. Vidal-Valverde, *Food Chem.*, 2008, **111**, 622.
77. R. Fernandez-Orozco, J. Frias, R. Muñoz, H. Zielinski, M. K. Piskula, H. Kozlowska and C. Vidal-Valverde, *Eur. Food Res. Technol.*, 2008, **227**, 979.
78. A. O. Ademiluyi and G. Oboh, *Asian Pac. J. Trop. Biomed.*, 2012, **5**, 692.
79. G. Kapravelou, R. Martinez, A. M. Andrade, C. Lopez Chaves, M. Lopez-Jurado, P. Aranda, F. Arrebola, F. J. Cañizares and M. Galisteo, *J. Sci. Food Agric.*, 2015, **95**, 1207.
80. M. Gabriele, L. Pucci, M. La Marca, D. Lucchesi, C. M. Della Croce, V. Longo and V. Lubrano, *Cell. Mol. Biol. Lett.*, 2016, **21**, 10.
81. S. K. Yeap, B. K. Beh, N. M. Ali, H. M. Yusof, W. Y. Ho, S. P. Koh, N. B. Alitheen and K. Long, *BioMed Res. Int.*, 2012, 285430.
82. N. M. Ali, S. K. Yeap, H. M. Yusof, B. K. Beh, W. Y. Ho, S. P. Koh, M. P. Abdullah, N. B. Alitheen and K. Long, *J. Sci. Food Agric.*, 2016, **96**, 1648.
83. H. Wu, X. Rui, W. Li, X. Chen, M. Jiang and M. Dong, *Food Sci. Technol.*, 2015, **63**, 445.
84. E. Peñas, M. Diana, J. Frias, J. Quilez and C. Martinez-Villaluenga, *Plant Foods Hum. Nutr.*, 2015, **70**, 97.
85. C. G. Rizzello, L. Nionelli, R. Coda, R. Di Cagno and M. Gobbetti, *Eur. Food Res. Technol.*, 2010, **230**, 645.

86. M. Duranti, *Fitoterapia*, 2006, **77**, 67.
87. C. G. Rizzello, L. Nionelli, R. Coda, M. De Angelis and M. Gobbetti, *Food Chem.*, 2010, **119**, 1079.
88. M. Gobbetti, C. G. Rizzello, R. Di Cagno and M. De Angelis, *Food Microbiol.*, 2014, **37**, 30.
89. J. A. Curiel, R. Coda, I. Centomani, C. Summo, M. Gobbetti and C. G. Rizzello, *Int. J. Food Microbiol.*, 2015, **196**, 51.
90. R. Coda, L. Melama, C. G. Rizzello, J. A. Curiel, J. Sibakov, U. Holopainen, M. Pulkkinen and N. Sozer, *Int. J. Food Microbiol.*, 2015, **193**, 34.
91. R. Coda, C. G. Rizzello and M. Gobbetti, *Int. J. Food Microbiol.*, 2010, **137**, 236.
92. M. Verni, C. Wang, M. Montemurro, M. De Angelis, K. Katina, C. G. Rizzello and R. Coda, *Frontiers Microbiol.*, 2017, **8**, 2461.
93. C. G. Rizzello, B. Hernandez-Ledesma, S. Fernandez-Tome, J. A. Curiel, D. Pinto, B. Marzani, R. Coda and M. Gobbetti, *Microb. Cell Fact.*, 2015, **14**, 168.
94. G. C. Rizzello, L. Nionelli, R. coda and M. Gobetti, *Nutr. Cancer*, 2012, **64**, 111.
95. C. C. Hsieh, C. Martinez-Villaluenga, B. O. Lumen and B. Hernandez-Ledesma, *J. Sci. Food Agric.*, 2017, **98**, 2070.
96. M. I. Torino, R. I. Limon, C. Martinez-Villaluenga, S. Makinen, A. Pihlanto, C. Vidal-Valverde and J. Frias, *Food Chem.*, 2013, **136**, 1030.
97. P. García-Mora, E. Peñas, J. Frias and C. Martínez-Villaluenga, *J. Agric. Food Chem.*, 2014, **62**, 4166.
98. P. García-Mora, J. Frias, E. Peñas, H. Zielinski, J. A. Gimenez-Bastida, W. Wiczkowski and C. Martínez-Villaluenga, *J. Funct. Foods*, 2015, **18**, 319.
99. M. Kojic, D. Fira, B. Bojovic, A. Banina and L. Topisirovic, *J. Appl. Bacteriol.*, 1995, **J79**, 61.
100. T. R. Klaenhammer, R. Barragou, B. L. Buck, M. A. Azcarate-Peril and E. Altermann, *FEMS Microbiol. Rev.*, 2005, **29**, 393.
101. S. Bautista-Exposito, E. Peñas, J. M. Silvan, J. Frias and C. Martínez-Villaluenga, *Food Chem.*, 2018, **248**, 262.
102. M. Estaban-Torres, J. M. Landere, I. Reverón, L. Santamaría, B. de las Rivas and R. Muñoz, *Appl. Environ. Microbiol.*, 2015, **81**, 3235.
103. R. Muñoz, B. de las Rivas, F. López de Felipe, I. Reverón, L. Santamaría, M. Estaban-Torres, J. A. Curiel, H. Rodriguez and J. M. Landete, Biotransformation of phenolic compounds by *Lactobacillus plantarum* in fermented foods, in *Fermented Foods in Health and Disease Prevention*, ed. J. Frias, C. Martínez-Villaluenga and E. Peñas, Elsevier, 2017, p. 63.
104. K. Savijoki, H. Ingmer and P. Varmane, *Appl. Microbiol. Biotechnol.*, 2006, **71**, 394.
105. Z. Bubelová, D. Sumczynski and R. N. Salek, *J. Food Process. Preserv.*, 2018, **42**, e13388.
106. E. W. Kwaw, Y. Ma, W. Tchabo, M. T. Apaliya, M. Wu, A. S. Sackey, L. Xiao and H. E. Tahir, *Food Chem.*, 2018, **250**, 148.

107. P. Filannino, R. Di Cargo and M. Gobetti, *Curr. Opin. Biotechnol.*, 2018, **49**, 64.
108. S. Bautista-Exposito, E. Peñas, M. Dueñas, J. M. Silvan, J. Frias and C. Martínez-Villaluenga, *Food Chem.*, 2018, **257**, 341.
109. B. D. Oomah, F. Caspar, L. J. Malcolmson and A.-S. Bellido, *Food Res. Int.*, 2011, **44**, 436.
110. B. Zhang, Z. Deng, D. D. Ramdath, Y. Tang, P. X. Chen, R. Liu, Q. Liu and R. Tsao, *Food Chem.*, 2015, **172**, 862.
111. C. L. Xuan, F. R. Yao, L. R. Guo, Q. Liu, S. K. Chang, K. X. Liu and C. W. Sun, *Eur. Rev. Med. Pharmacol. Sci.*, 2013, **17**, 2644.
112. S. Y. Lee and S. J. Hur, *Food Chem.*, 2017, **228**, 506.
113. A. Jakubczyk, M. Karas, U. Zlotek and U. Szymanowska, *Food Res. Int.*, 2017, **100**, 486–496.
114. M. Simsek, R. Quezada-Calvillo, M. G. Feruzzi, B. L. Nichols and B. Hamaker, *J. Agric. Food Chem.*, 2015, **63**, 3873.
115. N. Finer, W. P. James, P. G. Kopelman, M. E. Lean and G. Williams, *Int. J. Obesity Related Metabolic Disorders*, 2000, **24**, 306.
116. B. P. Singh and S. Vij, *LWT--Food Sci. Technol.*, 2017, **86**, 293.
117. P. García-Mora, M. Martin-Martinez, M. A. Bonache, R. Gonzalez-Muñiz, E. Peñas, J. Frias and C. Martínez-Villaluenga, *Food Chem.*, 2017, **221**, 464.
118. Y. W. Li and B. Li, *J. Theor. Biol.*, 2013, **318**, 29.
119. M. A. Ibrahim, M. J. Bester, A. W. H. Neitz and A. R. M. Gaspar, *Chem. Biol. Drug Des.*, 2018, **91**, 370.
120. H. Kumar, I.-S. Kim, S. V. More, B.-W. Kim and D.-K. Choi, *Nat. Prod. Rep.*, 2014, **31**, 109.
121. Y. Tan, S. K. C. Chang and Y. Zhang, *Food Chem.*, 2017, **214**, 259.
122. T. Bucholz and M. F. Melzig, *Planta Med.*, 2015, **81**, 771.

CHAPTER 10

Thermal Processing of Legumes

M. REBOLLO-HERNANZ*[a,b], M. A. MARTÍN-CABREJAS[a,b] AND Y. AGUILERA[a,b]

[a]Institute of Food Science Research (CIAL, UAM-CSIC), Department of Production and Characterization of Novel Foods, Nicolás Cabrera, 9, 28049, Madrid, Spain; [b]Universidad Autónoma de Madrid, Department of Agricultural Chemistry and Food Sciences, Francisco Tomás y Valiente, 7, 28049, Madrid, Spain
*E-mail: miguel.rebollo@uam.es

10.1 Introduction

As an essential part of the diet, legumes are good and inexpensive sources of macronutrients, *i.e.* proteins, lipids, carbohydrates. Additionally, legumes are a source of dietary fibre and micronutrients, including soluble vitamins from the B-group (thiamine, riboflavin and niacin), minerals, such as K, Ca, Mg, Cu, Fe and Zn, and also different phytochemicals as phenolic compounds.[1] Because of their exceptional nutritional value, they are used worldwide in both human and animal nutrition, especially in developing countries.[2] Most legumes contain non-nutritional factors, for instance, protease inhibitors, which can be either partially or entirely removed using different processing methods.[3] Legumes are commonly treated under thermal and non-thermal processes to improve their quality, preservation, safety, and suitability for specific product applications. The grade of processing affects the digestibility, solubility, and other associated parameters.[4] The

Food Chemistry, Function and Analysis No. 8
Legumes: Nutritional Quality, Processing and Potential Health Benefits
Edited by María Ángeles Martín-Cabrejas

Published by the Royal Society of Chemistry, www.rsc.org

digestibility, functionality, and suitability of legumes as ingredients in food applications can be improved through modifications of the physicochemical, structural, functional and thermal properties of whole pulses, flours or fractions.[5,6]

There is an extensive variety of treatments that can be applied to food legumes, ranging from essential, conventional and low-input technologies to advanced and more specialised techniques. Among these techniques, we can include wet and dry methods, thermal and non-thermal processing technologies such as milling, cooking/boiling, fermentation, germination, roasting, autoclaving, fractionation, extrusion, high-pressure cooking, and infrared radiation (micronisation).[7,8] Thermal processing, in general, improves the nutritional value of legumes, decreasing the content of non-nutritional factors and also improving the flavour and the overall acceptability of the legumes prepared. Nevertheless, heat treatments can also destroy some nutritional components of legumes or modify them, or affect the techno-functional and physiological properties of legumes, depending on several factors, such as the temperatures and times employed.[9]

Non-thermal processing such as soaking, germination and fermentation are relatively simple procedures which have been shown to be effective in reducing the content of non-nutritional factors and improving the composition and digestibility of legumes.[10,11] Other emerging non-thermal food processing technology, such as high hydrostatic pressure, which involves the application of hydrostatic pressure of several hundred MPa on foods,[12] may induce alterations in the native structure of legume macromolecules such as starch and protein, causing protein unfolding, enzyme inactivation, starch gelatinization, and texture and digestibility changes similar to thermally-induced modifications.[4] This chapter will describe the thermal treatments applied during the processing of legumes, presenting their impacts on legume composition and nutritional value, as well as on the techno-functional properties and health benefits of legumes.

10.2 Thermal Treatments

There is an extensive variety of thermal treatments available for food legume processing. Common thermal treatments are boiling, cooking, microwaving, roasting, autoclaving, and extruding (Figure 10.1).

Thermal processing involves the controlled use of temperature to produce modifications of raw food materials. According to the temperature and the type of treatment, the effect of thermal processing on whole seeds, flours, concentrates, or isolates may range from mild to severe. Furthermore, thermal processing may affect various components to different extents. Moisture decreases in dry thermal processing. High-temperature treatments may break bonds in starch and dietary fibre and affect their solubilisation, and the solubility of protein can also be reduced under such conditions due to denaturation and aggregation.[11,13]

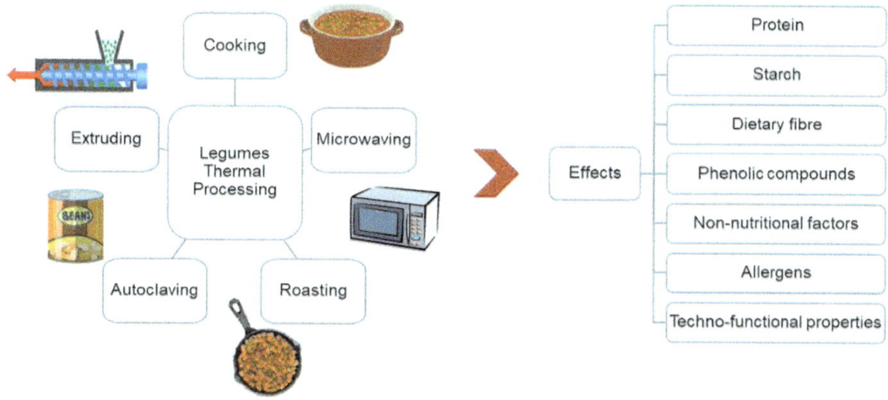

Figure 10.1 The main thermal treatments commonly employed in the processing of legumes and potential components and properties that can be affected by their use.

10.2.1 Cooking

Cooking might be the most ancient treatment for the culinary preparation of edible legumes. It usually comprises a previous soaking of the pulses, followed by cooking in boiling water, which lasts between 30 and 90 min, to achieve a softened product. Even if cooking methods are known to reduce the concentration of non-nutritional factors in legumes, and thus, improve their nutritional value, long cooking times may degrade some nutritive compounds of legumes, as compared with shorter treatments. Cooking quality of legume seeds appears to depend, at least in part, on the sequence of coagulation of storage proteins and swelling of starch granules during heating. Moreover, the addition of salts to the cooking media can shorten cooking times.[14]

Legumes are frequently prepared in a pressure cooker for decreased cooking times.[15] This process involves the use of both high temperature and pressure treatment for a short duration and results in a considerably enhanced digestibility and palatability of legumes. Pre-soaking legumes before pressure cooking is another strategy for reducing the processing time and also for obtaining a tender product.[3,8]

At this point, it is necessary to emphasise the environmental factors affecting cooking quality. Legumes when stored at high temperature and high moisture, develop the hard-to-cook (HTC) defect which has been observed for decades. The seed coats appear to be impermeable to water and show granular or sandy texture due to the physicochemical changes that occur during storage. Cooked HTC legumes are characterized by limited cell separation and restricted starch gelatinization. Several factors have been postulated to be involved in the formation of HTC legumes, including the cell wall components, structure and interactions, and the effect of environmental conditions.[16] The interaction of phytates, phenolic compounds, and divalent cations (mainly Ca and Mg) with pectic substances and proteins could be the central mechanisms involved. There are many events involved with ageing and soaking such as free radical formation, lipid peroxidation, acid formation, membrane deterioration, and protein denaturation, whereas pectin decomposition and solubilization, protein coagulation, and starch gelatinization are events that occur during cooking.[17]

10.2.2 Roasting

The roasting process sometimes denoted as parching, is used to prepare some specific recipes from pulses, legumes, and grains.[18] The basis of this method is the exposition of usually whole seed, non-dehulled grains to dry heat. Roasting is cooking food in a pan without water until obtaining a well-browned exterior and a moister cooked interior.[19] The traditional Turkish household practice for roasting legumes comprises a first sprinkling of the legumes with water followed by a mixing of the legumes with preheated sand or salt in a roasting pan. Then legumes are maintained at a temperature from 200 to 250 °C, depending on the legume.[18] Roasting whole grains directly on a hot metal surface is also a common traditional process.[18,20]

Traditional roasting of legumes is used primarily to enhance flavour, but other benefits include reductions of non-nutritional factors and extension of the product shelf-life.[21] The roasting process renders the grain digestible, without the loss of nutritious components and hence such grains are consumed throughout the world.[22] The products obtained by the roasting of legumes are generally wholly cooked and served as ready-to-eat snack foods. During pulse roasting, starch gelatinizes entirely, triggering an enhancement of the digestibility of the seed and lowering the amount of resistant starch.[23] Flours milled from roasted legumes show increased water retention capacity and viscosity.[21]

10.2.3 Autoclaving

Thermal treatment is an important stage during industrial legume processing, which often involves cooking and sterilisation, thus deactivating microorganisms and guaranteeing food safety.[24] This stage is usually performed in an industrial autoclave, which applies liquid water, steam water or a mixture of both, thus administering heat to the inside of the product.[25] The temperature of the process ranges from 110 to 135 °C, depending on the heating medium (steam, water, or mixture), and the pressure used in the process is typically 10^5 Pa.[26] Hence, autoclaving or sterilisation processes use higher temperatures (121 °C) and shorter times (between 15 and 45 min) than other thermal treatments, enhancing the nutritive value and enabling an extension of the shelf-life.[27,28] The time required for cooking and sterilisation of the product depends on some factors including the resistance of microorganisms or enzymes to heat, heating conditions, food pH, and packaging used.[29,30]

10.2.4 Microwave Cooking

Microwave cooking is an accessible means of cooking in the urban areas for saving both energy and time. This form of cooking legumes is even faster than pressure-cooking. Water is added to pre-soaked seeds that are then cooked in a microwave oven for 4–10 min or more until the legume has become tender.[15] Microwave cooking involves substantial decreases in fat, total ash, soluble carbohydrate fractions, non-nutritional factors, minerals, and B-group vitamins. The losses of micronutrients, B-group vitamins and minerals, in legumes prepared by microwave cooking are smaller than those cooked by traditional methods such as boiling and pressure-cooking.[31] Cooking processes decrease the concentrations of lysine, tryptophan, total aromatic and sulphur containing amino acids. *In vitro* protein digestibility, protein efficiency ratio and essential amino acid index are improved, however.[31,32] Thus, microwave cooking is presented as a useful cooking method of legumes, not only for enhancing their nutritional quality (by decreasing the content of non-nutritional factors as well as increasing protein digestibility and maintaining both B-vitamins and minerals) but also for reducing the cooking times.[32]

10.2.5 Extrusion

Food extrusion processing is a fully automated technology that allows more efficient use of energy during cooking and better process control.[33] The decrease in food nutritional quality is a challenging issue in conventional thermal processing when involving high temperatures.[34] Extrusion cooking is considered an HTST (High-Temperature Short Time) process, and as such, it is highly recommended for the processing of high protein content materials such as legumes.[35] In this regard, extrusion processing is preferable to achieve higher nutrient retention being generally operated at high temperatures but short times.[36,37] Although the material is exposed to higher temperatures, in comparison with other methods, the thermo-labile nutrients suffer less damage due to the short-term heat exposure since the food material has a short residence time in the extruder.[34] Extrusion combines both temperature, mechanical stress, and short times to shape food materials. Generally, legume flour is introduced into a heated barrel with a rotating screw. The screw forces the material through a die plate that is responsible for shaping the product.[38] Extrusion can physically deform proteins. These modifications in the structure (both tertiary and quaternary) of trypsin inhibitors is sufficient for their denaturation. The disadvantage is that the nutritious proteins can also be affected.[39] The intense mechanical shear during this small period in which the legume is inside the extruder during the process involves the breakdown of the covalent bonds in macromolecules and an increase of structural disruption that helps to texturise the final products.[34,36] This is translated into different techno-functional properties such as enhanced solubility but decreased emulsifying activity. The extrusion process has both positive and negative effects on the techno-functional properties of the final product.

Forty years ago, the application of extrusion processing in food technology was limited to mixing and forming of macaroni and ready-to-eat cereals.[40–43] However, this processing technique has been popularised and utilised for other purposes, *e.g.* from the production of ready-to-eat cereals to salty and sweet snacks and co-extruded snacks.[39]

10.3 Effect of Thermal Processing on the Nutritional Quality of Legumes

Legumes have been linked with good nutrition practices and the promotion of health from several years.[44] Although legumes can be associated with healthy food patterns, their attractiveness is sometimes impeded by their long preparation time, low protein and starch digestibility and the attendant discomfort that sometimes follows their consumption.[45] Processing affects the composition of macronutrients in legumes in numerous ways.[8,22,46–48] Specific changes to the macronutrient composition can influence the behaviour of legumes or legume-based ingredients in food and during digestion.[49] The processing of pulses might enlarge their markets and enhance their utilisation by making legumes more convenient and sometimes more nutritionally and functionally efficient.[50,51] The effects of thermal processing on the components of legumes are described below.

10.3.1 Impact on Protein

Thermal processing can induce denaturation and greater unfolding, that may expose side chains, increase chain flexibility and digestibility and protein-starch cross-linking, promoting aggregation and insolubility.[52,53] The solubility of the proteins from legumes has nutritional, health and functional implications.[54,55] Relatively insoluble proteins are slowly and poorly digested, and their disappearance from the gastrointestinal tract may be impaired with potentially adverse effects, especially for people with limited adsorption capacity.[56]

The literature shows increases in the protein contents of cooked legumes that may be attributed to the loss of soluble solids during cooking.[48,57–60] These increases not only appear in protein levels but also in non-protein nitrogen contents ranging from 22.6 to 26.2% and 2.3 to 2.7% in the legume cultivars, respectively.[59] However, the soluble protein fraction is reduced a 3- to 4-fold by thermal processing.[58] In this sense, the globulin fractions which are present in significant amounts (79–85%) significantly decreased after the heat treatment, accompanied by a significant increase in the glutelin-3 fractions.[59] The prolamin contents did not vary significantly after processing. In addition, SDS-PAGE profiles of numerous legumes exhibited a wide variation in electrophoresis pattern after heat processing. Therefore, thermal treatment appears to increase the insoluble protein fractions and eliminate the minor polypeptides.[59]

The severity of thermal processing affects the apparent recovery of most of the essential amino acids, depending on the legume seeds involved. It was detected that increased cooking times (30, 60 and 120 min) caused significant declines in amino acid contents, leading to a lower apparent recovery of methionine (28–31%) and cysteine (17–19%) in most of legumes; although the total essential amino acid levels in cooked seeds remained higher than that of the FAO/WHO requirement, except for methionine and cysteine in the majority of legumes.[46,48]

In vitro protein digestibility (IVPD) of lentils, chickpeas, peas, and soybeans after thermal treatment using the three-enzyme method (trypsin, chymotrypsin, and peptidase) was increases (2–13%),[61] while the effects of applying ultrasound or high hydrostatic pressure on the IVPD of legumes were mostly inconsistent or insignificant.

10.3.2 Impact on the Carbohydrate Fraction (Starch)

Thermal processing alters the content, composition, and quality of the carbohydrate fraction of legumes differently. Processing treatment and conditions, among which moisture content, the strength and extent of the application of heat and temperature, can significantly affect starch quality, its digestion and nutritive value.[6,62] At low to moderate moisture levels, thermal treatment causes the melting of legume starch, and disorganisation of starch molecules, without affecting the shape of the granule.[63] In the presence of excess water, gelatinization occurs. Gelatinization includes the disruption of starch molecular structure, its loss of birefringence, and the formation of a honeycomb-like structure. With the increase of temperature, the granular structure is destroyed, and starch is fragmented. These structural changes of the starch granules increase its availability for digestion by amylolytic enzymes,[53,62] and thus, increase the glycaemic index of the corresponding legume or legume-based product. The type of starch present also influences its response to heating at different moistures.[64,65] In legumes with high amylose content, the granules swell slower than starches rich in amylopectin.[66,67]

Several studies have demonstrated that different changes occurred in starch content after cooking that may be due not only to the type and physical characteristics of legumes but also to the influence of previous treatment and different conditions.[68,69] Aguilera *et al.*[13] indicated total starch content was not affected in chickpeas, whereas decreases (21%) were detected in lentils compared to raw legumes. It was observed that, as a result of cooking, available starch contents were further increased in these legumes (12–21%), and resistant starch levels decreased significantly (65% in chickpea and 49% in lentil with respect to raw seeds). Therefore, part of the resistant starch is modified by the cooking process and converted into digestible starch. A similar trend was observed in other conventionally processed legumes,[70,71] enhancing starch digestibility in legumes. Thus, processing treatments such as soaking, cooking, dehydration and autoclaving of legumes reduce the levels of starch but to lesser extents than non-nutritional factors (phenolic compounds, phytate, amylase inhibitor) and other soluble components.[68]

It is known that after soaking and thermal treatments an increase in starch digestibility occurs, due to gelatinization of starch granules and, probably, to the decreasing levels of antinutritional factors in the seeds. It seems a partial removal of tannins and phytic acid probably creates a large space within the matrix, which increases the susceptibility to enzymatic attack and consequently improves the digestibility of protein and starch after thermal processing.[72] Literature has also shown that thermal processing induces an increase in resistant starch values, suggesting amylose retrogradation as the primary mechanism behind the reduction in digestibility.[23,73] The significant variations in the results for resistant starch may be due to the use of several *in vitro* procedures for its quantification and consequently the resistant starch content will vary according to the procedure used.[74,75]

Commonly, legumes are considered as low glycaemic index foods that are therefore suitable for people with diabetes, as a replacement for foods that are more rapidly digested.[76] Nevertheless, the absorption of degraded carbohydrates from starch might elevate the blood glucose concentration.[77]

10.3.3 Impact on Dietary Fibre

The effects of thermal processing on the content and composition of the dietary fibre of legumes have been largely investigated with regard to the impacts of different heat processing, including cooking, roasting, extrusion, and dehydration.[78] Relevant modifications of polysaccharides of the soluble and insoluble dietary fibre fractions take place during processing, promoting changes in the structures and state of plant cell wall material.[69,78] Several studies confirmed that thermal processing causes increases in the soluble dietary fibre (SDF) fraction while decreasing the insoluble dietary fibre (IDF) fraction.[74,79] Boiling and cooking of legumes, can degrade the pectic polysaccharides that are involved in cell to cell adhesion in the middle lamella of the cell walls, resulting in softening of plant tissues.[78,80,81]

Cooking legumes, both boiling or microwaving them has been demonstrated not to modify the content of total non-starch polysaccharides in common legumes like beans and chickpeas, but increase the soluble fibre fraction and decrease the insoluble fraction.[71] These modifications could be due to partial solubilization and depolymerisation of hemicellulose and other insoluble pectic substances. Wang *et al.*[46] found significant increases in the content of total dietary fibre (TDF) in 7 different varieties of beans and 2 varieties of chickpeas. These increases included the enhancement in the concentration of both SDF and IDF fractions together with a decrease in the content of resistant starch in beans. Other studies have shown a similar trend in other legumes and cultivars including peas, common beans, chickpeas, and lentils.[74] Both conventional and microwave cooking have been shown to diminish the content of oligosaccharides in lentils with non-significant effects on the TDF.[7] After cooking chickpeas, their content of the different components constituting the dietary fibre was modified.[82] Whereas the levels of hemicellulose decreased, those of cellulose and lignin increased, and pectin content was maintained.

The use of autoclave processing has also shown to cause changes in the TDF of common beans, showing decreases in the SDF fraction, due to the loss of uronic acids from pectic substances, accompanied by increases in the IDF.[83] In lentils, the impact of autoclaving on dietary fibre was only remarkable in terms of concentration of oligosaccharides.[7] However, in chickpeas, the effect of the treatment was to decrease the content of TDF, due to high reductions in the IDF fraction, although the content of resistant starch was considerably increased.[23]

Extrusion in beans also caused a considerable redistribution of IDF to SDF, but the total dietary fibre content was not affected. Changes in solubility involved mainly the pectic polysaccharides, arabinose and uronic acid.[84] Similar results were shown after extrusion of peas, due to the solubilization of neutral pectic substances or pectic side chains rather than the rhamnogalacturonans or hemicellulose polymers;[85] in addition, the cellulosic backbone was not modified.[86] Owing to the high temperature and screw speed of extrusion, Berrios *et al.*[87] found decreases in the oligosaccharide concentrations in different pulses subjected to extrusion processing; the content of TDF in these flours was increased while the IDF decreased.

10.3.4 Impact on Phenolic Compounds

Thermal processing can modify the chemical form of phenolic compounds present in legumes.[88] Processing of legumes has commonly been reported to lead to either an increase, decrease, or no changes in the phenolic concentration.[89] A decrease in phenolics may be due to a processing method that allows movement of phenolics out of the system, possibly due to their dissolution into the processing water. There may also be chemical degradation or transformation of phenolics into other forms in which they are not assayable as phenolics. The processing method may also induce interactions between phenolic compounds and the other components of the legume seed that reduce their extractability and ability to be measured. Alternatively, an increase in the content of phenolic compounds could be due to the release of bound phenolics during thermal processing, rendering them more extractable and assayable.[88,89]

Several authors have described decreases in the total phenolic content (TPC) of bean after conventional cooking. These reductions were accompanied by reductions in the content of tannins and *in vitro* antioxidant capacity.[90,91] A very similar effect was presented by Maheshu *et al.*[92] in Dolichos beans when studying the impact of both conventional and pressure cooking. Nevertheless, Martín-Cabrejas *et al.*[11] found decreases in the TPC but increases in the content of catechins, in chickpeas and beans. Likewise, the dehydration process brought about higher increases in the catechin concentration. TPC and tannin content also increased in underutilised legumes following different thermal treatments.[93]

The phenolic profile of common brown beans was also modified by cooking, resulting in increased concentration of phenolic acids (vanillic, gallic, chlorogenic, and sinapic) and flavonols (kaempferol, catechin, quercetin, and their glycosides).[94] Nonetheless, Dueñas *et al.*[95] detected low decreases in most individual phenolic compounds from lentils and beans after cooking, even though a large part of those phenolics suffered redistribution from IDF to SDF. Aguilera *et al.*[96–98] observed decreased concentrations of the different families of phenolic compounds from diverse varieties of lentils, beans, and chickpeas; only slight increases in the content of hydroxybenzoic acids, the simple phenolics, were found after the legumes had been cooked.

10.3.5 Impact of Non-nutritional Factors

The impact of non-nutritional factors on nutrient metabolism and nutrition have been previously described.[99–102] Most of the so-called anti-nutrients, anti-nutritional factors, or non-nutritional factors may make positive contributions to human health and, as such, methods employed to reduce or remove these components from seeds may then impact negatively on the health benefits of food legumes.[103] Hence, tannins, oligosaccharides, and α-galactosides, have been included in previous sections, namely phenolic compounds and dietary fibre. Some of the health benefits of the non-nutritional factor from legumes include the anti-carcinogenic properties of trypsin/chymotrypsin inhibitors and saponins[104] or the benefits of oligosaccharides in gut microflora.[105] Certain classes of non-nutritional factors in legumes impair protein digestibility by inhibiting enzyme activities, in some cases leading to the formation of protein-non-nutritional factors complexes.[106] While it may be desirable to retain some of these non-nutritional factors in processed foods for their bioactive properties, the elimination or diminution of others during processing will improve legume nutritional quality.[107]

Thermal treatment can remove, reduce, inactivate or maintain the non-nutritional factors that are inherently present in legumes or that are generated during post-harvest processing and storage.[5,11,47] Thermal treatments significantly improve the nutritional quality and digestibility of pulses by inactivating or eliminating heat-labile non-nutritional factors.[100] The proportion of heat-stable and heat-labile non-nutritional factors varies among legume species and cultivars.[11] Thermal processes are usually able to remove, reduce, or inactivate the non-nutritional factors present in legumes, *i.e.* tannins and protease inhibitors, which can cause considerable reductions in the digestibility of protein and starch, the bioavailability of amino acids and nutrients and the overall quality of protein and starch.[108]

Cooking of pulses and pressure-cooking are the commonly used processes, and many researchers have evaluated the effects on pulse non-nutritional factors. Martín-Cabrejas *et al.*[11] described reductions in the content of phytic acid, amylase, trypsin and chymotrypsin inhibitors in common legumes including chickpeas, beans and lentils subjected to conventional cooking. Phytic acid, due to its heat-stable nature, was more resistant to the thermal treatment while enzyme inhibitor content was reduced further to negligible concentrations; compared to the raw seeds, almost total reductions in the level of these non-nutritional antinutrients were observed. Substantial reductions in the content of lectin were also observed.

Microwave cooking had a higher impact on the reduction of phytic acid in different legume seeds than both other thermal and non-thermal treatments including cooking and autoclaving.[109] Literature has shown that both microwaving, boiling, and autoclaving are able to destroy trypsin inhibitors in multiple kinds of legumes reaching 100% elimination of the inhibitory activity of these compounds.[7,8,47]

Autoclaving has also been shown to decrease the content of phytic acid and enzymes inhibitors (amylase and proteases inhibitors) in a great variety of legume seeds.[57,110] The effects of this treatment seemed not to be significantly different from those of conventional cooking (boiling). Autoclaving also had an impact on the content of lectins and saponins, reducing their concentration between 34 to 59%.[93]

10.3.6 Impact on Allergens

Legume allergies nowadays affect a noteworthy proportion of the population.[111–115] Consumption of legumes may cause to minor to severe anaphylactic symptoms in those sensitised individuals.[113,116,117] The allergenic potential of many legumes, such as lupins, peanuts, lentils, chickpeas, and soybeans, is well established.[113] Because of the increasing incidence of legume allergies, numerous efforts have been made to reduce the allergenicity of legume proteins.[113,118,119] Several effective methods can be implemented to minimise the risk of legume allergy.[120,121] In this sense, thermal processing of legumes has been observed as one of the most effective methods to decrease the allergenic problems associated with legume composition.[120,122]

Heating induces protein denaturation, aggregation, and the disruption of its structure and, consequently, it can modify allergenic properties of legume proteins.[123–127] The biochemical foundation of the changes in allergenic activity is the inactivation or destruction of epitope structures, the formation of new epitopes, or the enhancement of the access to cryptic epitopes by denaturation of the native allergen.[128,129] Thus, the temperature can impact both the physical and immunological properties of food proteins, modifying not only their structure, but their function, solubility, digestibility, and their IgE-binding epitopes and T-cell responses.[130,131]

Heat treatment might increase the digestibility of protein, increasing after that its absorption in the gastrointestinal tract, and hence, it may reduce the possibility of an allergenic protein eliciting an allergenic response.[132,133] In other cases, thermal processing may reduce the digestibility of a particular allergen or induce the formation of new antigens in the proteins, which were not present in its original form.[134] One of the main factors that induce the formation/degradation of antigens during legume processing is the Maillard reaction. This reaction may sometimes be beneficial by destroying the epitope of allergens and reducing allergenic potential, while it can also exacerbate allergic reactions by developing changes in the motifs of epitopes or generating new allergens.[135]

Nevertheless, the effects on legume allergens may be determined by the inherent characteristics of the treatment, *i.e.* the temperature, the duration of the process, the kind of heat treatment method used (dry or wet) along with the interaction with the food matrix.[136] It has been demonstrated that the allergenicity of protein can be reduced by one thermal treatment method but not by another.[137,138] The extent of thermal processing and its temperature have different effects on legume antigenicity depending on the protein.[123,139] All the treatments described in this chapter (cooking, microwaving, roasting, and autoclaving) have been shown to decrease the allergenic potential of legumes.[120] Literature reveals that, in a few cases, autoclaving has been more effective than boiling in decreasing the IgE-binding capacity. Clearly, microwave heating has been shown to be more effective than other thermal treatment methods. Roasting peanuts could even increase their allergenicity.[140,141]

10.4 Impact of Thermal Processing on the Techno-functional Properties of Legumes

To identify and improve the specific techno-functional properties of legume flours it is essential to determine the possible uses of legumes as food ingredients. The techno-functional properties provide information on how an ingredient would behave in the food matrix. These properties play a prominent role during food preparation, processing and storage and are correlated with sensory properties. These properties could be considered a result of different conformational changes or interactions that take place between the food components, such as the interactions between proteins, polysaccharides, lipids, phenolic compounds and phytic acid.[142,143]

Recent studies have revealed that different processing can change the structural characteristics and hydrophilicity : hydrophobicity ratio of legume components, mainly proteins. In this sense, these modifications might control the extent of protein unfolding and the ability of proteins to form films around dispersed oil.[144–146] Depending on the extent and type of processing, techno-functional properties show clear differences.[13,147] Hence, emulsification properties (emulsifying activity and stability) are essential in meat products, desserts and baked goods. The property of foaming is essential in products like whipped creams, frozen desserts and confectionery and occurs when films entrap air bubbles. Legume proteins are the components that can interact and also entrap air bubbles, migrate to the interface and keep the formed foam from collapsing.[145,147] Processing can affect the emulsion and foaming properties of legume products.[144] Equally, the property of water holding capacity in legume-based foods is modified during processing, the carbohydrate fraction being responsible for this property.[6,13,147] As an example, in the preparation of comminuted products, this techno-functional property is required to produce legume flours and fractions, which are responsible for imbibing, entrapping, and retaining more water. Similarly, processing may result in increases in fat absorption capacity with the exposure and modification of non-polar amino acid residues.[144,145,147] Nowadays, processing has been used to produce legume products with higher fat binding and retention properties, which may offer opportunities to use these products as ingredients for flavour retention or to mask undesirable flavours in foods. On the other hand, solubility and gelling properties of legumes are essential in many food applications (*e.g.* jellies, puddings, sauces and beverages). Developing products with improved gelling capacity[145,147] and solubility through processing is a trend in the food industry.[6,144,145,147] For example, recent studies revealed the impact of cooking and dehydration in the processing of flours (chickpeas, lentils, and beans).[13,97] The capacity of absorption and retention of water and swelling of these flours showed marked increases due to the increase in the availability of the starch fraction. The emulsifying, foaming and gelling properties are reduced because of cooking and dehydration. A decrease in protein level was the main change exhibited in all the legumes that were subjected to a dehydration process. Hence, it can be concluded that these dehydrated legume flours can be used as functional ingredients in food systems such as bakery products, dressings and sausages. Therefore, the processing step may disrupt legume protein and starch structures, increasing their bioaccessibility, and affecting the functional and end-use properties of the products.

Processed legumes with enhanced nutritional value and appealing texture and taste need to be developed in the market to provide functional food products with beneficial effects for human health. Further information on how to incorporate legume-based products into the food industry and design innovative and attractive food products with improved quality characteristics will be described in Chapter 11.

10.5 Concluding Remarks

Legumes have been presented as a very nutritious food, forming the basis of the diet of a large part of the world's population, which needs to be processed before consumption. Legume processing includes thermal and non-thermal treatments. The changes that take place during the thermal treatment of legumes depend not only on the processing carried out but also on the type of seed processed. This chapter has presented the different thermal processes for legumes and their effects on their nutritional and functional quality. Thermal treatments have been shown to improve the nutritional value of legumes, reducing the content of non-nutritional factors, but also, the use of temperature can remove beneficial components from legumes or modify them, thus affecting their techno-functional properties. Legumes are also a source of bioactive compounds with potential health benefits. As will be discussed in subsequent chapters (Chapters 12 to 15), the consumption of legumes has been associated with the prevention of several to chronic diseases. Diverse epidemiological and clinical studies have shown that the intake of legumes as part of a regular diet may limit the development of cancer, cardiovascular disease, and diabetes. However, thermal processes significantly influence not only the phytochemical profiles of legumes but also the potential beneficial biological effects (antioxidant, anti-inflammatory, antiproliferative, anti-hypertensive, anti-diabetic, anti-obesity) associated with these compounds. Since legumes must be cooked before consumption, the selection of the thermal cooking method to employ becomes vital to maintain or even increase the potential benefits of legumes.

Acknowledgements

M. Rebollo-Hernanz thank the FPU program of the Ministry of Education, Culture and Sports (FPU15/04238).

References

1. C. J. Rebello, F. L. Greenway and J. W. Finley, *J. Agric. Food Chem.*, 2014, **62**, 7029.
2. R. Tharanathan and S. Mahadevamma, *Trends Food Sci. Technol.*, 2003, **14**, 507.
3. S. Satya, G. Kaushik and S. N. Naik, *Med. J. Nutr. Metab.*, 2010, **3**, 183.
4. M. C. Vaz Patto, R. Amarowicz, A. N. A. Aryee, J. I. Boye, H.-J. Chung, M. A. Martín-Cabrejas and C. Domoney, *Crit. Rev. Plant Sci.*, 2015, **34**, 105.
5. M. Mondor, S. Aksay, H. Drolet, S. Roufik, E. Farnworth and J. I. Boye, *Innov. Food Sci. Emerg. Technol.*, 2009, **10**, 342.
6. P. J. M. Pelgrom, A. M. Vissers, R. M. Boom and M. A. I. Schutyser, *Food Res. Int.*, 2013, **53**, 232.
7. T. H. Hefnawy, *Ann. Agric. Sci.*, 2011, **56**, 57.
8. R. Y. Khattab, S. D. Arntfield and C. M. Nyachoti, *LWT - Food Sci. Technol.*, 2009, **42**, 1107.

9. N. Nidhina and S. P. Muthukumar, *Food Chem.*, 2015, **173**, 920.
10. Y. Aguilera, M. F. Díaz, T. Jiménez, V. Benítez, T. Herrera, C. Cuadrado, M. Martín-Pedrosa and M. A. Martín-Cabrejas, *J. Agric. Food Chem.*, 2013, **61**, 8120.
11. M. A. Martín-Cabrejas, Y. Aguilera, M. M. Pedrosa, C. Cuadrado, T. Hernández, S. Díaz and R. M. Esteban, *Food Chem.*, 2009, **114**, 1063.
12. D. Knorr, A. Froehling, H. Jaeger, K. Reineke, O. Schlueter and K. Schoessler, *Annu. Rev. Food Sci. Technol.*, 2011, **2**, 203.
13. Y. Aguilera, R. M. Esteban, V. Benítez, E. Mollá and M. A. Martín-Cabrejas, *J. Agric. Food Chem.*, 2009, **57**, 10682.
14. M. Rondanelli, M. Daglia, S. Meneghini, A. Di Lorenzo, G. Peroni, M. A. Faliva and S. Perna, *Food Sci. Nutr.*, 2017, **5**, 827.
15. Z. Rehinan, M. Rashid and W. H. Shah, *Food Chem.*, 2004, **85**, 245.
16. K. Liu, in *Food Technology (USA)*, Institute of Food Technologists, 1997, p. 58.
17. M. Pirhayati, N. Soltanizadeh and M. Kadivar, *Int. J. Food Sci. Technol.*, 2011, **46**, 1884.
18. Y. Coşkuner and E. Karababa, *Food Rev. Int.*, 2004, **20**, 257.
19. O. T. Olanipekun, E. C. Omenna, O. A. Olapade, P. Suleiman and O. G. Omodara, *Sky J. Food Sci.*, 2015, **4**, 24.
20. S. Bhattacharya, *Conventional and Advanced Food Processing Technologies*, John Wiley & Sons, 2015.
21. L. D. Griffith and M. E. Castell-Perez, *Cereal Chem. J.*, 1998, **75**, 780.
22. M. Kaur, N. Singh and N. S. Sodhi, *J. Food Eng.*, 2005, **69**, 511.
23. S. Mahadevamma and R. N. Tharanathan, *J. Food Qual.*, 2004, **27**, 289.
24. V. Schoeninger, S. R. M. Coelho, P. Z. Bassinello, V. Schoeninger, S. R. M. Coelho and P. Z. Bassinello, *Ciênc. Rural*, 2017, **47**, DOI: 10.1590/0103-8478cr20160672.
25. A. P. Singh, A. Singh and H. S. Ramaswamy, *Crit. Rev. Food Sci. Nutr.*, 2017, **57**, 1350.
26. A. Montanari, C. Barone, M. Barone and A. Santangelo, in *Thermal Treatments of Canned Foods*, eds. A. Montanari, C. Barone, M. Barone and A. Santangelo, Springer, Cham, 2018, p. 1.
27. N. N. Misra, M. Koubaa, S. Roohinejad, P. Juliano, H. Alpas, R. S. Inácio, J. A. Saraiva and F. J. Barba, *Food Res. Int.*, 2017, **97**, 318.
28. P. Puligundla, S. A. Abdullah, W. Choi, S. Jun, S.-E. Oh and S. Ko, *J. Food Process. Technol.*, 2013, **4**, 278.
29. G. B. Awuah, H. S. Ramaswamy and A. Economides, *Chem. Eng. Process. Process Intensif.*, 2007, **46**, 584.
30. V. A. Vaclavik and E. W. Christian, in *Essentials of Food Science*, Springer-Verlag, New York, NY, 2014, p. 323.
31. T. A. El-Adawy, *Plant Foods Hum. Nutr.*, 2002, **57**, 83.
32. S. A. Alajaji and T. A. El-Adawy, *J. Food Comp. Anal.*, 2006, **19**, 806.
33. R. M. Alonso-Salces, F. Serra, F. Reniero and K. Héberger, *J. Agric. Food Chem.*, 2009, **57**, 4224.
34. S. Singh, S. Gamlath and L. Wakeling, *Int. J. Food Sci. Technol.*, 2007, **42**, 916.

35. E. Pastor-Cavada, S. R. Drago, R. J. González, R. Juan, J. E. Pastor, M. Alaiz and J. Vioque, *Int. J. Food Sci. Technol.*, 2013, DOI: 10.1111/ijfs.12175.
36. C. W. P. Carvalho and J. R. Mitchell, *Int. J. Food Sci. Technol.*, 2000, **35**, 569.
37. N. Nikmaram, S. Y. Leong, M. Koubaa, Z. Zhu, F. J. Barba, R. Greiner, I. Oey and S. Roohinejad, *Food Control*, 2017, **79**, 62.
38. H. Akdogan, *Int. J. Food Sci. Technol.*, 2001, DOI: 10.1046/j.1365-2621.1999.00256.x.
39. M. S. Alam, J. Kaur, H. Khaira and K. Gupta, *Crit. Rev. Food Sci. Nutr.*, 2016, **56**, 445.
40. M. N. Riaz, *Qual. Assur. Saf. Crops Foods*, 2012, **4**, 156.
41. J. Milán-Carrillo, R. Gutiérrez-Dorado, J. X. K. Perales-Sánchez, E. O. Cuevas-Rodríguez, B. Ramírez-Wong and C. Reyes-Moreno, *Int. J. Food Sci. Technol.*, 2006, **41**, 727.
42. J. Eastman, F. Orthoefer and S. Solorio, *Cereal Foods World*, 2001, **46**, 468.
43. V. Offiah, V. Kontogiorgos and K. O. Falade, *Crit. Rev. Food Sci. Nutr.*, 2018, DOI: 10.1080/10408398.2018.1480007.
44. R. Polak, E. M. Phillips and A. Campbell, *Clin. Diabetes*, 2015, **33**, 198.
45. C. Ojiewo, D. J. D. H. Keatinge, J. Hughes, A. Tenkouano, R. Nair, R. Varshney, M. Siambi, E. Monyo, N. Ganga-Rao and S. Silim, *World Med. Health Policy*, 2015, **7**, 187.
46. N. Wang, D. W. Hatcher, R. T. Tyler, R. Toews and E. J. Gawalko, *Food Res. Int.*, 2010, **43**, 589.
47. R. Y. Khattab and S. D. Arntfield, *LWT - Food Sci. Technol.*, 2009, **42**, 1113.
48. N. Wang, D. W. Hatcher, R. Toews and E. J. Gawalko, *LWT - Food Sci. Technol.*, 2009, **42**, 842.
49. C. L. Nalle, G. Ravindran and V. Ravindran, *J. Sci. Food Agric.*, 2010, **90**, 1227.
50. R. Tyler, N. Wang and J. Han, *Cereal Chem. J.*, 2017, **94**, 1.
51. M. A. McCrory, B. R. Hamaker, J. C. Lovejoy and P. E. Eichelsdoerfer, *Adv. Nutr.*, 2010, **1**, 17.
52. M. G. Nosworthy, A. Franczyk, A. Zimoch-Korzycka, P. Appah, A. Utioh, J. Neufeld and J. D. House, *J. Agric. Food Chem.*, 2017, **65**, 3919.
53. T. Tinus, M. Damour, V. van Riel and P. A. Sopade, *J. Food Eng.*, 2012, **113**, 254.
54. C. J. Rebello, F. L. Greenway and J. W. Finley, *Obes. Rev.*, 2014, **15**, 392.
55. M. Duranti, *Fitoterapia*, 2006, **77**, 67.
56. M. Potier and D. Tomé, *J. AOAC Int.*, 2008, **91**, 1002.
57. C.-F. Chau and P. C.-K. Cheung, *J. Agric. Food Chem.*, 1997, **45**, 2500.
58. Z. Rehman and W. H. Shah, *Food Chem.*, 2005, **91**, 327.
59. V. B. Sashikala, Y. N. Sreerama, V. M. Pratape and H. V. Narasimha, *J. Food Sci. Technol.*, 2015, **52**, 1552.
60. N. Q. C. Melo, R. S. dos R. Moreira-Araújo, M. A. da M. Araújo, M. de M. Rocha, N. Q. C. Melo, R. S. dos R. Moreira-Araújo, M. A. da M. Araújo and M. de M. Rocha, *Rev. Ciênc. Agron.*, 2017, **48**, 811.
61. I. H. Han, B. G. Swanson and B.-K. Baik, *Cereal Chem. J.*, 2007, **84**, 518.

62. K. Mahasukhonthachat, P. A. Sopade and M. J. Gidley, *J. Food Eng.*, 2010, **96**, 18.
63. G. Annison and D. L. Topping, *Ann. Rev. Nutr.*, 1994, **14**, 297.
64. Z. Zhu and W. Guo, *Sci. Rep.*, 2017, **7**, 9311.
65. K. Muramatsu, A. Maiorka, F. Dahlke, A. Lopes and M. Pasche, *Rev. Bras. Ciência Avícola*, 2014, **16**, 367.
66. H. Fredrikssona, J. Silveriob, R. Andemon, A.-C. Eliassonb and P. Amanta, *Carbohydr. Polym.*, 1998, **35**, 9.
67. A. Lovegrove, C. H. Edwards, I. De Noni, H. Patel, S. N. El, T. Grassby, C. Zielke, M. Ulmius, L. Nilsson, P. J. Butterworth, P. R. Ellis and P. R. Shewry, *Crit. Rev. Food Sci. Nutr.*, 2017, **57**, 237.
68. Zia-ur-Rehman, *Food Chem.*, 2007, **100**, 764.
69. Y. Aguilera, M. A. Martín-Cabrejas, V. Benítez, E. Mollá, F. J. López-Andréu and R. M. Esteban, *J. Food Compos. Anal.*, 2009, **22**, 678.
70. M. Pujolà, A. Farreras and F. Casañas, *Food Chem.*, 2007, **102**, 1034.
71. E. Marconi, S. Ruggeri, M. Cappelloni, D. Leonardi and E. Carnovale, *J. Agric. Food Chem.*, 2000, **48**, 5986.
72. A. Negi, P. Boora and N. Khetarpaul, *Nahrung/Food*, 2001, **45**, 251.
73. P. Osorio-Díaz, L. A. Bello-Pérez, S. G. Sáyago-Ayerdi, M. del P. Benítez-Reyes, J. Tovar and O. Paredes-López, *J. Sci. Food Agric.*, 2003, **83**, 1283.
74. G. E. de Almeida Costa, K. da Silva Queiroz-Monici, S. M. Pissini Machado Reis and A. C. de Oliveira, *Food Chem.*, 2006, **94**, 327.
75. R. Hoover and Y. Zhou, *Carbohydr. Polym.*, 2003, **54**, 401.
76. S. W. Rizkalla, F. Bellisle and G. Slama, *Br. J. Nutr.*, 2002, **88**, 255.
77. J. M. W. Wong and D. J. A. Jenkins, *J. Nutr.*, 2007, **137**, 2539S.
78. S. M. Tosh and S. Yada, *Food Res. Int.*, 2010, **43**, 450.
79. T. Kutoš, T. Golob, M. Kač and A. Plestenjak, *Food Chem.*, 2003, **80**, 231.
80. M. J. Peña, C. E. Vergara and N. C. Carpita, in *Advanced Dietary Fibre Technology*, ed. B. V. McCleary and L. Prosky, Blackwell Science Ltd, Oxford, UK, 2008, p. 42.
81. K. W. Waldron, M. L. Parker and A. C. Smith, *Compr. Rev. Food Sci. Food Saf.*, 2003, **2**, 128.
82. H. Vasishtha and R. P. Srivastava, *J. Food Sci. Technol.*, 2013, **50**, 579.
83. M. A. Martín-Cabrejas, B. Sanfiz, A. Vidal, E. Mollá, R. Esteban and F. J. López-Andréu, *J. Agric. Food Chem.*, 2004, **52**, 261.
84. M. A. Martín-Cabrejas, L. Jaime, C. Karanja, A. J. Downie, M. L. Parker, F. J. Lopez-Andreu, G. Maina, R. M. Esteban, A. C. Smith and K. W. Waldron, *J. Agric. Food Chem.*, 1999, **47**, 1174.
85. M.-C. Ralet, G. Della Valle and J.-F. Thibault, *Carbohydr. Polym.*, 1993, **20**, 17.
86. M.-C. Ralet, L. Saulnier and J.-F. Thibault, *Carbohydr. Polym.*, 1993, **20**, 25.
87. J. D. J. Berrios, P. Morales, M. Cámara and M. C. Sánchez-Mata, *Food Res. Int.*, 2010, **43**, 531.
88. B. Xu and S. K. C. Chang, *J. Agric. Food Chem.*, 2009, **57**, 4754.

89. K. G. Duodu, *Cereal Foods World*, 2014, **59**, 64.
90. M. O. Silva, P. Brigide, N. M. V. de Toledo, S. G. Canniatti-Brazaca, M. O. Silva, P. Brigide, N. M. V. de Toledo and S. G. Canniatti-Brazaca, *Braz. J. Food Technol.*, 2018, **21**, e2016072.
91. L. G. Ranilla, Y.-I. Kwon, M. I. Genovese, F. M. Lajolo and K. Shetty, *J. Food Biochem.*, 2010, **34**, 329355.
92. V. Maheshu, D. T. Priyadarsini and J. M. Sasikumar, *J. Food Sci. Technol.*, 2013, **50**, 731.
93. A. Sathya and P. Siddhuraju, *J. Food Sci. Technol.*, 2015, **52**, 6157.
94. K. Huber, P. Brigide, E. B. Bretas and S. G. Canniatti-Brazaca, *J. Nutr. Food Sci.*, 2016, **6**, 1.
95. M. Dueñas, T. Sarmento, Y. Aguilera, V. Benitez, E. Mollá, R. M. Esteban and M. A. Martín-Cabrejas, *LWT - Food Sci. Technol.*, 2016, **66**, 72.
96. Y. Aguilera, M. Dueñas, I. Estrella, T. Hernández, V. Benitez, R. M. Esteban and M. A. Martín-Cabrejas, *J. Agric. Food Chem.*, 2010, **58**, 10101.
97. Y. Aguilera, I. Estrella, V. Benitez, R. M. Esteban and M. A. Martín-Cabrejas, *Food Res. Int.*, 2011, **44**, 774.
98. Y. Aguilera, M. Dueñas, I. Estrella, T. Hernández, V. Benitez, R. M. Esteban and M. A. Martín-Cabrejas, *Plant Foods Hum. Nutr.*, 2011, **66**, 187.
99. G. S. Gilani, K. A. Cockell and E. Sepehr, *J. AOAC Int.*, 2005, **88**, 967.
100. G. Sarwar Gilani, C. Wu Xiao and K. A. Cockell, *Br. J. Nutr.*, 2012, **108**, S315.
101. K. O. Soetan and O. E. Oyewole, *Afr. J. Food Sci.*, 2009, **3**, 223.
102. P. Thangaraj, in *Pharmacological Assays of Plant-based Natural Products*, Progress in Drug Research, Springer, Cham, vol. 71, 2016, p. 43.
103. S. Multari, D. Stewart and W. R. Russell, *Compr. Rev. Food Sci. Food Saf.*, 2015, **14**, 511.
104. S. Srikanth and Z. Chen, *Front. Pharmacol.*, 2016, **7**, 470.
105. W. Cheng, J. Lu, B. Li, W. Lin, Z. Zhang, X. Wei, C. Sun, M. Chi, W. Bi, B. Yang, A. Jiang and J. Yuan, *Front. Microbiol.*, 2017, **8**, 1750.
106. A. Clemente, M. C. Arques, M. Dalmais, C. Le Signor, C. Chinoy, R. Olias, T. Rayner, P. G. Isaac, D. M. Lawson, A. Bendahmane and C. Domoney, *PLoS One*, 2015, **10**, e0134634.
107. A. Clemente, G. Sonnante and C. Domoney, *Curr. Protein Pept. Sci.*, 2011, **12**, 358.
108. D. C. Dallas, M. R. Sanctuary, Y. Qu, S. H. Khajavi, A. E. Van Zandt, M. Dyandra, S. A. Frese, D. Barile and J. B. German, *Crit. Rev. Food Sci. Nutr.*, 2017, **57**, 3313.
109. U. Boettler, K. Sommerfeld, N. Volz, G. Pahlke, N. Teller, V. Somoza, R. Lang, T. Hofmann and D. Marko, *J. Nutr. Biochem.*, 2011, **22**, 426.
110. R. Borchers and C. W. Ackerson, *J. Nutr.*, 1950, **41**, 339.
111. H. A. Sampson, *N. Engl. J. Med.*, 2002, **346**, 1294.
112. S. L. Prescott, R. Pawankar, K. J. Allen, D. E. Campbell, J. K. Sinn, A. Fiocchi, M. Ebisawa, H. A. Sampson, K. Beyer and B.-W. Lee, *World Allergy Organ. J.*, 2013, **6**, 21.

113. A. K. Verma, S. Kumar, M. Das and P. D. Dwivedi, *Clin. Rev. Allergy Immunol.*, 2013, **45**, 30–46.
114. S. H. Sicherer, *J. Allergy Clin. Immunol.*, 2011, **127**, 594.
115. B. Cabanillas, U. Jappe and N. Novak, *Mol. Nutr. Food Res.*, 2018, **62**, 1700446.
116. https://www.anaphylaxis.org.uk(last accessed June 2018).
117. A. Cianferoni and A. Muraro, *Immunol. Allergy Clin. North Am.*, 2012, **32**, 165.
118. S. H. Sicherer and H. A. Sampson, *J. Allergy Clin. Immunol.*, 2018, **141**, 41.
119. J. M. Lomas and K. M. Järvinen, *J. Asthma Allergy*, 2015, **8**, 115.
120. A. K. Verma, S. Kumar, M. Das and P. D. Dwivedi, *Plant Foods Hum. Nutr.*, 2012, **67**, 430.
121. H.-W. Huang, C.-P. Hsu, B. B. Yang and C.-Y. Wang, *Compr. Rev. Food Sci. Food Sci.*, 2013, DOI: 10.1111/1541-4337.12045.
122. J. J. Riascos, A. K. Weissinger, S. M. Weissinger and A. W. Burks, *J. Agric. Food Chem.*, 2010, **58**, 20.
123. P. J. Davis and S. C. Williams, *Allergy*, 1998, **53**, 102.
124. C. L. Okolie, A. N. A. Aryee and C. C. Udenigwe, in *Proteins in Food Processing*, ed. R. Y. Yada, Elsevier Ltd, 2018, p. 367.
125. Q. Xu, J. Shi, M. Yao, M. Jiang and Y. Luo, *J. Food Agric. Immunol.*, 2016, **27**, 401.
126. E. N. C. Mills, A. I. Sancho, N. Rigby, J. A. Jenkins and A. R. Mackie, *Mol. Nutr. Food Res.*, 2009, **53**, 963.
127. J. Xu, Z. Chen, D. Han, Y. Li, X. Sun, Z. Wang and H. Jin, *Molecules*, 2017, **22**, 1893.
128. C. Koch, S. S. Jensen, A. Oster and G. Houen, *APMIS*, 1996, **104**, 115.
129. M. Besler, H. Steinhart and A. Paschke, *J. Chromatogr. B: Biomed. Sci. Appl.*, 2001, **756**, 207.
130. A. N. A. Aryee, D. Agyei and C. C. Udenigwe, in *Proteins in Food Processing*, ed. R. Y. Yada, Elsevier Ltd, 2018, p. 27.
131. S. K. Sathe and G. M. Sharma, *Mol. Nutr. Food Res.*, 2009, **53**, 970.
132. M. Teodorowicz, J. van Neerven and H. Savelkoul, *Nutrients*, 2017, **4**, E835.
133. S. Benedé, I. López-Expósito, E. Molina and R. López-Fandiño, *Food Funct.*, 2015, **6**, 694.
134. J. Pekar, D. Ret and E. Untersmayr, *Mol. Immunol.*, 2018, **100**, 14.
135. P. Singh, R. H. Jayaramaiah, S. B. Agawane, G. Vannuruswamy, A. M. Korwar, A. Anand, V. S. Dhaygude, M. L. Shaikh, R. S. Joshi, R. Boppana, M. J. Kulkarni, H. V. Thulasiram, A. P. Giri, F. A. Van De Laar, A. J. Scheen, N. Fukaya, K. Matsumoto, M. Brownlee, M. Brownlee, M. Luthra, D. Balasubramanian, M. J. Kulkarni, A. M. Korwar, S. Mary, H. S. Bhonsle, A. P. Giri, J. A. Vinson, T. B. Howard, P. Urios, A. M. Grigorova-Borsos, M. Sternberg, D. Ruggiero-Lopez, S. K. Kesavan, S. Vasan, K. Mizutani, K. Ikeda, Y. Yamori, Y. Tang, P. Raskin, K. Matsumoto, J. L. Chiasson,

L. Naditch, P. Prakash, N. Gupta, F. L. Hakkim, G. Arivazhagan, R. Boopathy, K. P. Bhargava, N. Singh., K. Karthikeyan, P. Gunasekaran, N. Ramamurthy, S. Govindasamy, P. U. Devi, A. Ganasoundari, B. Vrinda, K. K. Srinivasan, M. K. Unnikrishnan, A. Khan, A. Ahmad, N. Manzoor, L. A. Khan, P. Singh, R. R. Chattopadhyay, J. M. Hannan, W. Bors, C. Michel, M. Zhang, W. Xu, Y. Deng, K. Yanagisawa, Y. M. Kolekar, B. Elya, H. S. Bhonsle, F. Y. Cheng, K. Blackburn, Y. M. Lin, M. B. Goshe, J. D. Williamson, H. S. Bhonsle, A. M. Korwar and K. A. Jones, *Sci. Rep.*, 2016, **6**, 18798.

136. K. C. M. Verhoeckx, Y. M. Vissers, J. L. Baumert, R. Faludi, M. Feys, S. Flanagan, C. Herouet-Guicheney, T. Holzhauser, R. Shimojo, N. van der Bolt, H. Wichers and I. Kimber, *Food Chem. Toxicol.*, 2015, **80**, 223.
137. R. Jiménez-Saiz, S. Benedé, E. Molina and I. López-Expósito, *Crit. Rev. Food Sci. Nutr.*, 2015, **55**, 1902.
138. A. Paschke, *Mol. Nutr. Food Res.*, 2009, **53**, 959.
139. M. Besler, H. Steinhart, A. Paschke and M. Besler, *Internet Symp. Food Allergens*, 2001, **3**, 119.
140. K. Beyer, E. Morrowa, X.-M. Li, L. Bardina, G. A. Bannon, A. W. Burks and H. A. Sampson, *J. Allergy Clin. Immunol.*, 2001, **107**, 1077.
141. C. Cuadrado, B. Cabanillas, M. M. Pedrosa, A. Varela, E. Guillamón, M. Muzquiz, J. F. Crespo, J. Rodriguez and C. Burbano, *Mol. Nutr. Food Res.*, 2009, **53**, 1462.
142. M. A. H. Ismond, E. D. Murray and S. D. Arntfield, *Food Chem.*, 1986, **21**, 27.
143. C. Patanè, E. Iacoponi and S. A. Raccuia, *Int. J. Food Sci. Nutr.*, 2004, **55**, 547.
144. J. I. Boye, S. Aksay, S. Roufik, S. Ribéreau, M. Mondor, E. Farnworth and S. H. Rajamohamed, *Food Res. Int.*, 2010, **43**, 537.
145. F. Zare and A. Pletch, *Food Res. Int.*, 2010, **43**, 414.
146. N. A. Avramenko, N. H. Low and M. T. Nickerson, *Food Res. Int.*, 2013, **51**, 162.
147. M. Kaur and N. Singh, *Food Chem.*, 2007, **102**, 366.

CHAPTER 11

Innovative Legume Foods

J. PINELA[a], S. A. PETROPOULOS*[b] AND I. C. F. R. FERREIRA*[a]

[a]Centro de Investigação de Montanha (CIMO), Instituto Politécnico de Bragança, Campus de Santa Apolónia, 5300-253 Bragança, Portugal; [b]Department of Agriculture, Crop Production and Rural Environment, University of Thessaly, 38446 N. Ionia, Magnissia, Greece
*E-mail: spetropoulos@uth.gr, iferreira@ipb.pt

11.1 Significance of Legume Foods in Human Diet

Leguminous crops belong to the Fabaceae or Leguminosae family and their fruit and seeds (dry or fresh) are essential in human diet, since they contribute significantly to dietary needs of various macro- and micro-nutrients on a daily basis. There are many economically important cultivated species in this family including soybeans (*Glycine max*), chickpeas (*Cicer arietinum*), cowpeas (*Vigna unguiculata*), peas (*Pisum* sp.), broad beans (*Vicia faba*), common beans (*Phaseolus vulgaris*), lentils (*Lens culinaris*), grass peas (*Lathyrus sativus*), sweet peas (*Lathyrus odoratus*) and lupins (*Lupinus* sp.), that have been used by humans throughout the centuries. Other species of minor or regional importance include *Vigna* species such as mung bean (*V. radiata*), black gram (*V. mungo*) and Bambara groundnut (*V. subterranea*), while pigeon pea (*Cajanus cajan*) and vetch (*Vicia sativa*) have also been used for human consumption from time to time. The most common use of the above mentioned species is the consumption of their fresh fruit (pods) which can be consumed either as raw or processed vegetables, as well as the consumption

Food Chemistry, Function and Analysis No. 8
Legumes: Nutritional Quality, Processing and Potential Health Benefits
Edited by María Ángeles Martín-Cabrejas

Published by the Royal Society of Chemistry, www.rsc.org

of their naturally dried pods and grains (pulses) which are ingredients in many dishes throughout the world.[1] Depending on their end-use (vegetables or pulses), the nutritional value and chemical composition of legume products varies significantly, with pulses being excellent sources of proteins, carbohydrates, dietary fibres, vitamins (mainly B-complex vitamins) and minerals, while fresh pods contain significant amounts of vitamins, macro- and micro-nutrients.[2] Moreover, pulses are considered as gluten-free products which is a very important quality feature for the food and baking products industry in particular, considering the recent market trends and consumer preference for such products.[3–5]

As shown in the previous chapters, legumes are a very rich protein source, with dried grains (pulses) containing more than 20% of protein, while they are also a good source of vitamins, minerals and energy without containing high amounts of fat.[6] Despite the health benefits of pulses, annual consumption per capita showed a gradual decrease during the last decades in both developed and developing countries which could be attributed to the modern way of living and the consumers' preference for more expensive protein sources such as dairy and meat products.[7] However, recently consumption of pulses shows an increasing trend throughout the world, where in undeveloped and developing countries they are the main protein source, while in developed countries consumers appreciate their health benefits and their importance for a balanced and healthy diet. Recent studies have confirmed the importance of legume consumption in the fight against coronary heart diseases and diabetes, since they contain proteins, dietary fibres and antidiabetic substances which are correlated with reduced serum cholesterol and glucose absorption,[8–10] while other studies report high polyphenol content and bioactive properties of legume seeds and sprouts.[11]

Legume products are usually consumed in raw or minimally processed form (cooked, canned or frozen food products); however, there is an increasing trend and research activity for incorporating legumes in innovative food products and formulations that will help towards increasing legume consumption and improving human diet throughout the world.[12] Although legumes contain several compounds beneficial to human health, there is great concern regarding the antinutrient and even toxic and allergenic factors they may also contain (tannins, trypsin inhibitors, saponins, raffinose, pyrimidine glycosides, oxalates *etc.*), that needs to be further investigated before the food industry follows the trend and legume-based food products become widely accepted.[6,13–17]

Grain legumes can be included in various food products and formulations, thus increasing not only the nutritional value but also the functional properties of the final products, and indeed there are many products based on legumes available in the market already, including breads and baked products, milk and dairy products substitutes, pasta *etc.*[14] Incorporation of legume proteins, isolates and concentrates has been described to affect various functional and physicochemical properties of food matrices, including solubility of proteins, water and fat binding capacity, and emulsification, gelation and foaming properties, all of which are highly appreciated in the food industry.[14] Moreover, there is great potential for creating innovative legume-fortified cereal products, especially in terms of essential amino acids and minerals contents.[18] In order to achieve this, the complex nature of the nutritional properties of legumes and the effect of processing techniques on the final food matrices has to be further elucidated.[12] Legumes may also help towards increasing dietary fibre consumption in human diets.[19] The interaction between the various components of primary matrixes and legume ingredients has needs to be studied, since such interactions (*e.g.* proteins, polysaccharides, metal ions, lipids and enzymes) may affect not only the visual appearance and texture but also the nutritional value and chemical composition of the final product.[20] In addition, significant variation exists among cultivars of the same species regarding chemical composition and nutritional value that has to be further elucidated and explored.[21] Finally, apart from the cultivated conventional legume species, there are many unexploited and underutilized wild species of regional interest with important nutritional features that could be studied and incorporated into extensive breeding programs in order to improve existing germplasm, with a view to designing and creating innovative legume-based food products.[13,22]

11.2 Chemical Composition of the Most Important Legume Crops

The proximate composition of important legume crops, namely common bean, broad bean, chickpea, cowpea, pea (*Pisum sativum*), grass pea, soybean, lentil, white lupin (*Lupinus albus*) and mung bean is shown in Table 11.1. The mean values of nutrient content and the range of variability in the literature data are presented. These differences between reported values can be attributed to the cultivars studied, the edaphoclimatic conditions of the cultivation sites and also to the analytical techniques used.[23,24] These legume seeds contain high amounts of protein ranging from 20.4 g/100 g in common bean[14,25–27] to 38.6 g/100 g in soybean[28–33] and white lupin.[29,34–37] In addition to providing essential amino acids and bioactive peptides for human nutrition, legume proteins possess functional properties such as water holding, fat binding and gelation, which promote their use in the development of a wide range of food products.[14] The nutritional quality of proteins is largely determined by their essential amino acid composition and digestibility. Pulse proteins are generally high in glutamic acid, arginine, aspartic acid, lysine and leucine content, but deficient in methionine, cysteine and tryptophan.[14,34,38] Proteins with higher digestibility have a better nutritional value. *In vitro* protein digestibility of values of 63.8%, 64.2–74.0%, 65.6%, 68.1–70.6%, 70.8–75.4% and 74.5% have been reported for raw cowpeas,[39] chickpeas,[39,40] lentils,[39] kidney beans,[26,40] broad beans,[26,40] and peas, respectively.[40]

The protein digestibility also varies depending on the legume variety and can be improved by processing methods such as sacking, dehulling, cooking and extrusion;[39,40] however, it can be adversely affected by the presence of trypsin inhibitors, which must be inactivated during processing in order to increase protein bioavailability upon digestion.

Legumes are a good source of structural carbohydrates mainly due to their high content of dietary fibre. This complex mixture of indigestible components can be classified according to its solubility as soluble and insoluble dietary fibres, which have dissimilar nutritional benefits and health-promoting effects. As described in Chapter 4, insoluble dietary fibre (*e.g.* cellulose and hemicellulose) is fermented in the large intestine, supporting the growth of intestinal microflora, including probiotic species, while soluble dietary fibre (*e.g.* oligosaccharides and pectins) helps to lower blood cholesterol and regulate blood glucose levels.[50] Legume-based flours and fibre-rich fractions have been incorporated into food products to increase the dietary fibre content and/or serve as a functional ingredient (*e.g.* used to enhance or modify the food texture).[51] As showed in Table 11.1, mean values of total dietary fibre in legume seeds range from 5.5 to 24.6 g/100 g, with the highest levels found in common bean (22.2 g/100 g)[14,25,27] and white lupin (24.4 g/100 g).[29,34–37] In addition, the levels of insoluble dietary fibre in legume seeds have been reported to be higher than those of soluble dietary fibre.[33,51] For example, in the case of yellow and green soybeans insoluble fibre corresponds to 74–78% of the total fibre fraction.[33] It is also known that dietary fibre characteristics may be altered by processing methods such as grinding or extrusion. Furthermore, cooking and heat treatments can decrease the insoluble fraction and increase the soluble one mainly by solubilisation of hemicelluloses and degradation of pectic polysaccharides.[52]

Starch is generally the most abundant carbohydrate found in legumes, reaching up to ~50 g/100 g in mung bean (Table 11.2).[53–55] On the other hand, soybean and white lupin have the lowest values (5.6 and 6.3 g/100 g, respectively),[24,32,37] but with a high protein and fat content, with fat appearing to replace starch as the main source of energy, since mean fat values reach ~8.5 g/100 g in these seeds (Table 11.1).[29,30,33,34,37] In the food industry, starch has been used as a raw material to impart specific functions to food products such as thickening, gelling, coating and encapsulation.[56] However, the swelling, gelatinization and pasting properties and the *in vitro* starch digestibility may differ among the cultivars of the same species (*e.g.* desi and kabuli chickpeas).[21] As reported for proteins, the starch digestibility can also be improved by processing techniques. For example, cooking in boiling water may improve the starch digestibility of chickpeas, lentils and kidney beans by 84.0–90.4%.[57] Moreover, extrusion can be more effective in improving starch digestibility than dehulling or soaking.[26] In turn, Lu *et al.*[58] demonstrated that, although roasted pea starch had a low *in vitro* digestibility, the slowly digestible starch and resistant starch fractions became rapidly digestible after baking.

Table 11.1 Proximate composition of seeds of the most important legume crops. The mean values (g/100 g of dry mater)[a] and the range of variability (within parenthesis) of the literature data are presented.

	Proteins	Carbohydrates	Dietary fibre	Fat	Ash	References
Common bean	20.4 (16.9–26.4)	63.7 (56.6–69.3)	22.2 (6.2–30.3)	2.4 (0.8–3.6)	3.5 (1.1–4.5)	14 and 25–27
Broad bean	30.1 (26.9–31.9)	48.0 (47.7–48.3)	9.1 (8.7–9.9)	1.5 (1.1–2.0)	3.0 (3.0–3.1)	26, 34, 41 and 42
Chickpea	22.0 (15.4–29.0)	62.5 (48.0–71.2)	11.4 (3.0–27.2)	5.6 (2.8–7.3)	3.2 (2.2–5.7)	14, 23, 25, 30, 34, 39 and 43–45
Cowpea	24.2 (22.4–25.7)	63.5 (63.3–63.6)	14.4 (1.6–27.4)	2.5 (1.2–4.8)	3.6 (2.9–4.2)	38, 39 and 46
Pea	24.3 (19.4–34.7)	57.7 (40.5–65.7)	13.4 (4.3–35.3)	3.1 (1.5–12.1)	3.5 (2.4–4.6)	14, 21, 23, 25, 29 and 46
Grass pea	27.6 (24.1–35.9)	62.2 (60.5–64.2)	12.1 (5.3–31.7)	1.3 (0.5–2.3)	3.5	25, 34 and 44
Soybean	38.6 (20.7–52.1)	25.5 (9.9–38.6)	14.8 (4.2–23.6)	8.5 (0.9–18.6)	5.4 (3.4–7.0)	28–33
Lentil	25.8 (18.8–31.5)	63.7 (62.9–69.0)	16.6 (13.1–24.7)	1.8 (0.5–3.5)	2.6 (2.0–4.16)	14, 21, 23, 25, 39 and 47
White lupin	38.3 (30.6–52.9)	45	24.6 (13.7–39.4)	8.4 (1.4–14.6)	3.8 (2.7–5.0)	29 and 34–37
Mung bean	23.9 (20.0–30.6)	65.1 (52.1–91.3)	5.5 (4.2–8.5)	1.6 (1.2–2.6)	3.5 (2.8–4.6)	48 and 49

[a]Some values were converted from other units, based on the originally reported data.

Table 11.2 Starch, sucrose and α-galactosides (raffinose, stachyose and verbascose) contents in seeds of the most important legume crops. The mean values (g/100 g of dry mater)[a] and the range of variability (within parenthesis) of the literature data are presented.

	Starch	Sucrose	Raffinose	Stachyose	Verbascose	References
Common bean	37.1 (36.9–37.2)	2.6 (1.7–3.7)	0.4 (0.2–0.6)	1.9 (1.2–2.8)	0.2	27 and 62
Broad bean	42.9 (41.2–44.3)	2.5 (1.2–3.8)	0.5 (0.3–1.3)	1.9 (0.9–2.5)	2.4 (0.7–5.0)	61 and 64
Chickpea	44.3 (42.4–46.3)	2.0 (0.8–3.3)	0.8 (0.6–0.9)	1.9 (1.4–2.3)	1.0 (0.7–1.3)	19, 21, 44, 45 and 57
Cowpea	34.3 (30.5–38.1)	3.0 (2.3–3.7)	1.0 (0.7–1.2)	2.4 (1.8–3.4)	0.8 (0.4–1.2)	38, 39, 45, 63 and 65
Pea	45.8 (39.7–49.4)	0.5 (0.4–0.6)	0.9 (0.3–1.6)	1.4 (0.7–2.0)	0.8	19, 21 and 66
Grass pea	41.4 (40.5–42.3)	2.5	0.4 (0–0.9)	0.4 (0–0.9)	nd	44, 67 and 68
Soybean	5.6 (4.9–6.3)	6.0 (3.2–9.5)	0.7 (0.3–1.0)	3.5 (0.2–4.5)	0.1 (0.0–0.2)	31, 32 and 60
Lentil	44.6 (28.7–49.7)	2.3 (0.7–3.2)	0.6 (0.5–1.2)	2.7 (2.3–3.0)	0.8 (0.7–1.0)	19, 21, 39, 47 and 57
White lupin	6.3 (3.3–9.3)	3.2 (2.5–4.1)	0.5 (0.3–0.6)	6.5 (5.7–7.2)	1.1 (0.9–1.2)	24, 36 and 37
Mung bean	50.5 (46.1–54.9)	1.4	0.4 (0.3–0.4)	1.5 (1.5–1.6)	1.0	53–55

[a]Some values were converted from other units, based on the originally reported data; nd: not detected.

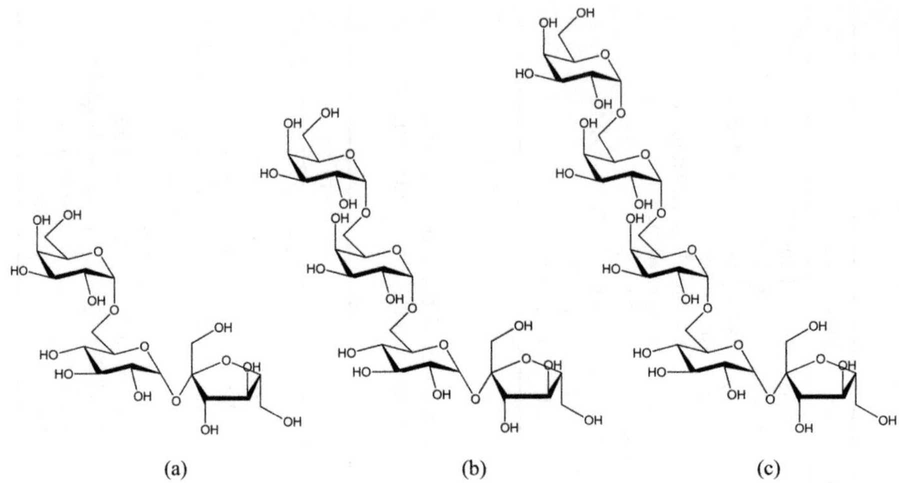

Figure 11.1 Chemical structures of (a) raffinose, (b) stachyose, and (c) verbascose.

As shown in Table 11.2, α-galactosides of sucrose, namely raffinose, stachyose and verbascose (Figure 11.1) can be found (some of them in appreciable concentrations) in seeds of the most important legume crops. Due to the absence of α-galactosidase activity in the gut, these antinutritional factors escape digestion in the upper part of the intestinal tract and pass into the large intestine where they are fermented by the intestinal microbiota.[59] High amounts of sucrose and stachyose have been reported for soybeans (6.0 and 3.5 g/100 g, respectively).[31,32,60] White lupin also contains high levels of stachyose ranging from 5.7 to 7.2 g/100 g.[37] In turn, verbascose is particularly abundant in broad bean (2.4 g/100 g).[61] The lowest levels of raffinose were described in common bean, grass pea and mung bean, while the lowest levels of stachyose were reported in grass pea, in which verbascose was not detected (Table 11.2). In fact, stachyose generally predominates over raffinose in legumes, except for the case of grass pea where equal amounts have been reported. These flatulence-causing oligosaccharides can be removed from legume seeds using different processing methods.[37,59,62,63] However, in moderate doses, they can act as prebiotics through the selective stimulation of bifidobacteria growth, thus being promising functional ingredients in food products.[59]

As shown in Chapter 7, the nutritional quality of legumes is affected by the presence of antinutritional factors such as phytic acid, tannins and trypsin inhibitors. Phytic acid influences the absorption of essential minerals (*e.g.* Fe and Zn) by turning them into insoluble forms unavailable to the organism, while it also decreases the digestibility of proteins.[69] In legume seeds, phytic acid levels ranging from 0.75 g/100 g (in mung bean)[49] to 1.79 g/100 g (in common bean)[26,57,62] have been reported (Table 11.3). Tannins are particularly abundant in cowpea, with levels range from 0.38 to 12.31 g/100 g.[39,63,70] These antinutrients can form complexes with enzymes and

Table 11.3 Phytic acid, tannin and trypsin inhibitors content in seeds of important legume crops.[a] The mean values and the range of variability (within parenthesis) of the literature data are presented.

	Phytic acid	Tannins	Trypsin inhibitors	References
Common bean	1.79 (1.23–2.41)	1.45 (0.54–2.88)	13.6 (3.1–29.3)	26, 27, 57, 62 and 75
Broad bean	0.89 (0.12–2.17)	0.22 (0.20–0.23)	7.2 (2.9–14.1)	26, 42 and 74
Chickpea	0.89 (0.48–1.21)	0.49 (0.18–0.77)	14.5 (6.5–20.9)	39, 43, 45, 57, 75 and 76
Cowpea	0.80 (0.40–1.40)	4.75 (0.38–12.31)	7.0	39, 45, 63 and 70
Pea	1.02 (0.72–1.23)	0.15 (1.07–2.37)	9.5 (5.8–12.6)	75, 77 and 78
Soybean	0.89 (0.21–1.20)	2.69 (1.76–3.60)	58.1 (43.6–71.6)	28, 31, 32 and 73
Lentil	0.76 (0.19–1.25)	0.54 (0.34–0.92)	2.8 (2.7–2.9)	39, 47, 57 and 79
White lupin	1.38 (0.76–1.63)	—	—	77
Mung bean	0.75 (0.75–0.77)	0.25 (0.19–0.31)	16.2 (15.8–16.5)*	49, 54, 55 and 76

[a]Results in g/100 g of dry material for phytic acid and tannins and in TIU (trypsin inhibitor units)/mg of sample (or protein*) for trypsin inhibitors. Some values were converted from other units, based on the originally reported data.

alter the absorption and utilization of proteins, carbohydrates and minerals.[71] Among the antinutritional factors, trypsin inhibitors (a group of serine protease enzymes) are of high importance as they can inhibit the proteolytic activity of the digestive enzymes trypsin and chymotrypsin, which may lead to reduced availability of amino acids.[72] Levels of this antinutrient are particularly high in soybean (43.6–71.6 Trypsin Inhibitor Units (TIU)/mg).[28,73] In general, raw legume seeds contain higher amounts of antinutritional factors than their processed forms,[26,40,54,57,62,63,74] which indicates the necessity of processing these raw materials before incorporating them into food products.

11.3 New Legume-based Food Products

There is a great potential for using legumes in the food industry, not only as an alternative to meat protein sources, but also as stand-alone innovative food formulations that can satisfy the consumers' need for healthy, balanced and varied diets.[14] Apart from the usual consumption in raw form (fresh or dried pods and seeds), there are various applications of legumes in the food industry so far (Table 11.4), including:

- baby and weaning foods (use of mung beans, cowpeas and kidney beans in mixes with cereals or vegetables)
- imitation milks (mainly from soybeans and soy protein isolates)
- bean curds (mainly from soybeans and limited use of other pulses)
- meat products (incorporation of pulses flours and protein isolates in meat products)
- meat extenders (pulse flours used in comminuted meat products)

Table 11.4 Applications of legume crops in the food industry.

Food product	Legume crop	References
Baby foods	Mung beans	80
Baby foods	Mung beans and cowpeas	81
Baby foods	Kidney beans	82
Baby foods	Cowpeas	83
Ready-to eat food for the elderly	Soybean	84
Imitation milks	Soybeans	14
Imitation milks	Lentils, lupin, peanut, cowpeas and peas	85 and 86
Bean curd	Chickpea, faba bean, lentil, mung bean, smooth pea, pea, pinto bean, winged bean	87–89
	Field pea	90
Meat products	Soy, common bean, cowpea, black gram and green gram	14
Meat analogues and extenders	Pulses flours and isolates, texturized vegetable proteins	91–94
Baked products	Pulses flours and fractions, legume hydrolysates, legume sprouts	95–98
Extruded products (pastas, noodles and puffed products)	Pulse flours and fractions	99–101
Functional beverages	Soybeans	102
Functional foods	Germinated seeds of beans, lentils and peas	103

- baked products (breads, sweets, cookies, pastes and glazes from pulses flours, powders and isolates)
- extruded products (pasta noodles and puffed products from pulse flours and fractions)
- functional food products and beverages (soybeans and sprouts of various legume species)

Proteins are important macronutrients, usually ingested from meat and meat products, since on one hand they provide the with amino acids essential for human body functions, while on the other hand they contribute to specific functions and physicochemical properties of the final food products.[104] More specifically, meat proteins are responsible for visual quality and texture of meat and meat products, while in processed meat products they contribute to gelation, emulsification and water holding capacity.[104,105] However, considering the great concerns from consumers of developed and developing countries regarding the health side-effects of meat and meat products consumption, there is an increasing trend for meat proteins replacement in the food industry. This becomes even more important when considering that specific parts of world population do not consume or do not have access to meat and meat products for various reasons (*e.g.* religious issues where consumption of all meat products or meat from specific animals is prohibited, veganism, limited supply of meat, *etc.*).[104] Apart from protein deficiency, malnutrition is highly correlated with various minerals deficiencies which may severely affect functions of human body systems.[106] Although many minerals and nutrient supplements are available in the market to support human diets, biofortification of food products seems to be a promising and highly sought alternative in the food industry as more and more consumers turn to healthy diets and take up functional food product consumption.[107] Such an approach may be achieved through integrated plant breeding programs where traditional and modern breeding techniques are essential cost-effective tools for the creation of new elite genotypes with enhanced nutritional value.[108–110] In addition, cultivation practices may improve the nutritional value of legume crops, mostly through the application of targeted fertilization regimes which help towards mineral and vitamin biofortification of the final product.[107] So far the efforts are limited to biofortification of common beans in terms of zinc and iron content, since this species is the most important legume to be consumed in raw form and zinc and iron deficiencies affect a large part of world population.[111–113] Fractionation of legume products (*i.e.* flours of legume grains) into their main ingredients is another important aspect for the production of legume-based food products, since it may be achieved with various methods (*e.g.* air classification, isoelectric precipitation, milling, ultrafiltration/diafiltration, and wet extraction) and serve towards several applications in the food industry for the design of customised nutritionally enhanced and functional food products.[23,43,114]

Proteins of plant origin have long been considered to have great potential as an alternative to meat proteins, with legumes being the most important candidates due to their high protein content.[104] Moreover, texturized vegetable proteins are currently being used as ingredients in meat analogue products, to increase protein content, as meat extenders to enhance yield and water binding properties of comminuted meat products, or as standalone meat analogue products with low fat content, improved functional properties and texture and visual appearance that are similar to meat.[94,115] Despite their nutritional value, the application of legume proteins in food products is limited mostly to soybeans protein, and although other legumes species are less studied, there are many promising reports which suggest their incorporation in novel food products with functional properties.[116–118] Meat analogue and meat extender products are also very promising considering the shortage of animal protein throughout the world, the various animal diseases outbreaks that come under the spotlight from time to time, the increasing incidence of food allergies, and consumer's health concerns in general.[94,104,119] Although these products are usually intended for vegetarians and people who exclude meat products from their diets for religious purposes, they have gained great consumer interest since they combine high nutritional value with health beneficial properties against various chronic diseases.[104]

Among vegetable proteins, soybeans protein is the most popular legume protein, and is used in various forms throughout the world such as soymilk, tofu, fermented products, protein isolates, defatted flour and protein concentrates,[120] while several raw materials from soybeans can be used in texturized form and further incorporated in various food products.[115,121] According to Granito *et al.*,[27] Luo *et al.*[122] and Curiel *et al.*,[25] fermentation of various legume seed flours with lactic acid bacteria (sourdough fermentation) can be implemented for designing novel food products, since bacterial inoculation significantly increased all nutritional parameters of seed flours, as well as antioxidant properties and phytase activity.[123] An example of such food products is a bread with functional properties produced by a mixture of fermented flours of legumes and pseudo-cereals seeds that has increased gamma-aminobutyric acid content.[30] Moreover, fortification of rice porridge extrudates with soybean and mung bean powders altered significantly the physicochemical and pastry properties of porridge powder.[124] Maziya-Dixon *et al.*[125] have also suggested the use of cassava, soybean and cowpea flour in food preparation of snacks with high protein content.

Apart from soybeans, chickpeas show also great potential for the production of novel food products, since although they contain moderately high protein content compared to other legumes, they present high amounts of starch that allows for better extrusion, pasting, foaming and emulsification properties in food formulations.[1,19] Bhat and Bhattacharya[126] have suggested the use of chickpea flours for preparation of snacks and sweets, while Cardoso-Santiago and Arêas[127] found that snacks of chickpeas and bovine lung blends have similar nutritional value to commercially available snacks. Moreover, Gómez *et al.*[128] studied the quality of cakes produced with partial substitution of wheat by chickpea flours and found that the chickpea genotype has a significant effect on the physical properties of cakes, depending on the proportion of wheat and chickpea in the flour blend. Chickpea flour has been also reported to have ideal flow properties for imitation milk and formulations production,[129] while Chandra-Hioe *et al.*[130] have reported that fermentation of chickpea flour increased protein digestibility and reduced trypsin inhibitor activity compared to raw (unfermented) flour.

Green beans are considered very rich sources of Ca and Fe (15% and 45% of recommended daily allowance, RDA) and contain low amounts of antinutrient factors such as raffinose and trypsin inhibitors, features that can find practical uses in food biofortification.[1] Randhir and Shetty[131] have recommended the use of green beans in combination with other food products in order to reduce glycaemic index and increase antioxidant properties. Other uses of green beans include wheat flour substitutes in crackers, breads, pizzas, milk substitutes, noodles, pasta, jellies and so forth.[132–134] According to Huerta-Abrego *et al.*,[135] the addition of lima bean protein concentrates to starches of corn, cassava and lima bean significantly alters the functional properties of blends in terms of gelatinization temperature, swelling power, water absorption capacity and maximum viscosity.[135] Pinto dry beans also show great potential for the food industry and the design of functional food products, since they contain numerous bioactive compounds that have beneficial effects against oxidative stress, cardiovascular diseases, cancer, diabetes and various chronic diseases.[89,136] Broad beans isolates, concentrates and seed flours have been incorporated in various gluten-free bakery products to date in order to produce functional foods.[137]

Lentils are an important ingredient in various dishes throughout the world and considered amongst the most widely used legumes. Apart from consumption in their raw form, alternative uses of lentils have been also suggested, including tasty snacks and cakes, suitable for vegetarians and people with diabetes problems, while industrial uses of lentil starch and proteins have been implemented in the food industry.[19,138]

Although cowpeas are indigenous to Africa and represent one of the major sources of protein in African countries since they are commonly used in various forms in many traditional dishes, they are also used in novel food products (baby foods, snacks, breakfast products, *etc.*) in other parts of the world.[1] Akinyele *et al.*[139] and Jirapa *et al.*[38] have suggested the industrial use of cowpeas flours in weaning products, while Rangel *et al.*[140] and Phillips *et al.*[66] have highlighted the potential of using cowpeas in baked products and extruded snacks due to their excellent emulsifying properties. However, excessive amounts of cowpea flour in blends with cereal flours may result in hard texture and dark colour of baked products which may reduce consumers' acceptance.[141,142] In addition, fermentation and thermal processing of cowpea flour resulted in higher antioxidant properties, in both *in vitro* and *in vivo* tests, which is of great practical interest for the design of functional and healthy food products.[70,143] Modification of cowpea protein isolates through thermal processing or high hydrostatic pressure may also be implemented in novel foodstuffs, beverages and desserts due to the high solubility and water holding capacity of the processed materials.[144]

Mungo beans are consumed for their raw green pods as a vegetable or for their ground dry seeds in various Indian and other tropical countries' dishes. Their consumption has beneficial effects on human health (mainly activities against cholesterol and cardiac diseases); however their high content in antinutrient factors reduces protein bioavailability[8,145] and further studies regarding the effect of processing techniques and incorporation in food products need to be carried out prior to industrial use.

Pigeon peas are a very promising source of protein and may have a major contribution to make against malnutrition and diet deficiencies in undeveloped and developing countries. Protein of pigeon peas may provide up to 80% of total protein requirements in the less developed countries of the world, while fermentation may further improve bioavailability and digestibility of pigeon peas incorporated into pasta products.[101] Pigeon pea seed flour (5%) has also been mixed with cassava flour in various baked products and extrudates with promising results.[146]

Green peas are another legume crop, usually consumed for their fresh pods or fresh and dried seeds. Their consumption is correlated with hypocholesterolemic properties due to increased excretion of faecal and biliary bile acids.[147] Seed flour is a rich source of protein and can be mixed with wheat flour in novel foodstuff products, while seed powder has been used as a supplement in soups, and in baked and puffed products to improve emulsifying, gelling and water holding properties.[1] According to Belghith-Fendri *et al.*,[141] dried pea flour may substitute for wheat flour by up to 15% without affecting the quality of baked products, while at the same time protein and dietary fibre contents is significantly increased. Recently, Brennan *et al.*[148] studied the effect of adding green and yellow peas in barley-oat breakfast cereal products, and they reported that the presence of peas affected the colour features of the products (greenness, redness, browning index and chroma values), while starch digestibility remained unaffected.

The addition of fenugreek (*Trigonella foenum-graecum*) flour (raw and debittered flour) in chickpeas and rice flour blends has been reported to significantly lower the glycaemic index of snack products intended for diabetics, although amounts higher than 15% may severely affect the physical and sensory properties of the snacks produced.[149] Other applications of fenugreek include its use as a stabilizing, emulsifying and adhesion agent in food products, in baked products, as well as in extruded products.[150]

All these examples show the great interest in and potential of legume-based food products. However, further studies need to be conducted in order to alleviate problems with antinutritional and allergenic factors and thus help towards the wider acceptance of legumes and legume-based products. Moreover, the food industry sector has to listen to market and consumers' demands and invest in designing innovative and attractive food products with functional and health properties.

11.4 Processing Techniques in the Food Industry Regarding Legume-based Foods

Legumes are often processed into semi-finished and finished food products in order to transform raw forms into edible and novel food products, reduce or eliminate antinutritional factors, improve the nutritional quality and digestibility, and improve customer's acceptance.[39,40,63,151] In the first instance, the highly fibrous hull fraction of legume seeds can be removed (dehulling) to improve palatability, digestibility and overall nutritional value.[26,29,39] Whole, dehulled and defatted legume seeds can be wet- or dry-milled to produce a paste or flour, respectively, whose particle size is obtained as a function of the sieves or using air classification techniques. This pre-processing/processing step may disrupt protein and starch structures, increasing their bioaccessibility, and affect the functional and end-use properties of the products.[14] Whole seeds, flours or isolate fractions can also be subjected to secondary processing using traditional or innovative techniques such as soaking, debittering, fermentation, roasting, puffing, cooking, high-temperature pressure cooking (HTPC), ultra-high pressure homogenization (UHPH), extrusion cooking and couette cell in order to develop novel legume-based food products.[152]

Debittering is a procedure widely used for washing out bitter components of whole lupin seeds, including toxic alkaloids such as sparteine and lupanine.[153] The process includes a soaking step to increase the water content of seeds and facilitate the subsequent extraction of alkaloids. Cooking is then performed to inactivate enzymes (*e.g.* lipase and lipoxygenase) and germination capacity, eliminate microorganisms, reduce protein losses, and facilitate the leaching of alkaloids. Subsequently, the removal of alkaloids is achieved by biological, chemical or aqueous processing.[154,155] Aqueous debittering processes are used at a household and commercial scale. Carvajal-Larenas *et al.*[154] established optimum processing conditions for debittering of lupin seeds consisting of 18 h of soaking, 1 h of cooking and a washing step including 3 water changes/day and 22 h of agitation/day. Ertaş and Bilgiçli[156] studied different debittering processes and demonstrated that lupin seeds soaked in 0.5% $NaHCO_3$ solution for 144 h had better sensory quality than seeds submitted to other soaking treatments. Nevertheless, it would be interesting to develop novel processing techniques with soaking and cooking steps assisted by ultrasound and microwaves, respectively, in order to reduce the time needed for debittering.

Legumes are fermented with the aim of obtaining novel food products with improved nutritional and sensory characteristics as are described in Chapter 9.[27,152,157] There is a large number of fermented legume foods, most of them obtained from soybeans. Tempeh is a traditional fermented food made from soaked and cooked soybeans inoculated with strains of *Rhizopus* species.[158] It consists of soybeans bound together into a compact cake by a dense mycelium with a pleasant mushroom-like aroma and nutty flavour, which makes it a good alternative to meat products. During the fermentation process, soybean stachyose and raffinose are hydrolysed to digestive sugars and antinutritional constituents can also be decreased or eliminated, thus improving the nutritional quality of the product.[158] Other legumes, such as cowpeas and chickpeas can also be used for tempeh preparation.[152] Soy sauce (*shoyu*) is another fermented food product prepared from soybeans and wheat, first fermented with *Aspergillus oryzae* (*koji* mold) and then with yeast and lactic acid bacteria, after the addition of brine (~8–17%). This process takes about 1 to 6 months and, consequently, it provides products with different organoleptic properties.[159] This soy sauce is free of soybean and wheat allergens, promotes digestion, and has antimicrobial, anticarcinogenic and antihypertensive properties.[160]

Puffing is a processing method used to produce light, porous and crispy food products. Puffed legumes are prepared by soaking seeds in water for 15–20 min, mixing with sand that has been heated to 200–250 °C, and then toasting for about 15–25 s. After removal of the sand, the expanded grains are dehusked between a hot plate and a roller. The texture and organoleptic properties of the grains are thus improved. These products are eaten as a snack or can be milled into flour. This traditional method is commonly applied to peas and chickpeas, which retains the nutritional quality and improves the protein and carbohydrate digestibility.[152,161] Legumes can also be roasted to improve their sensory characteristics. Lu *et al.*[58] developed new pea ingredients with enhanced slowly digestible starch and resistant starch content by roasting and encapsulation techniques, and used this novel ingredient to make gluten-free pea breads, which had an appearance, crumb structure and texture, and palatability comparable to wheat-based white bread. In fact, roasting legumes prior to their use in baking may be a more appropriate processing technique than cooking or fermentation, since dough handling properties are improved, a greater volume of bread with roasted flavour is produced, and nutritional quality attributes are promoted.[162]

Soymilk is a homogeneous protein colloid obtained from a series of processing techniques such as soaking, grinding, cooking and filtration that is consumed worldwide as a nutritious protein beverage and is used to produce soy gel foods, such as tofu and soy yogurt. When the raw soymilk is heated, the soy protein subunits are denatured, dissociated and subsequently aggregated to form particles with different size and composition, causing soymilk gelation.[163] In industrial-scale production of soymilk products, traditional cooking methods normally result in the non-uniform heating of soymilk which influences the yield and quality of the final product.[164,165] HTPC promotes heat transfer efficiency by applying pressure and increasing the cooking temperature above the boiling point of water. Feng *et al.*[165] compared the effects of HTPC (105–125 °C, 0.12–0.235 MPa) and traditional cooking (97 °C, 0.1 MPa) on soymilk protein particles. The content and average size of these particles was increased by HTPC, a treatment that affected soymilk protein denaturation. HTPC soymilk at 115 °C for 10 min had a homogeneous, smooth and creamy texture and greater acceptability. A similar study[164] demonstrated that the increased protein particle content promoted by HTPC increases the tofu yield and contributes to the formation of a dense network of tofu gel, which further improves texture properties and allows the incorporation of more water or dry matters into the tofu matrix.

UHPH is a continuous non-thermal food processing technology that can also be used for preparing soymilk and tofu. While 20–50 MPa are applied in conventional homogenization processes, this innovative technique operates at pressures up to 350 MPa. Conventional homogenization is used to form stable fine emulsions and to disperse, mix and process colloidal products. The use of UHPH brings several advantages, since it promotes deflocculation of clusters and uniform dispersion of agglomerates, changes in protein conformation, increases of emulsion viscosity and stability, and inactivation of microorganisms and enzymes. Liu *et al.*[166] compared the quality of tofu prepared by both traditional thermal treatment and UHPH and found that tofu made from 20% soy flour suspension with thermal treatment had poor quality, while tofu made from the same flour percentage with 150 MPa had the best quality and water holding capacity during storage. Furthermore, UHPH reduced the soy flour particle size, denatured the soy proteins and provided tofu with a high level of functional ingredients such as dietary fibre. In another study, Ferragut *et al.*[167] investigated the effects of UHPH (300 MPa, 15 s) on soymilk for producing soy yogurts. The yogurts developed presented improved mechanical properties and water holding capacity when compared with yogurts made from soymilk treated at 95 °C for 15 min, while the yogurts prepared were stable for at least 28 days of refrigerated storage. The suitability of UHPH (200 and 300 MPa) for processing soymilk was also shown by Cruz *et al.*[168] The process produced more stable soymilk than an ultra-high temperature (142 °C, 2 s) treatment and ensured microbiological safety. However, aggregation of soymilk components was detected at 300 MPa.

Extrusion cooking is a high-temperature short-time processing technique that has been used to create novel legume-based food products including pasta, breakfast cereals, baby foods, snack foods and soy meat analogues, among many other products.[14,51] For extruding, the legume flour is initially mixed with other ingredients in an extruder barrel and preconditioned with water or steam. This mixture (extrudate) is then passed through the extruder while being forced through an orifice or a die with the help of a piston or a large, tight-fitting rotating screw within the barrel, which conveys, compresses and kneads the extrudate into plasticized mass. During this step, the extrudate is subjected to high-temperature short-time cooking and then pushed through a small die with the desired shape and cut to obtain the required size.[14,169] This is a thermal process that also promotes the inactivation of microorganisms and enzymes responsible for spoilage and reduces the water activity of the final product.[170] Alonso *et al.*[26] showed that extruded broad beans and kidney beans have better protein and starch digestibility than dehulled, soaked or germinated beans, as well as reduced protein antinutritional factors. In another study, Giménez *et al.*[171] concluded that extrusion cooking at 100 °C and 28% moisture was appropriate to obtain corn-broad beans spaghetti type pasta with adequate physicochemical and textural characteristics and enhanced protein and dietary fibre contents. Comparable results were achieved by Wójtowicz and Mościcki,[172] who verified that extrusion cooking can be successfully applied in the manufacture of precooked pasta products from common wheat flour enriched with white bean, split yellow pea and lentil flours. The quality of this precooked pasta was directly related to the type and amount of legume in the recipe and its hardness decreased with increasing addition of legumes. Additionally, pastas containing up to 30 g/100 g of legume flour had an acceptable sensory quality. Extrusion cooking also promotes the reduction of aflatoxins that usually requires severe conditions of shear and temperature.[173] Regarding textured vegetable protein, Osen *et al.*[174] have shown that pea protein isolates are suitable for the production of fibrous meat alternatives using high moisture extrusion cooking, thus opening up a wide range of food products for different consumer requirements.

Highly structured meat replacers can also be produced using the concept of shear-induced structuring through simultaneous application of simple shear flow and heat under mild process conditions in a couette cell.

Unlike with extrusion, the deformation inside the device is well defined and constant upon processing. The number of studies regarding the use of this novel technology is limited. Nevertheless, fibre-structured meat analogues with high anisotropy indices have already been successfully produced from soy protein isolate (90% protein) and vital wheat gluten (81% protein).[175]

11.5 Practices that Reduce Antinutrient Contents to Provide Safe Legume-based Food Production

Legumes may contain several species specific compounds that present antinutrient and/or toxic effects when consumed by humans, including compounds of various chemical classes such as secondary metabolites (saponins, alkaloids, tannins, pyrimidine glycosides) and oligosaccharides.[12] The effects of these compounds may be alleviated by various processing techniques, with domestic cooking being the most common and easily applicable one, as well as growing conditions and plant breeding tools, thus rendering legume-based and legume-incorporated food products safe for human consumption.[157,176–178] Apart from cooking, other processes have been confirmed to increase nutritional value and decrease antinutrient factors contents in legumes, such as dehulling, autoclaving, hydration, roasting, micronization and microwaving.[28,63,151] According to Ghavidel and Prakash[39] and El-Hady and Habiba,[40] soaking, germination and dehulling of various legume seeds (green grams, cowpeas, lentils and chickpeas) prior to consumption resulted in a significant increase in digestible proteins and starch contents, while it also increased bioavailability of proteins, thiamine, iron and calcium. Moreover, dehulling of seeds helped towards a further decrease in antinutrients such as phytic acid and tannins. Similar results have been reported by Alonso *et al.*[26] for faba and kidney beans, while extrusion cooking was the most efficient method for antinutrient factors elimination. Germination and heat treatment have been reported to eliminate undesirable flavours in soybean protein, which is the primary concern of using this high quality protein in the food industry and especially in imitation milk and dairy products.[179]

Faba beans is an underutilized legume crop and not widely consumed throughout the world, since its consumption is implicated in haemolytic anaemia incidents (favism) for consumers that carry a genetic deficiency of the erythrocyte-located glucose-6-phosphate dehydrogenase.[180] The compounds responsible for this anaemia are pyrimidine glucosidic compounds, such as vicine and convicine, divicine and isouramil, whose content varies among the various faba beans genotypes;[12,137] however, processing techniques, such as fermentation and dehulling, have been reported to significantly reduce these antinutrients content by up to 91%, thus eliminating haemolytic anaemia risks.[30,176]

Grass pea is another unappreciated legume crop, due to its association with lathyrism and odoratism, diseases which are associated with consumption of *Lathyrus sativus* and *L. odoratus* seeds, respectively.[75] The culprit for these diseases are a non-protein amino acid (β-*N*-oxalyl-$_L$-α,β-diaminopropionic acid; β-ODAP) in the case of lathyrism, while for odoratism β-aminopropionitrile is to be blamed.[75] However, thermal processing techniques may reduce the contents of these antinutrients and make further exploitation of these species from the food industry feasible.[181]

Although processing is a useful means towards elimination of antinutrient factors, many aspects have to be considered, since removal of these compounds with processing treatments may alter digestibility of food products and bioavailability and quality of other nutrients such as protein and starch.[47,182] Moreover, some of these antinutrient compounds may also have effects that are beneficial to health in low amounts, therefore processing treatments have to be applied with caution and reasonably in order to retain specific food qualities and health properties.[12]

11.6 Future Prospects

Recent trends in the food industry indicate that there is an urgent need in the market for functional food products, with enhanced nutritional value and appealing texture and taste. Moreover, there is an ever-growing global demand for a greater supply of food, especially for undeveloped and developing countries, that has to be satisfied within the next decades. In this framework, legume crops have to be further utilized, aiming not only for higher production but also for improvement in nutritional value and effects that are beneficial to human health. Limiting factors such as antinutrient compounds and lack of integration of novel processing techniques in the food industry are the major drawbacks that hinder exploiting the full potential of legume crops. Furthermore, underutilized legume species have to be further investigated and incorporated into the food industry so that innovative and attractive food products with enhanced quality features can be introduced. Breeding programs and field experiments will be very helpful towards identifying elite genotypes and best practices that increase nutritional value and bioactive properties, while at the same time minimize the content of antinutrient factors.

11.7 Conclusion

Legume-based and legume-incorporating food products are promising ingredients for the design and production of innovative products. Moreover, their high nutritional value renders them promising tools to ensure food security and prevent global malnutrition, especially in the undeveloped countries of the world, while their health promoting effects make them attractive to consumers who seek functional and healthy foods.

Acknowledgements

The authors are grateful to the Foundation for Science and Technology (FCT, Portugal) and FEDER under Programme PT2020 for financial support to CIMO (UID/AGR/00690/2013).

References

1. M. Asif, L. W. Rooney, R. Ali and M. N. Riaz, *Crit. Rev. Food Sci. Nutr.*, 2013, **53**, 1168–1179.
2. US Department of Agriculture (USDA), Agricultural Research Service, Nutrient Data Laboratory. *USDA National Nutrient Database for Standard Reference, Legacy*. Version Current: April 2018, Internet: http://www.ars.usda.gov/nutrientdata.
3. P. Watts, in *Pulse Foods: Processing, Quality and Nutraceutical Applications*, eds. B. K. Tiwari, A. Gowen and B. McKenna, Elsevier, 2011, pp. 437–464.
4. F. Melini, V. Melini, F. Luziatelli and M. Ruzzi, *Compr. Rev. Food Sci. Food Saf.*, 2017, **16**, 1101–1122.
5. R. Schoenlechner, *Bodenkultur*, 2016, **67**, 239–248.
6. J. J. Riascos, A. K. Weissinger, S. M. Weissinger and A. Wesley Burks, *J. Agric. Food Chem.*, 2010, **58**, 20–27.
7. S. Akibode and M. Maredia, *Global and Regional Trends in Production, Trade and Consumption of Food Legume Crops*, Department of Agricultural, Food, and Resource Economics, Michigan State University, 2012.
8. L. A. Bazzano, J. He, L. G. Ogden, C. Loria, S. Vupputuri, L. Myers and P. K. Whelton, *Arch. Intern. Med.*, 2001, **161**, 2573.
9. I. Flight and P. Clifton, *Eur. J. Clin. Nutr.*, 2006, **60**, 1145–1159.
10. M. Gwtek, N. Czech, M. Muc-Wierzgon, E. Grochowska-Niedworok, T. Kokot and E. Nowakowska-Zajdel, *J. Evidence-Based Complementary Altern. Med.*, 2014, **2014**, 12.
11. S.-U. Chon, *Curr. Pharm. Des.*, 2013, **19**, 6112–6124.
12. M. C. V. Patto, R. Amarowicz, A. N. A. Aryee, J. I. Boye, H. Chung, M. A. Martín-cabrejas and C. Domoney, *Crit. Rev. Plant Sci.*, 2015, **34**, 105.
13. L. Díaz-Batalla, J. M. Widholm, G. C. Fahey, E. Castaño-Tostado and O. Paredes-López, *J. Agric. Food Chem.*, 2006, **54**, 2045–2052.
14. J. Boye, F. Zare and A. Pletch, *Food Res. Int.*, 2010, **43**, 414–431.
15. M. Radcliffe, G. Scadding and H. M. Brown, *Lancet*, 2005, **365**, 1360.
16. J. A. Martin, J. A. Compaired, B. de la Hoz, S. Quirce, M. D. Alonso, J. M. Igea and E. Losada, *Allergy*, 1992, **47**, 185–187.
17. L. K. Massey, R. G. Palmer and H. T. Horner, *J. Agric. Food Chem.*, 2001, **49**, 4262–4266.
18. Z. Farooq and J. Boye, in *Pulse Foods: Processing, Quality and Nutraceutical Applications*, eds. B. K. Tiwari, A. Gowen and B. McKenna, Elsevier, 2011, pp. 283–234.

19. J. D. J. Berrios, P. Morales, M. Cámara and M. C. Sánchez-Mata, *Food Res. Int.*, 2010, **43**, 531–536.
20. C. Lv, G. Zhao and Y. Ning, *Crit. Rev. Food Sci. Nutr.*, 2017, **57**, 1718–1728.
21. H. J. Chung, Q. Liu, R. Hoover, T. D. Warkentin and B. Vandenberg, *Food Chem.*, 2008, **111**, 316–321.
22. R. Bhat and A. A. Karim, *Compr. Rev. Food Sci. Food Saf.*, 2009, **8**, 305–331.
23. J. I. Boye, S. Aksay, S. Roufik, S. Ribéreau, M. Mondor, E. Farnworth and S. H. Rajamohamed, *Food Res. Int.*, 2010, **43**, 537–546.
24. S. Calabrò, M. I. Cutrignelli, V. Lo Presti, R. Tudisco, V. Chiofalo, M. Grossi, F. Infascelli and B. Chiofalo, *J. Sci. Food Agric.*, 2015, **95**, 3127–3136.
25. J. A. Curiel, R. Coda, I. Centomani, C. Summo, M. Gobbetti and C. G. Rizzello, *Int. J. Food Microbiol.*, 2015, **196**, 51–61.
26. R. Alonso, A. Aguirre and F. Marzo, *Food Chem.*, 2000, **68**, 159–165.
27. M. Granito, J. Frias, R. Doblado, M. Guerra, M. Champ and C. Vidal-Valverde, *Eur. Food Res. Technol.*, 2002, **214**, 226–231.
28. P. Siddhuraju, H. P. S. Makkar and K. Becker, *Food Chem.*, 2002, **78**, 187–205.
29. M. Bähr, A. Fechner, K. Hasenkopf, S. Mittermaier and G. Jahreis, *LWT - Food Sci. Technol.*, 2014, **59**, 587–590.
30. R. Coda, C. G. Rizzello and M. Gobbetti, *Int. J. Food Microbiol.*, 2010, **137**, 236–245.
31. Z. Jiao, X. Si, Z. Zhang, G. Li and Z. Cai, *Food Chem.*, 2012, **135**, 285–291.
32. S. Sharma, M. Kaur, R. Goyal and B. S. Gill, *J. Food Sci. Technol.*, 2014, **51**, 551–557.
33. A. Redondo-Cuenca, M. J. Villanueva-Suárez, M. D. Rodríguez-Sevilla and I. Mateos-Aparicio, *Food Chem.*, 2007, **101**, 1216–1222.
34. Z. Y. Yan, P. S. Spencer, Z. X. Li, Y. M. Liang, Y. F. Wang, C. Y. Wang and F. M. Li, *Phytochemistry*, 2006, **67**, 107–121.
35. A. Sujak, A. Kotlarz and W. Strobel, *Food Chem.*, 2006, **98**, 711–719.
36. M. Erbas, M. Certel and M. K. Uslu, *Food Chem.*, 2005, **89**, 341–345.
37. C. Martínez-Villaluenga, J. Frías and C. Vidal-Valverde, *Food Chem.*, 2006, **98**, 291–299.
38. P. Jirapa, H. Normah, M. M. Zamaliah, R. Asmah and K. Mohamad, *Plant Foods Hum. Nutr.*, 2001, **56**, 203–216.
39. R. A. Ghavidel and J. Prakash, *LWT -- Food Sci. Technol.*, 2007, **40**, 1292–1299.
40. E. A. A. El-Hady and R. A. Habiba, *LWT -- Food Sci. Technol.*, 2003, **36**, 285–293.
41. K. Crépon, P. Marget, C. Peyronnet, B. Carrouée, P. Arese and G. Duc, *Field Crop. Res.*, 2010, **115**, 329–339.
42. E. A. E. Elsheikh, A. H. El Tinay and I. A. Fadul, *Food Chem.*, 1999, **67**, 379–383.
43. M. Mondor, S. Aksay, H. Drolet, S. Roufik, E. Farnworth and J. I. Boye, *Innovative Food Sci. Emerging Technol.*, 2009, **10**, 342–347.

44. A. Sarmento, L. Barros, Â. Fernandes, A. M. Carvalho and I. C. Ferreira, *J. Sci. Food Agric.*, 2015, **95**, 179–185.
45. Y. N. Sreerama, V. B. Sashikala, V. M. Pratape and V. Singh, *Food Chem.*, 2012, **131**, 462–468.
46. A. Iqbal, I. A. Khalil, N. Ateeq and M. Sayyar Khan, *Food Chem.*, 2006, **97**, 331–335.
47. N. Wang, D. W. Hatcher, R. Toews and E. J. Gawalko, *LWT -- Food Sci. Technol.*, 2009, **42**, 842–848.
48. S. D. Chandrasiri, R. Liyanage, J. K. Vidanarachchi, P. Weththasinghe and B. C. Jayawardana, *Procedia Food Sci.*, 2016, **6**, 352–355.
49. P. K. Dahiya, A. R. Linnemann, M. J. R. Nout, M. A. J. S. van Boekel and R. B. Grewal, *LWT -- Food Sci. Technol.*, 2013, **54**, 249–256.
50. C. W. C. Kendall, A. Esfahani and D. J. A. Jenkins, *Food Hydrocolloids*, 2010, **24**, 42–48.
51. S. M. Tosh and S. Yada, *Food Res. Int.*, 2010, **43**, 450–460.
52. E. Marconi, S. Ruggeri, M. Cappelloni, D. Leonardi and E. Carnovale, *J. Agric. Food Chem.*, 2000, **48**, 5986–5994.
53. A. Kataria and B. M. Chauhan, *Plant Foods Hum. Nutr.*, 1988, **38**, 51–59.
54. A. E. Mubarak, *Food Chem.*, 2005, **89**, 489–495.
55. T. A. El-Adawy, E. H. Rahma, A. A. El-Bedawey and A. E. El-Beltagy, *Plant Foods Hum. Nutr.*, 2003, **58**, 1–13.
56. R. N. Tharanathan and S. Mahadevamma, *Trends Food Sci. Technol.*, 2003, **14**, 507–518.
57. Zia-ur-Rehman and A. M. Salariya, *Int. J. Food Sci. Technol.*, 2005, **40**, 695–700.
58. Z.-H. Lu, E. Donner and Q. Liu, *Food Chem.*, 2018, **245**, 71–78.
59. C. Martínez-Villaluenga, J. Frias and C. Vidal-Valverde, *Crit. Rev. Food Sci. Nutr.*, 2008, **48**, 301–316.
60. X. Saldivar, Y.-J. Wang, P. Chen and A. Hou, *Food Chem.*, 2011, **124**, 1369–1375.
61. E. J. Landry, S. J. Fuchs and J. Hu, *J. Food Compos. Anal.*, 2016, **50**, 55–60.
62. E. A. Shimelis and S. K. Rakshit, *Food Chem.*, 2007, **103**, 161–172.
63. V. Kalpanadevi and V. R. Mohan, *LWT -- Food Sci. Technol.*, 2013, **51**, 455–461.
64. K. Crépon, P. Marget, C. Peyronnet, B. Carroué, P. Arese and G. Rard Duc, *Field Crop. Res.*, 1010, **115**, 329–339.
65. W. Prinyawiwatkul, L. Beuchat, K. McWatters and R. Phillips, *Food Chem.*, 1996, **57**, 405–413.
66. R. D. Phillips, K. H. McWatters, M. S. Chinnan, Y. C. Hung, L. R. Beuchat, S. Sefa-Dedeh, E. Sakyi-Dawson, P. Ngoddy, D. Nnanyelugo, J. Enwere, N. S. Komey, K. Liu, Y. Mensa-Wilmot, I. A. Nnanna, C. Okeke, W. Prinyawiwatkul and F. K. Saalia, *Field Crop. Res.*, 2003, **82**, 193–213.
67. A. B. M. Moslehuddin and I. Chowdhury, *Bangladesh J. Nutr.*, 1989, **2**, 14–16.
68. G. Akalu, F. Tufvesson, C. Jonsson and B. M. Nair, *Starch-Starke*, 1998, **50**, 374–382.

69. A.-S. Sandberg, *Br. J. Nutr.*, 2002, **88**, 281.
70. P. Siddhuraju and K. Becker, *Food Chem.*, 2007, **101**, 10–19.
71. M. C. Chang, J. W. Bailey and J. L. Collins, *J. Nutr.*, 1994, **124**, 283–288.
72. B. H. Vagadia, S. K. Vanga and V. Raghavan, *Trends Food Sci. Technol.*, 2017, **64**, 115–125.
73. T. C. de Toledo, S. G. Canniatti-Brazaca, V. Arthur and S. M. S. Piedade, *Radiat. Phys. Chem.*, 2007, **76**, 1653–1656.
74. A. Sharma and S. Sehgal, *Food Chem.*, 1992, **43**, 383–385.
75. M. Muzquiz, A. Varela, C. Burbano, C. Cuadrado, E. Guillamón and M. M. Pedrosa, *Phytochem. Rev.*, 2012, **11**, 227–244.
76. V. R. Mohan, P. S. Tresina and E. D. Daffodil, in *Encyclopedia of Food and Health*, Elsevier, 2016, pp. 211–220.
77. M. Hídvégi and R. Lásztity, *Period. Polytech. Chem. Eng.*, 2002, **46**, 59–64.
78. Z. Ma, J. I. Boye and X. Hu, *Food Res. Int.*, 2017, **92**, 64–78.
79. T. H. Hefnawy, *Ann. Agric. Sci.*, 2011, **56**, 57–61.
80. U. Singh and B. Singh, *Econ. Bot.*, 1992, **46**, 310–321.
81. L. M. Marero, E. M. Payumo, A. R. Aguinaldo, I. Matsumoto and S. Homma, *Leb. und-Technologie*, 1991, **24**, 177–181.
82. Y. Abebe, B. J. Stoecker, M. J. Hinds and G. E. Gates, *Afr. J. Food, Agric., Nutr. Dev.*, 2006, **6**, 1–19.
83. H. H. Laminu, M. Sheriff, B. Bintu and A. Muhammad, *Sch. J. Agric. Sci.*, 2014, **4**, 536–542.
84. P. Satusap, V. Chavasit, W. Kriengsinyos and K. Judprasong, *Springerplus*, 2014, **3**, 451.
85. B. G. Swanson, *J. Am. Oil Chem. Soc.*, 1990, **67**, 276–280.
86. S. Sethi, S. K. Tyagi and R. K. Anurag, *J. Food Sci. Technol.*, 2016, **53**, 3408–3423.
87. R. Cai, B. Klamczynska and B.-K. Baik, *J. Agric. Food Chem.*, 2001, **49**, 3068–3073.
88. S. Sri Kantha, N. S. Hettiarachchy and J. W. Erdman Jr, *J. Food Sci.*, 1983, **48**, 441–444.
89. C. Câmara, C. Urrea and V. Schlegel, *Agriculture*, 2013, **3**, 90–111.
90. A. Gebre-Egziabher and A. K. Sumner, *J. Food Sci.*, 1983, **48**, 375–377.
91. T. Dzudie, J. Scher and J. Hardy, *J. Food Eng.*, 2002, **52**, 143–147.
92. M. Serdaroğlu, G. Yildiz-Turp and K. Abrodímov, *Meat Sci.*, 2005, **70**, 99–105.
93. V. K. Modi, N. S. Mahendrakar, D. Narasimha Rao and N. M. Sachindra, *Meat Sci.*, 2004, **66**, 143–149.
94. V. Joshi and S. Kumar, *Int. J. Food Ferment. Technol.*, 2015, **5**, 107.
95. E. Sironi, F. Sessa and M. Duranti, *Eur. Food Res. Technol.*, 2005, **221**, 145–150.
96. B. Klamczynska, Z. Czuchajowska and B. K. Baik, *Int. J. Food Sci. Technol.*, 2001, **36**, 563–572.
97. L. Menon, S. D. Majumdar and U. Ravi, *J. Food Sci. Technol.*, 2015, **52**, 4156–4165.

98. H. Franco-Miranda, L. Chel-Guerrero, S. Gallegos-Tintoré, A. Castellanos-Ruelas and D. Betancur-Ancona, *LWT -- Food Sci. Technol.*, 2017, **80**, 250–256.
99. K. B. Filli and I. Nkama, *Br. Food J.*, 2007, **109**, 68–80.
100. Y. Bahnassey, K. Khan and R. Harrold, *Cereal Chem.*, 1986, **63**, 210–215.
101. A. Torres, J. Frias, M. Granito and C. Vidal-Valverde, *J. Agric. Food Chem.*, 2006, **54**, 6685–6691.
102. M. R. Corbo, A. Bevilacqua, L. Petruzzi, F. P. Casanova and M. Sinigaglia, *Compr. Rev. Food Sci. Food Saf.*, 2014, **13**, 1192–1206.
103. C. Vidal-Valverde, J. Frias, I. Sierra, I. Blazquez, F. Lambein and Y. H. Kuo, *Eur. Food Res. Technol.*, 2002, **215**, 472–477.
104. M. A. Asgar, A. Fazilah, N. Huda, R. Bhat and A. A. Karim, *Compr. Rev. Food Sci. Food Saf.*, 2010, **9**, 513–529.
105. Y. Xiong, in *Proteins in Food Processing*, ed. R. Yada, Woodhead Publishing Limited, 2004, pp. 100–122.
106. K. Hirschi, *Annu. Rev. Nutr.*, 2009, **29**, 401–421.
107. P. J. White, P. J. White and M. R. Broadley, *New Phytol.*, 2009, **182**, 49–84.
108. S. L. Dwivedi, K. L. Sahrawat, K. N. Rai, M. W. Blair, M. S. Andersson and W. Pfeiffer, *Plant Breed. Rev.*, 2012, **36**, 169–262.
109. J. M. Al-Khayri, S. M. Jain and D. V. Johnson, *Advances in Plant Breeding Strategies: Agronomic, Abiotic and Biotic Stress Traits*, Springer, 2016, vol. 2.
110. M. W. Blair, *J. Agric. Food Chem.*, 2013, **61**, 8287–8294.
111. W. J. Broughton, G. Hernández, M. Blair, S. Beebe, P. Gepts and J. Vanderleyden, *Plant Soil*, 2003, **252**, 55–128.
112. H. E. Bouis, *Proc. Nutr. Soc*, 2003, **62**, 403–411.
113. H. E. Bouis and R. M. Welch, *Crop Sci.*, 2010, **50**, S20–S32.
114. C. Barbana and J. I. Boye, *Food Funct.*, 2013, **4**, 310–321.
115. M. Riaz, in *Proteins in Food Processing*, ed. R. Yada, Woodhead Publishing Limited, 2004, pp. 517–557.
116. E. J. Papavergou, J. G. Bloukas and G. Doxastakis, *Meat Sci.*, 1999, **52**, 421–427.
117. E. Makri, E. Papalamprou and G. Doxastakis, *Food Hydrocolloids*, 2005, **19**, 583–594.
118. G. Doxastakis, in *Novel Macromolecules in Food Systems*, eds. G. Doxastakis and V. Kiosseoglou, Elsevier, 2000, pp. 7–38.
119. R. Egbert and C. Borders, *Food Technol.*, 2006, **60**, 28–34.
120. D. Fukushima, in *Proteins in Food Processing*, ed. R. Y. Yada, Woodhead Publishing Limited, 2004, pp. 123–145.
121. A. U. Achidi, B. Tiencheu, N. Tenyang, H. M. Womeni, M. N. Moyeh, L. T. Ebini and F. Tatsinkou, *Int. J. Food Sci. Nutr. Eng.*, 2016, **6**, 21–31.
122. Y. Luo, Z. Gu, Y. Han and Z. Chen, *J. Sci. Food Agric.*, 2009, **89**, 861–866.
123. R. Y. Gan, N. P. Shah, M. F. Wang, W. Y. Lui and H. Corke, *Int. J. Food Sci. Technol.*, 2016, **51**, 875–884.
124. P. Mayachiew, C. Charunuch and S. Devahastin, *J. Food Sci.*, 2015, **80**, E2782–E2791.

125. B. Maziya-Dixon, E. O. Alamu, I. O. Popoola and M. Yomeni, *Food Sci. Nutr.*, 2017, **5**, 805–811.
126. K. K. Bhat and S. Bhattacharya, *Int. J. Food Sci. Technol.*, 2001, **36**, 499–507.
127. R. A. Cardoso-Santiago and J. A. G. Arêas, *Food Chem.*, 2001, **74**, 35–40.
128. M. Gómez, B. Oliete, C. M. Rosell, V. Pando and E. Fernández, *LWT - Food Sci. Technol.*, 2008, **41**, 1701–1709.
129. R. Ravi and S. Bhattacharya, *J. Food Eng.*, 2004, **65**, 619–624.
130. M. V. Chandra-Hioe, C. H. M. Wong and J. Arcot, *Plant Foods Hum. Nutr.*, 2016, **71**, 90–95.
131. R. Randhir and K. Shetty, *Innovative Food Sci. Emerging Technol.*, 2007, **8**, 197–204.
132. W. Liu and Q. Shen, *J. Food Eng.*, 2007, **79**, 1310–1314.
133. E. M. T. Mendoza, M. Adachi, A. E. N. Bernardo and S. Utsumi, *J. Agric. Food Chem.*, 2001, **49**, 1552–1558.
134. Z. Kohajdová, J. Karovičová and M. Magala, *Chem. Pap.*, 2013, **67**, 398–407.
135. A. Huerta-Abrego, M. Segura-Campos, L. Chel-Guerrero and D. Betancur-Ancona, *Food Technol. Biotechnol.*, 2010, **48**, 36–41.
136. G. Sikand, P. Kris-Etherton and N. M. Boulos, *Curr. Cardiol. Rep.*, 2015, **17**, 39.
137. S. Multari, D. Stewart and W. R. Russell, *Compr. Rev. Food Sci. Food Saf.*, 2015, **14**, 511–522.
138. M. A. Suliman, A. H. El Tinay, A. E. O. Elkhalifa, E. E. Babiker and E. A. I. Elkhalil, *Pak. J. Nutr.*, 2006, **5**, 589–593.
139. I. O. Akinyele, M. H. Love and M. Ringe, *Int. J. Food Sci. Technol.*, 1988, **23**, 297–301.
140. A. Rangel, G. B. Domont, C. Pedrosa and S. T. Ferreira, *J. Agric. Food Chem.*, 2003, **51**, 5792–5797.
141. L. Belghith-Fendri, F. Chaari, F. Kallel, S. Zouari-Ellouzi, R. Ghorbel, S. Besbes, S. Ellouz-Chaabouni and D. Ghribi-Aydi, *J. Food Sci.*, 2016, **81**, C2360–C2366.
142. E. Hallén, Ş. Ibanoğlu and P. Ainsworth, *J. Food Eng.*, 2004, **63**, 177–184.
143. G. Kapravelou, R. Martínez, A. M. Andrade, C. López, J. Cañizares and J. M. Porres, *J. Sci. Food Agric.*, 2015, **95**, 1207–1216.
144. F. Peyrano, F. Speroni and M. V. Avanza, *Innovative Food Sci. Emerging Technol.*, 2016, **33**, 38–46.
145. B. Niroula, *Our Nature*, 2003, **1**, 10–14.
146. R. Rampersad, N. Badrie and E. Comissiong, *J. Food Sci.*, 2003, **68**, 363–367.
147. J. M. Martins, M. Riottot, M. C. de Abreu, M. J. Lança, A. M. Viegas-Crespo, J. A. Almeida, J. B. Freire and O. P. Bento, *J. Nutr.*, 2004, **134**, 3305–3312.
148. M. A. Brennan, T. Lan and C. S. Brennan, *J. Food Process. Preserv.*, 2016, **40**, 405–413.
149. G. Shirani and R. Ganesharanee, *J. Food Eng.*, 2009, **90**, 44–52.

150. S. A. Wani and P. Kumar, *J. Saudi Soc. Agric. Sci.*, 2018, **17**, 97–106.
151. R. Y. Khattab and S. D. Arntfield, *LWT -- Food Sci. Technol.*, 2009, **42**, 1113–1118.
152. G. A. Annor, Z. Ma and J. I. Boye, in *Food Processing: Principles and Applications*, John Wiley & Sons, Ltd, Chichester, UK, 2014, pp. 305–337.
153. S. C. Q. Magalhães, F. Fernandes, A. R. J. Cabrita, A. J. M. Fonseca, P. Valentão and P. B. Andrade, *Ind. Crops Prod.*, 2017, **95**, 286–295.
154. F. E. Carvajal-Larenas, M. J. R. Nout, M. A. J. S. van Boekel, M. Koziol and A. R. Linnemann, *LWT -- Food Sci. Technol.*, 2013, **53**, 507–516.
155. M. Erbas, *J. Food Qual.*, 2010, **33**, 742–757.
156. N. Ertaş and N. Bilgiçli, *J. Food Sci. Technol.*, 2014, **51**, 3348–3354.
157. Y. H. Kuo, H. M. Bau, P. Rozan, B. Chowdhury and F. Lambein, *J. Sci. Food Agric.*, 2000, **80**, 2209–2215.
158. K. A. Hachmeister and D. Y. C. Fung, *Crit. Rev. Microbiol.*, 1993, **19**, 137–188.
159. R. Harada, M. Yuzuki, K. Ito, K. Shiga, T. Bamba and E. Fukusaki, *J. Biosci. Bioeng.*, 2017, **123**, 203–208.
160. S. Kataoka, *J. Biosci. Bioeng.*, 2005, **100**, 227–234.
161. S. Paul Mukhopadhyay, J. A. Wood, A. J. Saliba, C. L. Blanchard, B. T. Carr and P. D. Prenzler, *LWT -- Food Sci. Technol.*, 2015, **64**, 959–965.
162. B.-K. Baik and I. H. Han, *Cereal Chem. J.*, 2012, **89**, 269–275.
163. X. Peng, C. Ren and S. Guo, *Trends Food Sci. Technol.*, 2016, **54**, 138–147.
164. F. Zuo, Z. Chen, X. Shi, R. Wang and S. Guo, *Food Chem.*, 2016, **213**, 561–566.
165. F. Zuo, X. Peng, X. Shi and S. Guo, *Food Chem.*, 2016, **209**, 50–56.
166. H.-H. Liu, J.-T. Chien and M.-I. Kuo, *Food Hydrocolloids*, 2013, **32**, 278–285.
167. V. Ferragut, N. S. Cruz, A. Trujillo, B. Guamis and M. Capellas, *J. Food Eng.*, 2009, **92**, 63–69.
168. N. Cruz, M. Capellas, M. Hernández, A. J. Trujillo, B. Guamis and V. Ferragut, *Food Res. Int.*, 2007, **40**, 725–732.
169. N. Nikmaram, S. Y. Leong, M. Koubaa, Z. Zhu, F. J. Barba, R. Greiner, I. Oey and S. Roohinejad, *Food Control*, 2017, **79**, 62–73.
170. P. J. Fellows and P. J. Fellows, in *Food Processing Technology*, Elsevier, 2017, pp. 753–780.
171. M. A. Giménez, R. J. González, J. Wagner, R. Torres, M. O. Lobo and N. C. Samman, *Food Chem.*, 2013, **136**, 538–545.
172. A. Wójtowicz and L. Mościcki, *LWT -- Food Sci. Technol.*, 2014, **59**, 1175–1185.
173. F. K. Saalia and R. D. Phillips, *LWT -- Food Sci. Technol.*, 2011, **44**, 1496–1501.
174. R. Osen, S. Toelstede, F. Wild, P. Eisner and U. Schweiggert-Weisz, *J. Food Eng.*, 2014, **127**, 67–74.
175. G. A. Krintiras, J. Gadea Diaz, A. J. van der Goot, A. I. Stankiewicz and G. D. Stefanidis, *J. Food Eng.*, 2016, **169**, 205–213.

176. A. Gnanasambandam, J. Paull, A. Torres, S. Kaur, T. Leonforte, H. Li, X. Zong, T. Yang and M. Materne, *Agronomy*, 2012, **2**, 132–166.
177. Y. Takada, I. Tayama, T. Sayama, H. Sasama, M. Saruta, A. Kikuchi, M. Ishimoto and C. Tsukamoto, *Breed. Sci.*, 2012, **61**, 639–645.
178. S. Kumar, G. Bejiga, S. Ahmed, H. Nakkoul and A. Sarker, *Food Chem. Toxicol.*, 2011, **49**, 589–600.
179. F. Suberbie, D. Mendizábal and C. Mendizábal, *J. Am. Oil Chem. Soc.*, 1981, **58**, 192–194.
180. C. G. Rizzello, R. Coda and M. Gobbetti, *Fermented Foods Health Dis. Prev.*, 2016, 433–452.
181. K. M. Tarade, R. S. Singhal, R. V. Jayram and A. B. Pandit, *Food Chem.*, 2007, **104**, 643–649.
182. M. A. Martín-Cabrejas, Y. Aguilera, M. M. Pedrosa, C. Cuadrado, T. Hernández, S. Díaz and R. M. Esteban, *Food Chem.*, 2009, **114**, 1063–1068.

Part IV

Health Benefits Of Legumes

CHAPTER 12

Legumes and Oxidative Stress

D. GUARDADO FÉLIX[a,b] AND J. A. GUTIERREZ URIBE*[a,c]

[a]Tecnológico de Monterrey, Centro de Biotecnología-FEMSA, Av. Eugenio Garza Sada 2501 Sur, 64849, Monterrey, Mexico; [b]Programa Regional de Posgrado en Biotecnología, Facultad de Ciencias Químico-Biológicas, Universidad Autónoma de Sinaloa, FCQB-UAS, AP 1354, C.P. 80000 Culiacán, Sinaloa, Mexico; [c]Tecnológico de Monterrey, Escuela de Ingeniería y Ciencias, Campus Puebla, Vía Atlixcáyotl 2301, Reserva Territorial Atlixcáyotl, C.P. 72453 Puebla, Puebla, México
*E-mail: jagu@itesm.mx

12.1 Oxidative Stress

Oxidative stress is referred to as the imbalance between the production of reactive oxygen/nitrogen species (ROS/RNS) and the capacity of a biological system to counteract its action. This can occur due to excessive generation of ROS and RNS or a deficit of antioxidant protection systems.[1] ROS and RNS are generated naturally during cellular respiration and physical activity (Figure 12.1). At low concentrations, they play important roles in mechanisms of signaling, defense, proliferation and death of cells, due to the regulation of gene expression or diverse interactions with enzymes.[2] The exposure to environmental factors such as radiation, metals, pesticides, toxic substances and some medications increases the generation of ROS and RNS.[3]

ROS and RNS are highly reactive molecules represented by both free radicals and non-free radicals. The free radical contains one or more unpaired electrons and non-free radicals are formed when two free radicals share

Food Chemistry, Function and Analysis No. 8
Legumes: Nutritional Quality, Processing and Potential Health Benefits
Edited by María Ángeles Martín-Cabrejas

Published by the Royal Society of Chemistry, www.rsc.org

$$2O_2 + NADPH \xrightarrow{(oxidase)} 2O_2^{\bullet -} + NADP^+ + H^+ \quad (1)$$

$$2O_2^{\bullet -} + 2H^+ \xrightarrow{(SOD)} H_2O_2 + O_2 \quad (2)$$

$$Cl^- + H_2O_2 + H^+ \xrightarrow{(MPO)} HOCl + H_2O \quad (3)$$

$$H_2O_2 + Fe^{2+} \longrightarrow {}^{\bullet}OH + OH^- + Fe^{3+} \quad (4)$$

$$O_2^{\bullet -} + H_2O_2 \longrightarrow {}^{\bullet}OH + OH^- + O_2 \quad (5)$$

$$L\text{-}Arg + O_2 + NADPH \xrightarrow{(MPO)} NO^{\bullet} + citrulline \quad (6)$$

$$NO^{\bullet} + O_2^{\bullet -} \longrightarrow ONOO^- \quad (7)$$

Figure 12.1 Generation of reactive oxygen species (ROS) and reactive nitrogen species (RNS) during aerobic metabolism. (1) NADPH in the presence of molecular oxygen is oxidized to $NADP^+$ by the action of NADPH oxidase producing superoxide anion radical ($O_2^{\bullet -}$). (2) $O_2^{\bullet -}$ is converted into hydrogen peroxide (H_2O_2) by the action of superoxide dismutase (SOD). (3) Myeloperoxidase (MPO) in the presence of the chloride ion and H_2O_2 is converted into hypochlorous (HOCl). (4) H_2O_2 is catalyzed in the presence of iron generating the hydroxyl radical ($^{\bullet}OH$) by the Fenton reaction. (5) The Haber–Weiss reaction generates $^{\bullet}OH$ from H_2O_2 and $O_2^{\bullet -}$. (6) Nitric oxide synthase (NOS) produces nitric oxide ($NO^{\bullet}$) from arginine. (7) Peroxynitrite ($ONOO^-$) is produced by the reaction between $NO^{\bullet}$ and $O_2^{\bullet -}$.

their unpaired electrons.[1] Hydrogen peroxide (H_2O_2), superoxide ($O_2^{\bullet -}$) and hydroxyl radical ($^{\bullet}OH$) are the most important species. The $^{\bullet}OH$ is highly reactive and can cause extensive damage in cell structures of macromolecules such as nucleic acids, lipids and proteins, which results in the chemical modifications and alteration of their functions.[4,5] In the cells, around 4 million hydroxyl radicals can be generated per day.[6] The accumulation of $^{\bullet}OH$ in the cells promotes lipid peroxidation mainly of polyunsaturated fatty acids (PUFAs). Lipid peroxidation process involves three steps: initiation, propagation and termination. In the initiation step, $^{\bullet}OH$ reacts with the carbon=carbon double bond of PUFA, inducing the extraction of a hydrogen atom and formation of a lipid radical. In the propagation steps, the lipid radical reacts rapidly with molecular oxygen to form a lipid peroxyl radical that can initiate the oxidation of other PUFA molecules and form new lipid radicals and lipid hydroperoxides, this radical molecule can break and form more radicals species and aldehydes, which can attack other part of the cells and cause damage. In the termination steps, antioxidant compounds donate a hydrogen atom to the lipid peroxyl radical to form stable molecules[6,7] (Figure 12.2). High lipid peroxidation can induce cell death mechanisms of apoptosis and

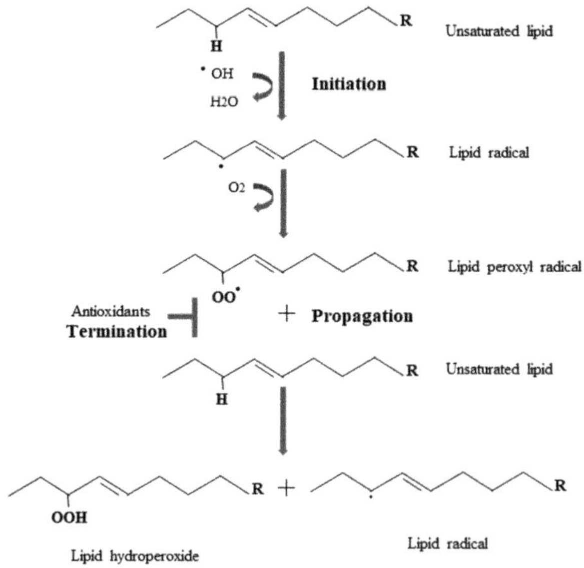

Figure 12.2 Lipid peroxidation process. In the initiation, hydroxyl radical (•OH) reacts with the carbon=carbon double bond lipid to form lipid radical. In the propagation, lipid radical reacts with oxygen to form lipid peroxyl radical, which reacts with new lipid generating lipid radical and lipid hydroperoxide. In the termination, antioxidants donate a hydrogen atom to lipid peroxyl radical to form stable molecules.

necrosis, as well as, decrease the capacity for cell repair in favor of the development of diseases and accelerated aging.[6]

Under oxidative stress conditions, the phospholipids of the cell membranes are attacked by ROS causing their oxidation and structural changes. Phospholipid oxidation causes the formation of pores in the cell membranes and increases intracellular diffusion of reactive oxygen/nitrogen species, resulting in oxidative damage of intracellular macromolecules such as DNA and proteins.[8] The •OH reacts with the nitrogenous bases and sugar molecules of DNA causing structural changes in their double helix by a variety of mechanisms.[7] ROS can react easily with the amino acids proline, arginine, lysine and histidine. These events alter the biological functions of proteins. Oxidative damage to DNA and protein molecules can cause genetic and epigenetic modifications that affect genetic regulation of oncogenes and tumor suppressor genes, which promotes the development of cancer.[9] It has been confirmed that there is a strong relationship between oxidative stress and the development of chronic diseases such as atherosclerosis, diabetes, inflammation, neurological malfunctioning and weakening of the immune system.[1]

To counteract the damage caused by oxidative stress, living organisms have antioxidant defense mechanisms that can be divided into enzymatic and non-enzymatic.[5] The main enzymatic antioxidant systems include superoxide dismutase (SOD) that catalyzes the conversion of superoxide to hydrogen, catalase (CAT) converts hydrogen peroxide (H_2O_2) to oxygen and water, and glutathione peroxidase (GPx) is responsible of the reduction of H_2O_2 and lipid hydroperoxides. Non-enzymatic antioxidant systems are composed of metabolic compounds (uric acid, bilirubin, glutathione, coenzyme Q, melatonin and alpha lipoic acid).[10] When endogenous antioxidant systems are limited, the dietary intake of antioxidants is a key step for the control of oxidative stress levels.

Plants contain a large number of antioxidant compounds that can be divided into phytochemicals like polyphenols, flavonoids and carotenoids, as well as, proteins, peptides, lipids, vitamins and minerals. The presence of these compounds in the diet provides a defense mechanism against the development of diseases associated with oxidative stress. It has been widely shown that the consumption of diets rich in antioxidant compounds decreases the risks of chronic diseases, such as cardiovascular diseases, arthritis, chronic inflammation and cancers.[11]

12.2 Antioxidant Potential of Phenolic Compounds in Legumes

Legumes represent an excellent source of exogenous antioxidants due to their high content of phenolic compounds such as phenolic acids, flavonoids, lignans and tannins.[12] The antioxidant capacity of legumes depends on the biological variety of the plant, concentration and chemical forms of the phenolic compounds. Food processing affects the availability and stability of these phenolic compounds.

Phenolic compounds are produced in plants as a defense mechanism against biotic and abiotic stress, diseases and insect attack, they are resistant to oxidation and protect against cell damage. The antioxidant activity of phenolic compounds is dependent of their chemical structure, degree of glycosylation and position of hydroxyl groups.[13] These compounds have the ability to inhibit lipid peroxidation and subsequent damage to DNA and protein structure due to their free-radical scavenging and metal ion-chelating activities.[7] Phenolic compounds can donate hydrogen atoms or an electron to free radicals to form stable intermediates. Previous epidemiological studies have shown that increased intake of legumes can reduce the risks of cancer and cardiovascular diseases, and these health benefits are attributed to the decrease of oxidative stress caused by antioxidant compounds.[14–16]

Soybeans (*Glycine max*), chickpeas (*Cicer arietinum* L.), common beans (*Phaseolus vulgaris* L.), lentils (*Lens culinaris*), peas (*Pisum sativum*), cowpeas (*Vigna unguiculata*), and peanuts (*Arachis hypogaea*) constitute important sources of antioxidant compounds in the human diet.[17] As shown in Chapter 3, flavonoids, phenolic acids and procyanidins are the main phenolic compounds present in legumes. The phenolic compounds such as anthocyanidins determine the color of the seeds and are concentrated in the coats of the legumes.[18] Once cooked, the concentration of anthocyanidins is reduced due to their chemical instability in response to heat.

The antioxidant activity and capacity of the seeds or their extracts have been evaluated by different *in vitro* methods, which determine the activity of the antioxidant against chromogenic substances of a radical nature such as 2,2-diphenyl-1-picrylhydrazyl (DPPH) and 2,2′-azino-bis(3-ethylbenzothiazoline-6-sulphonic acid) (ABTS). The loss of color of these compounds occurs proportionally with the concentration of the antioxidant.[7] Oxygen radical absorbance capacity (ORAC) evaluates the antioxidant capacity of compounds through a change in fluorescent molecules; antioxidants protect the fluorescent molecule from oxidative degeneration. The antioxidant protection is compared to the antioxidant Trolox (Vitamin E analog). The antioxidant potential of various legumes is shown in Table 12.1.

Colored beans showed higher antioxidant capacity than uncolored beans due to the presence of catechins and procyanidins.[13] Higher antioxidant potential was observed in black and red chickpeas compared to beige and cream chickpeas.[26] The total antioxidant capacity (TAC) of pinto bean, cowpea, baby lima bean, lentil, chickpea, small red bean, red kidney bean, black kidney bean, navy bean and mung bean extracts were reported to be 567, 222, 116, 721, 648, 622, 516, 602, 215 and 304 U/g, respectively.[27] It has been observed that lentils possess high antioxidant capacity, measured through the DPPH scavenging (38%) method, in comparison to other legumes.[27] From 88% to 96% of peanut seeds antioxidant capacity is attributed to phenolic content.[28] Lentil seeds contain high concentrations of flavonoids, catechin and epicatechin, which contribute to their high ORAC values (67 μmol trolox equivalents (TE) g^{-1} DW).[29,30] Faba beans contain a good concentration of phenolic compounds that enhance their antioxidant capacity. Ferric reducing antioxidant power (FRAP) value ≥1.157 mmol g^{-1} and DPPH radical-scavenging capacity >44% were observed in 13 faba bean genotypes, indicating their potential as natural sources of antioxidant foods.[31] Seed coats of black kidney beans showed higher antioxidant capacity (491 μmol TE g^{-1}) than white kidney beans (4.00 μmol TE g^{-1}).[32] The high antioxidant capacity of faba bean was correlated with the content of myricetin, quercetin, apigenin, kaempferol and luteolin free aglycones or glycosides.[33] As in many other plants, most of the flavonoids in legumes are naturally found conjugated with one or more sugar moieties.

In addition to their excellent antioxidant activity, phenolic compounds of legumes have been recognized for their anticancer activity through the modulation of oxidative stress. Genistein, a soy isoflavone, reduced the proliferation of MCF-7 breast cancer cells through modulation of the antioxidant system.[34] In another study, genistein administered orally for 24 weeks to diabetic rats resulted in a decrease in blood glucose, glycosylated hemoglobin and expression of tumor necrosis factor (TNF-α) and transforming growth factor (TGF-β1). In addition, genistein increased the total antioxidant reserve of the heart.[35] Genistein effectively protects cortical neurons against

Table 12.1 Antioxidant activity of extracts of legumes and their coats.[a]

Source	Activity (μmol TE g^{-1})	Assay	References
Broad bean	583	TAC	18
Black turtle bean	49.2	DPPH	19
Black soybean	23.7		
Black bean	5.9	DPPH	20
Kidney bean	5.2		
Mung bean	3.7		
Red bean	4.9		
Red kidney bean	5.5		
Soybean	3.2		
Chickpea	3.4		
Red sword bean coat	699	ABTS	21
Pinto bean coat	237		
Adzuki bean coat	95.3		
Broad bean coat	113		
Black hyacinth bean	55.5		
Black hyacinth bean coat	308		
Red kidney bean coat	212		
Violet red kidney bean	288		
Violet red kidney bean coat	282		
Big speckled kidney	71.1		
Bean coat	118		
Small speckled kidney	203		
Bean coat	37		
Panda bean coat	94.7		
Big runner bean coat	29.0		
Small runner bean	29.7		
Small runner bean coat	95.2		
Brown string bean	140		
Pea seed coat	35.5		
Red cowpea coat	112		
Mottled cowpea coat	211		
Pigeon pea coat	206		
White kidney bean	217.97	ORAC	22
Round purple bean	304.24		
Green lentil	136.1		23
Red lentil	134.8		
Yellow pea	10.36		
Yellow pea hull	22.89		
Black cowpeas	13.49	ORAC	24
Black cowpeas coat	15.47		
Black cowpeas cotyledons	13.94		
Horse gram	IC_{50} 22.9 μg ml^{-1}	DPPH	25
Cowpea flour	IC_{50} 48.7 μg ml^{-1}		
Chickpea	IC_{50} 31.4 μg ml^{-1}		

[a]TE, Trolox equivalents; TAC, total antioxidant activity.

oxidative stress *via* inactivation of nuclear factor-κB as well as mitogen-activated protein kinase signaling pathways.[36] These findings demonstrate the potential use of genistein in the treatment of chronic diseases such as diabetes, cancer, cerebrovascular and inflammation.

Crude water extract of lentils, black soybeans and black turtle beans with high phenolic contents and antioxidant activity showed potent angiotensin I-converting enzyme (ACE) inhibitory activity *in vitro*.[37] In a study with human and rat cardiomyocytes, a crude lentil phenolic extract attenuated angiotensin II-induced cardiomyocyte hypertrophy *via* decreasing intracellular ROS levels.[38] In another study, phenolic extract of azuki bean (*Vigna angularis*) seed coats containing polyphenols such as proanthocyanidins reduced blood pressure and intracellular ROS level in spontaneously hypertensive rats, suggesting that polyphenols attenuate vascular oxidative stress during the progression of hypertension.[39] Seed coats and different extracts of legumes are rich in phenolic compounds, which possess health benefits through oxidative stress modulation. These extracts can be used as ingredients in nutraceuticals and functional foods.

12.3 Antioxidant Potential of Legume Proteins

Legumes represent an excellent source of high quality proteins and they are an important source of these compounds in the human diet worldwide.[40] The protein concentration in legumes is in the range of 20–40%.[41,42] Previous studies have shown that the consumption of proteins from legumes decreases the risk of diseases associated with oxidative stress.[43] The antioxidant activity of horse gram (brown and black), and cowpea (brown) proteins has been positively compared to that of ascorbic acid. In addition, the high protein content was related to the increase in antioxidant activity in these legumes.[44]

During digestion of proteins, a variety of peptides with antioxidant activity are produced. The peptides contain from 2 to 20 amino acids with molecular masses less than 6.0 kDa.[45] The antioxidant properties of protein hydrolysates and peptides from legumes are related with their ability to inactivate reactive oxygen species, scavenge free radicals, chelate pro-oxidative transition metals and inhibit lipid peroxidation.[46] These properties are dependent of the structure, amino acid composition, peptide sequence and physicochemical properties of the C-terminal and N-terminal regions, which can be affected by the type of proteases used for hydrolysis, hydrolysis conditions, degree of hydrolysis and protein source.[47]

The presence of sulfur amino acids (Cys and Met), as well as aromatic amino acids (Phe, Trp and Typ), can enhance the activities of peptide antioxidants due to their capacity of donate hydrogen atoms. Additionally, basic amino acids possess chelating capacity for metallic ions. The hydrophobic amino acid residues at the C-terminal, Val and Leu increase their antioxidant potential due to their high lipid solubility.[48]

Peptides isolated from soybean hydrolysate containing histidine (VNPHDHQN, LVNPHDHQN and LLPHH, LLPHHADADY) have been recognized for their high antioxidant capacity.[49] In chickpea protein, histidine-rich peptides such as LLPH and AHH have been identified, as well as, the presence of aromatic amino acids, which, showed high antioxidant activity, as effective as the synthetic antioxidant butylated hydroxytoluene (BHT).[50] Three

Table 12.2 Bioactive peptides with antioxidant activity found in different legumes.

Source	Treatment	Peptide sequence	References
Lentil	Savinase	SDRFTYVAF	52
		DLPVLRWLKL	
		REQIEELRRL	
		QTIYENENGHIRLL	
		NSLTLPILRYL	53
		LLSGTQNQPSFLSGF	
		TLEPNSVFLPVLLH	
Chickpea	Alcalase	VGDI	54
		DHG	
	Pepsin/pancreatin	ALEPDHR	50
		TETWNPNHPEL	
		FVPH	
		SAEHGSLH	
	Alcalase/Flavorzyme	RQSHFANAQP	48
Pinto beans	Protamex	PPHMLP	55
		PPMHLP	
		PLPPHMLP	
		PLPLHMLP	
		ACSNHSPLGWRGH	
		LSSLQMGSLGALFVCM	

antioxidant peptides were isolated from peanut protein and the amino acid sequences of the peptides were identified as TPA, I/L-PS, and SP.[51]

Adzuki bean protein isolate with good essential amino acids content and high solubility exhibited high DPPH radical-scavenging capacity and reducing power compared to soybean, black soybean and mung bean protein isolates. Adzuki beans showed the highest DPPH radical-scavenging capacity, due to the presence of hydrophobic amino acid residues. Additionally, the strongest scavenging activity against hydroxyl radicals was observed in mung bean protein followed by adzuki bean protein.[40] Antioxidant peptides and their amino acid sequences identified in common legumes are shown in Table 12.2.

Several commercial enzymes, such as alcalase and flavourzyme, as well as enzymes implicated in the gastrointestinal digestion process such as pepsin and pancreatin are used in the production of protein hydrolysates, and their antioxidant potentials has been evaluated *in vitro* (Table 12.3). The combination of several enzymes during protein hydrolysis has served as a good strategy to increase the formation of lower molecular weight peptides with higher antioxidant capacity.[56,59] Antioxidant and metal chelating activities increased in phaseolin and lectin protein when they were hydrolyzed with pepsin and pancreatin.[64] On the other hand, high pressure (HP) and high temperature (HT) are used to increase the efficiency of enzymatic protein hydrolysis and structural changes that promote an increase in their antioxidant activity. Pea protein hydrolyzed with alcalase and pretreated at 400 MPa for 5 min at 24 °C

Table 12.3 Antioxidant hydrolysates of legume proteins.

Source	Treatment	Activity	Assay	References
Plus black	Alcalase/ Flavourzyme	4.34 mM of TE mg^{-1} protein	ABTS	56
Azufrado Higuera		4.82 mM of TE mg^{-1} protein		
Pinto Saltillo		25.96 mM of TE mg^{-1} protein		
Pinto	Alcalase	340 μmol TE mg^{-1} protein	ORAC	52
Black jamapa		8.1 mM of TE mg^{-1} protein	ABTS	57
	Alcalase/ Flavourzyme	6922 1 mM of TE mg^{-1} protein		58
Black bean	Pepsin/ pancreatin	1.7 μmol TE mg^{-1} protein	ORAC	59
	Alcalase	69%[a]	$ABTS^{•+}$	60
Lentil	Savinase	810 μmol of TE mg^{-1} protein	ORAC	61
Cowpea	Pepsin/ pancreatin	0.784 μmol TE mg^{-1} protein	ORAC	62
Soybean	Alcalase	IC_{50}[b] value of 4.22 mg mL^{-1} IC_{50}[b] value of 2.93 mg mL^{-1}	DPPH $ABTS^{•+}$	63

[a]Percentage of inhibition of the $ABTS^{•+}$.
[b]IC_{50} value (mg extract mL^{-1}) is the effective concentration at which the radicals were scavenged by 50%.

increased its antioxidant activity compared with when it was pretreated with HP of 200 and 600 MPa or HT of 100 °C.[65]

Protein hydrolysate and peptides with antioxidant potential derived from legumes offer a number of potential benefits. They can be used as a safe alternatives for the control of diseases related to oxidative stress since they are naturally produced by digestive enzymes after legume consumption. In an *in vitro* study, chickpea peptide RQSHFANAQP derived from chickpea albumin hydrolysate with high antioxidant activity inhibited the proliferation of breast cancer cells by increasing the level of p53.[66] Antioxidant peptides of pinto bean protein, obtained by hydrolysis with alcalase at 100 MPa, showed inhibitory activity against ACE, involved in peripheral hypertension and cardiovascular functioning.[67] Mung bean vicilin protein hydrolysate showed high antioxidant activity and ACE inhibitory activity, as well as, antiproliferative activity against breast cancer cells.[68] A soybean hydrolysate was reported to have antiglycation, antioxidant and antiproliferative activity.[69] Therefore protein hydrolysates and antioxidant peptides of legumes can be used in the formulation of functional foods for the control of diseases related to oxidative stress.

12.4 Antioxidant Potential of Legume Carotenoids and Tocopherols

Legumes are an important source of lipophilic antioxidants such as carotenoids and tocopherols.[70] These compounds protect the structures of cell membranes against the attack of free radicals, mainly through the inhibition of lipid peroxidation.[71,72] Several tocopherol isoforms (α, β, γ, and δ) have been identified in legumes, γ-tocopherol being the most abundant in legume seeds.[73] High tocopherol content has been identified in peanut, cowpea and soybean seeds, as well as high carotenoid content in chickpea, soybean and lentil seeds compared to other legumes.[74] The total tocopherols content in soybean can vary from 90.4 to 120.9 μg g^{-1} dry weight (DW), followed by black soybean from 66.1 to 100.8 μg g^{-1} DW. Tocopherols, sterols and tocotrienols have been identified in chickpea seeds. The α-tocopherol present in chickpeas is higher (8.2 mg 100 g^{-1} DW) than in other legumes such as lentils (4.9 mg 100 g^{-1} DW), green peas (1.3 mg 100 g^{-1} DW), red kidney beans (2.1 mg 100 g^{-1} DW) and mung beans (5.1 mg 100 g^{-1} DW).[75] In another study, the total tocopherol content in chickpea seeds (170 μg g^{-1} DW) was higher than the observed in pea (57 μg g^{-1} DW), lentil (46 μg g^{-1} DW) and bean (36 μg g^{-1} DW) seeds.[76]

The antioxidant activities of pea, lentil, bean and chickpea lipophilic extracts have been correlated with the total tocopherol content and not the total carotenoid content.[76] The carotenoids and tocopherols present in 20 lentil red and green cultivars were shown to have a synergic effect on the antioxidant activity of their lipophilic extracts.[70] γ-Tocopherol was found in cranberry bean cultivars in the range 16.6–27.5 μg g^{-1} DW and β-carotene was identified in the range 3.6–7.4 ng g^{-1}. Strong positive correlation was also found between γ-tocopherol and antioxidant activity of cranberry beans lipophilic extracts.[77] In Brazilian cowpea genotypes the δ-tocopherol content was strongly related to free radical scavenging capacity.[78]

Carotenoids are powerful lipid antioxidants and are responsible for the red, yellow and orange colors of different legumes.[79] The important carotenoids present in chickpeas include β-carotene, lutein, zeaxanthin, β-cryptoxanthin, lycopene and α-carotene.[80] The highest carotenoids contents were found in lentils from 4.5 to 21.3 μg g^{-1} DW and red kidney beans from 8.3 to 20.9 μg g^{-1} DW.[73] Daily consumption of a diet rich in β-cryptoxanthin, lycopene, lutein plus zeaxanthin, β-carotene and α-carotene decreased oxidative stress markers in middle-aged men.[81]

The intake of carotenoids and tocopherols in the diet provides additional health benefits mainly through the inhibition of lipid peroxidation. High intake of legumes rich in carotenoids and tocopherols could serve to counteract oxidative damage in cell membranes caused by reactive oxygen species. More importantly, in contrast to other foods, legumes retain their lipophilic antioxidants after cooking and therefore, retain the potential health benefits.[82] However, other processing methods to produce more convenient foods derived from legumes should be evaluated in terms of the bioaccessibility of carotenoids and tocopherols.

12.5 Non-digestible Fermentable Components and Oxidative Stress

Legumes contain resistant starch, soluble fiber and non-starch polysaccharides that after microbial metabolism improve the defense against oxidative stress. In particular, the consumption of cooked red kidney beans increased almost three times the antioxidant capacity of mice feces in comparison with those fed white beans.[83] Interestingly, the difference in the antioxidant capacity between the diets supplemented with white and dark red kidney beans was only 15.6% but metabolites of anthocyanidins in colored beans could be responsible for this difference.[83] In fact, natural products rich in anthocyanins not only alleviate oxidative stress processes but also have cytoprotective functions and regulate inflammatory processes.[84]

Legume phenolic compound consumption also alters microbiota composition to favor the growth of microorganisms responsible for producing more bioactive metabolites. Mice fed with cooked chickpea flour had an altered cecal and fecal microbiota community structure that, using predictive microbial functional outcomes, was shown to have a metagenome enriched in genes related to flavonoid biosynthesis.[85] In dogs that consumed cooked bean powders, shifts in the relative abundance of some taxa were observed, but in a healthy canine the intestinal microbiome was maintained.[86] It is also important to point out that legume dietary fibre may contain phenolic compounds and tannins bound to the sugar moieties of the polymer[87] and therefore, may be slowly released during digestion thus altering microbiota composition.

In overweight diabetic patients, a non-soya legume-based diet significantly decreased high-sensitivity C-reactive protein, IL-6 and TNF-α.[88] Additionally, during a median follow-up of 4.3 years, individuals in the highest quartile of total legume and lentil consumption had a lower risk of diabetes than those in the lowest quartile.[89] Another study that included 1026 pregnant women, aged 18–45 years, a higher consumption of legumes during reproductive age decreased the risk of gestational diabetes mellitus.[90]

Dietary fiber has also been related to a reduction in postprandial glycemia after the consumption of common beans.[91] The consumption of beans also affects starch digestibility at different levels depending on the content of phytic acid, tannins and α-amylase inhibitors.[92]

12.6 Legume Consumption and Gut Health

Ulcerative colitis is associated with elevated oxidative stress and legume consumption reduces the severity of this inflammatory response. The consumption of beans has been demonstrated to play an important role in mucosal protection and epithelial restitution in mice with colonic inflammation following dextran sodium sulphate (DSS) exposure.[85] Dietary fiber content of legumes has been associated with a reduction in inflammation as a result of decreased lipid oxidation. For instance, the replacement of two servings of red meat by non-soya legumes reduced the plasma concentrations of inflammatory markers among overweight diabetic patients.[88] Recently, it has been proved that pea seed albumin extract prevented microbiota dysbiosis in a mouse model of dextran sodium sulfate (DSS)-induced colitis.[93] Interestingly, even in the absence of polysaccharides, pea seed extract ameliorated DSS-induced colitis in mice.[94]

Inflammation in adipose tissue is also a consequence of obesity. The transference of non-esterified fatty acids from inflamed adipose tissue to the liver is one of the causes of non-alcoholic fatty liver disease (NAFLD). Exercise and germinated legume consumption reduce the risk of developing NAFLD. Particularly, peptides in germinated mung bean (*Vigna radiata*) have been correlated with this preventive effect.[95] Thus, legume proteins also have an impact on gut health due to the production of bioactive peptides, some of them related to antioxidant and anti-inflammatory effects.[59,96,97]

Gut microbiota affects the host health and, independently of the phenolic composition, beans consumption reduced the abundance of the Clostridiaceae, which includes the potentially harmful bacterium species *C. perfringens*.[98] Anti-microbial defenses were also observed when the diet of mice was supplemented with 20% cooked chickpea flour.[85]

Microbial dysbiosis is a dysfunctional interaction between a host and his microbiota that, among other negative affects, may elevate oxidative stress among. Microbiota may be altered by different factors besides diet. Navy or black bean supplementation prior to colitis onset attenuated the tissue damage and inflammatory response by altering the colonic microenvironment.[99] Previously, it was demonstrated that navy and black bean supplementation affects microbiota and host tissues to improve overall gut health and barrier integrity.[100]

Due to advances in food technology, legumes are transformed into diverse food products, rather than simply boiling them in water. The complexity of the interaction between polysaccharides and proteins and phytochemicals and the changes that occur during digestion alter gut microbiota. Additionally, the release of compounds with anti-inflammatory activity as a result of increased consumption of legumes means they provide a good food alternative to improve health.

12.7 Germination of Legumes and Antioxidant Activity

Germination of legumes enhances their nutritional value, antioxidant capacity and concentration of phenolic compounds.[101,102] The enzymatic activity of lipases, amylases and proteases increases during germination and the main macromolecules are hydrolyzed to their constituent molecules, in response to the nutritional and energy requirements of plant development.[103] Additionally, germination increases the synthesis of phenolic compounds in the seeds.[102,104] The phenolic compounds' content and composition profiles change during germination under stress conditions such as light, temperature and salts concentrations.[101,105,106] The phenolic compounds play a key role as defense mechanisms against oxidative stress.[107,108] In germinated legumes, the increase in the antioxidant capacity was directly correlated with the concentration of phenolic compounds. Phenolic compound content dramatically increased in beans, white cowpeas, black beans, adzuki beans, soybeans and peanuts after five days of germination compared to ungerminated seeds.[109] Germinated peanuts showed the highest content of total phenolic compounds and the strongest antioxidant capacity compared to other legumes. The content of sinapic acid, ferulic acid and cinnamic acid, were strongly related to DPPH radical scavenging activity, and ellagic acid, cinnamic acid and ferulic acid showed maximum correlation with reducing power.[109] In addition, it has been reported that during peanut germination the resveratrol content increases.[110] High flavonoids and epicatechins concentrations has been found in germinated peanuts.[111]

The germination of cowpea, jack bean, dolichos, and mucuna seeds for 4 days showed increased total phenolic compounds content, from 12% to 136%, accompanied by increased antioxidant activity, from 22% to 89%. The highest antioxidant activity was found in germinated mucuna (79.6 µmol of Trolox per gram of DW by FRAP and 10.70 µmol Trolox per gram of DW by DPPH), followed by cowpea and dolichos.[112] During the germination of chickpea, phenolic compounds and hydroxyl radical ($^{\bullet}OH$) scavenging activity were increased by 93% and 493%, respectively.[113] Methanolic extracts of germinated lentils showed increased phenolic compounds content and antioxidant activity by 47% and 38% at 5 days of germination compared to ungerminated lentils.[114] Phenolic compound content increased from 1341 mg 100 g^{-1} DW in lentil seeds to 1411, 1463, 1630 and 1510 at 3, 4, 5 and 6 days of germination. DPPH radical scavenging increased from 41% to 62% in 6 days germinated lentils compared to ungerminated lentils.[115]

The total phenolic compounds content and antioxidant capacity of black bean extracts were higher when they were obtained from seed coats, mainly from the third day of germination.[102] Compared to untreated sprouts, total isoflavonoid and antioxidant capacity showed remarkable increases of 83% and 33%, respectively, in germinated chickpeas that were treated with a high sodium selenite concentration (2 mg 100 g^{-1} seeds).[101] The germination of dark bean seeds increased their daidzein, genistein, and biochanin A content compared with bean seeds.[116] The germination of soybeans for 3 days increased their total aglycones content and DPPH free radical scavenging capacity by 147% and 220%, respectively, compared to ungerminated soybean.[117] Higher antioxidant capacities of soybean, chickpea, mung bean, red lentils and kidney sprouts were found when the seeds were germinated at 40 °C compared to those germinated at 30 °C, as a result of the increase in total polyphenols.[118] Germinated legumes can be used as a source of natural antioxidants in foods, but more importantly, a metabolic engineering technique may be designed to produce certain groups of metabolites, either phenolic compounds or bioactive peptides.

12.8 Conclusions and Perspectives

Legumes are staple foods of many cultures but innovative processing methods are required to increase their consumption, particularly in urban areas where cooking time is a major concern. In the past 15 years, many bioactive compounds have been validated *in vitro* or *in vivo* but their stability in processed foods has not always been tested. In addition to the antioxidant capacity exerted by phenolic compounds and certain peptides, regulation of gene expression has also been related to the prevention of oxidative stress. Seed coat antioxidants have a wide chemical diversity and potential to be used as bioactive ingredients in nutraceutical formulations. Legume proteins are also a reservoir of bioactive peptides and, as their release is naturally controlled depending on interaction with microbiota, they may be used to design personalized foods in the future.

References

1. A. M. Pisoschi and A. Pop, *Eur. J. Med. Chem.*, 2015, **97**, 55–74.
2. B. Poljsak, D. Suput and I. Milisav, *Oxid. Med. Cell Longev.*, 2013, DOI: 10.1155/2013/956792.
3. E. Birben, U. M. Sahiner, C. Sackesen, S. Erzurum and O. Kalayci, *World Allergy Organ J.*, 2012, **5**, 9.
4. M. P. Gomes and Q. S. Garcia, *Biologia*, 2013, **68**, 351.
5. M. Ristow, *Nat. Med.*, 2014, **20**, 709–711.
6. A. Ayala, M. F. Muñoz and S. Argüelles, *Oxid. Med. Cell Longev.*, 2014, DOI: 10.1155/2014/360438.
7. S. B. Nimse and D. Palb, *RSC Adv.*, 2015, **5**, 27986.
8. J. Van der Paal, E. C. Neyts, C. C. W. Verlackt and A. Bogaerts, *Chem. Sci.*, 2016, **7**, 489.
9. M. Murata, R. Thanan, N. Ma and S. Kawanishi, *J. Biomed. Biotechnol.*, 2012, DOI: 10.1155/2012/623019.
10. B. A. Aslani and S. Ghobadi, *Life Sci.*, 2016, **146**, 163.
11. M. Marrelli, B. Cristaldi, F. Menichini and F. Conforti, *Food Chem. Toxicol.*, 2015, **86**, 16.
12. S. C. Magalhães, M. Taveira, A. R. Cabrita, A. J. Fonseca, P. Valentão and P. B. Andrade, *Food Chem.*, 2017, **215**, 177.
13. Y. Aguilera, I. Estrella, V. Benitez, R. M. Esteban and M. A. Martín-Cabrejas, *Food Res. Int.*, 2011, **44**, 774.
14. J. S. Chia, J. L. Du, M. S. Wu, W. B. Hsu, C. P. Chiang, A. Sun, J. J. Y. Lu and W. B. Wang, *Integr. Cancer Ther.*, 2012, **12**, 248.
15. B. Xu and S. K. C. Chan, *Food Chem.*, 2012, **134**, 1287.
16. S. Marventano, M. Izquierdo-Pulido, C. Sánchez-González, J. Godos, A. Speciani, F. Galvano and G. Grosso, *Public Health Nutr.*, 2016, **20**, 245.
17. B. Singh, J. P. Singh, K. Shevkani, N. Singh and A. Kau, *J. Food Sci. Technol.*, 2017, **54**, 858.
18. R. Amarowicz and F. Shahidi, *J. Funct. Foods*, 2017, DOI: 10.1016/j.jff.2017.04.002.

19. Y. Tan, S. K. C. Chang and Y. Zhang, *Food Chem.*, 2017, **214**, 259.
20. Y. K. Wang, X. Zhang, G. L. Chen, J. Yu, L. Q. Yang and Y. Q. Gao, *J. Funct. Foods*, 2016, **24**, 359.
21. R. Y. Gan, Z. Q. Deng, A. X. Yan, N. P. Shah, W. Y. Lui, C. L. Chan and H. Corke, *LWT -- Food Sci. Technol.*, 2016, **73**, 168.
22. A. García-Lafuente, C. Moro, N. Manchón, A. Gonzalo-Ruiz, A. Villares, E. Guillamón and L. Mateo-Vivaracho, *Food Chem.*, 2014, **161**, 216.
23. B. D. Oomah, F. Caspar, L. J. Malcolmson and A. S. Bellido, *Food Res. Int.*, 2011, **44**, 436.
24. J. A. Gutiérrez-Uribe, I. Romo-López and S. O. Serna-Saldívar, *J. Funct. Foods*, 2011, **3**, 290.
25. Y. N. Sreerama, V. B. Sashikala and V. M. Pratape, *Food Chem.*, 2012, **133**, 156.
26. A. Segev, H. Badani, Y. Kapulnik, I. Shomer, M. Oren-Shamir and S. Galili, *J. Food Sci.*, 2010, **75**, S115.
27. Y. Zhao, S. K. Du, H. Wang and M. Cai, *Food Chem.*, 2014, **152**, 462.
28. R. Attree, B. Du and B. Xu, *Ind. Crops Prod.*, 2015, **67**, 448.
29. Y. Aguilera, M. Dueñas, I. Estrella, T. Hernández, V. Benitez, R. M. Esteban and M. A. Martín-Cabrejas, *J. Agric. Food Chem.*, 2010, **58**, 10101.
30. B. Xu and S. K. Chang, *J. Agric. Food Chem.*, 2010, **58**, 1509.
31. N. Chaieb, J. L. González, M. López-Mesas, M. Bouslama and M. Valiente, *Food Res. Int.*, 2011, **44**, 970.
32. L. Kan, S. Nie, J. Hu, S. Wang, S. W. Cui, Y. Li and M. Xie, *Food Chem. Toxicol.*, 2016, DOI: 10.1016/j.fct.2016.09.007.
33. S. J. Kwon, K. Dong-Gun, J. M. Kim, K. Kyung-Yun, L. Min-Kyu, H. Min Jeong, J. -K. Baek, E. Seok Hyun, K. Si-Yong Kang, H. Bo-Keun and R. Jaihyunk, *Plant Breed. Biotech.*, 2018, **6**, 65.
34. R. F. Prietsch, L. G. Monte, F. A. da Silva, F. T. Beira, F. A. B. Del Pino, V. F. Campos, T. Collares, L. S. Pinto, R. M. Spanevello, G. D. Gamaro and E. Braganhol, *Mol. Cell. Biochem.*, 2014, DOI: 10.1007/s11010-014-1974-x.
35. S. K. Gupta, S. Dongare, R. Mathur, R. Mathur, I. R. Mohanty, S. Srivastava, S. Mathur and T. C. Nag, *Mol. Cell. Biochem.*, 2015, **408**, 63.
36. Y. Qian, L. Cao, T. Guan, L. Chen, H. Xin, Y. Li, R. Zheng and D. Yu, *Pharm. Biol.*, 2015, **53**, 1124.
37. Y. Zhang, T. Pechan and S. K. C. Chang, *J. Funct. Foods*, 2018, **42**, 289.
38. F. R. Yao, C. W. Sun and S. K. C. Chang, *J. Agric. Food Chem.*, 2010, **58**, 10382.
39. Y. Mukai and S. Sato, *J. Nutr. Biochem.*, 2011, **22**, 16.
40. Z. Chen, J. Wang, W. Liu and H. Chen, *J. Food Sci. Technol.*, 2017, **54**, 964.
41. C. P. Marinangeli and P. J. Jones, *Br. J. Nutr.*, 2011, **105**, 110.
42. W. Li, C. Shu, S. Yan and Q. Shen, *J. Food Sci. Technol.*, 2010, **45**, 1205.
43. F. Roy, J. L. Boye and B. K. Simpson, *Food Res. Int.*, 2010, **43**, 432.
44. C. Petchiammal and W. Hopper, *Int. J. Pharm. Sci.*, 2014, **6**, 476.
45. B. Hernández-Ledesma, M. J. García-Nebot, S. Fernández-Tomé, L. Amigo and I. Recio, *Int. Dairy J.*, 2014, **38**, 82.
46. L. J. Zhu, J. Chen, X. Tang and Y. L. Xiong, *Agric. Food Chem.*, 2008, **56**, 2714.

47. C. F. Ajibola, J. B. Fashakin, T. N. Fagbemi and R. E. Al, *Int. J. Mol. Sci.*, 2011, **12**, 6685.
48. K. Kou, J. Gao, Z. Xue, Z. Zhang, H. Wang and X. Wang, *LWT -- Food Sci. Technol.*, 2013, **50**, 591.
49. H. M. Chen, K. Muramoto and F. Yamauchi, *J. Agric. Food Chem.*, 1995, **43**, 574.
50. C. Torres-Fuentes, M. M. Contreras, I. Recio, M. Alaiz and J. Vioque, *Food Chem.*, 2015, **180**, 194.
51. N. Ji, C. Sun, Y. Zhao, L. Xiong and Q. Su, *Food Chem.*, 2014, **161**, 148.
52. P. Garcia-Mora, J. Frias, E. Peñas, H. Zielinski, J. A. Giménez-Bastida and W. Wiczkowski, *J. Funct. Foods*, 2015, **18**, 319.
53. P. García-Mora, M. Martín-Martínez, M. Angeles-Bonache, R. González-Múniz, E. Peñas, J. Frias and C. Martinez-Villaluenga, *Food Chem.*, 2017, **221**, 464.
54. A. Mokni Ghribi, I. Maklouf Gafsi, A. Sila, C. Blecker, C. Danthine, H. Attia, A. Bougatef and S. Besbes, *Food Chem.*, 2015, **187**, 322.
55. Y. Y. Ngoh and C. Y. Gan, *Food Chem.*, 2016, **190**, 331–337.
56. T. J. Ariza-Ortega, E. Y. Zenón-Briones, J. L. Castrejón-Flores, J. Yáñez-Fernández, Y. M. Gómez-Gómez and M. C. Oliver-Salvador, *Eur. Food Res. Technol.*, 2014, **239**, 785.
57. D. Betancur-Ancona, T. Sosa-Espinoza, J. Ruiz-Ruiz, M. Segura-Campos and L. Chel-Guerrero, *Int. J. Food Sci. Technol.*, 2014, **49**, 2.
58. J. Ruiz-Ruiz, G. Dávila-Ortíz, L. Chel-Guerrero and D. Betancur-Ancona, *J. Food Biochem.*, 2013, **37**, 26.
59. L. López-Barrios, M. Antunes-Ricardo and J. A. Gutiérrez-Uribe, *Food Chem.*, 2016, **203**, 417.
60. J. A. Evangelho, N. L. Vanier, V. A. Pinto, J. J. Berrio, A. R. Guerra Dias and E. Zavareze, *Food Chem.*, 2017, **214**, 460.
61. P. Garcia-Mora, E. Peña, J. Frias and C. Martínez-Villaluenga, *J. Agric. Food Chem.*, 2014, **62**, 4166.
62. M. Rodrigues Marques, R. A. M. Soares, A. C. Corrêa Carlos, E. S. Siguemoto, G. Guadagnucci Fontanari and J. A. Gomes Arêas, *Food Chem.*, 2015, **168**, 288.
63. Q. Zhang, X. Tong, B. Qi, Z. Wang, Y. Li, X. Sui and L. Jiang, *J. Funct. Foods*, 2018, **42**, 298.
64. J. Carrasco-Castilla, A. J. Hernández-Álvarez, C. Jiménez-Martínez, C. Jacinto-Hernández, M. Alaiz and J. Girón-Calle, *J. Agric. Food Chem.*, 2012, **135**, 1789.
65. A. T. Girgih, D. Chao, L. Lin, R. He, S. Jung and R. E. Aluk, *Food Chem.*, 2015, **188**, 510.
66. Z. Xue, H. Wen, L. Zhaia, Y. Yu, Y. Li, W. Yu, A. Cheng, C. Wang and X. Kou, *Food Res. Int.*, 2015, **77**, 75.
67. P. Garcia-Mora, E. Peñas, J. Frias, H. Zieliński, W. Wiczkowski, D. Zielińska and C. Martínez-Villaluenga, *J. Agric. Food Chem.*, 2016, **64**, 1730.
68. N. Gupta, N. Srivastava and S. S. Bhagyawant, *PLoS One*, 2018, DOI: 10.1371/journal.pone.0191265.

69. A. Kuerban, S. S. Moselhy, T. A. Kumosani, O. A. S. Baothman, M. Zeyadi, N. Helmi, T. M. Alghamdi, M. A. Zamzami, M. B. Mirza, F. Q. Syed, F. A. Mohammed, A. L. AL-Malki and K. O. Abulnaja, *J. Pharm. Res. Int.*, 2017, **19**, 1.
70. B. Zhang, Z. Deng, Y. Tang, P. Chen, R. Liu, D. D. Ramdath, Q. Liu, M. Hernandez and R. Tsao, *Food Chem.*, 2014, **161**, 296.
71. S. E. Sattler, L. U. Gilliland, M. Magallanes-Lundback, M. Pollard and D. DellaPenna, *Plant Cell*, 2014, **16**, 1419.
72. C. E. Seal, R. Zammit, P. Scott, T. J. Flowers and I. Kranner, *Sci. Res.*, 2010, **20**, 47.
73. A. Elgersma, K. Søegaard and S. K. Jensen, *J. Agric. Food Chem.*, 2013, **61**, 11913.
74. B. Fernández-Marín, R. Milla, N. Martín-Robles, E. Arc, I. Kranner, J. M. Becerril and J. I. García-Plazaola, *BMC Plant Biol.*, 2014, **20**(14), 1599.
75. A. K. Jukanti, P. M. Gaur, C. L. L. Gowda and R. N. Chibbar, *Br. J. Nutr.*, 2012, **108**, S11.
76. E. M. T. Padhi, R. Liu, M. Hernandez, R. Tsao and D. D. Ramdath, *J. Funct. Foods*, 2017, **38**, 602.
77. X. P. Chen, Y. Tang, M. F. Marcone, P. K. Pauls, B. Zhang, R. Liu and R. Tsao, *Food Chem.*, 2015, **185**, 298.
78. A. F. U. Carvalho, N. M. Sousa, D. F. Farias, L. C. B. Rocha-Bezerra, R. M. P. Silva and M. P. Viana, *J. Food Comp. Anal.*, 2012, **26**, 81.
79. G. E. Bartley and P. A. Scolnik, *Plant Cell*, 1995, **7**, 1027.
80. K. Ashokkumar, B. Tar'an, M. Diapari, G. Arganosa and T. D. Warkentin, *Crop Sci.*, 2014, **54**, 2225.
81. P. Cocate, A. Natali, R. Alfenas, A. De Oliveira, E. Dos Santos and H. Hermsdorff, *Nutrition*, 2015, **114**, 257.
82. B. Zhang, Z. Deng, Y. Tang, P. X. Chen, R. Liu, D. D. Ramdath, Q. Liu, M. Hernandez and R. Tsao, *J. Agric. Food Chem.*, 2014, **62**, 12585.
83. J. M. Monk, C. P. Zhang, W. Wu, L. Zarepoor, J. T. Lu, R. Li and K. A. Power, *J. Nutr. Biochem.*, 2015, **26**, 752.
84. H. R. Sodagari, M. H. Farzaei, R. Bahramsoltani, A. H. Abdolghaffari, M. Mahmoudi and N. Rezaei, *Expert Rev. Gastroenterol. Hepatol.*, 2015, **9**, 807.
85. J. M. Monk, D. Lepp, W. Wu, D. Graf, L. H. McGillis, A. Hussain and Y. Brummer, *J. Funct. Foods*, 2017, **38**, 663.
86. A. N. Beloshapka, G. M. Forster, H. D. Holscher, K. S. Swanson and E. P. Ryan, *J. Vet. Sci. Technol.*, 2016, **7**, 2.
87. L. X. López-Martínez, N. Leyva-López, E. P. Gutiérrez-Grijalva and J. B. Heredia, *J. Funct. Foods*, 2017, **38**, 624.
88. S. Hosseinpour-Niazi, P. Mirmiran, A. Fallah-Ghohroudi and F. Azizi, *Br. J. Nutr.*, 2015, **114**, 213.
89. N. Becerra-Tomás, A. Díaz-López, N. Rosique-Esteban, E. Ros, P. Buil-Cosiales, D. Corella and R. M. Lamuela-Raventós, *Clin. Nutr.*, 2017, **37**, 906.
90. A. Goshtasebi, S. Hosseinpour-Niazi, P. Mirmiran, M. Lamyian, L. M. Banaem and F. Azizi, *Diabetes Res. Clin. Pract.*, 2018, **139**, 131.

91. D. M. Winham, A. M. Hutchins and S. V. Thompson, *Nutrients*, 2017, **9**, 1095.
92. C. D. Ferreira, V. Ziegler, S. L. M. El Halal, N. L. Vanier, E. R. Zavareze and M. de Oliveira, *Starch/Staerke*, 2017, **69**, 1.
93. I. Aranda-Olmedo, R. Ruiz, M. J. Peinado and L. A. Rubio, *J. Funct. Foods*, 2017, **35**, 279.
94. M. Utrilla, M. Peinado, R. Ruiz, A. Rodriguez-Nogales, F. Algieri, M. Rodriguez-Cabezas and L. A. Rubio, *Mol. Nutr. Food Res.*, 2015, **59**, 807.
95. G. Kapravelou, R. Martínez, E. Nebot, M. López-Jurado, P. Aranda, F. Arrebola and J. M. Porres, *Nutrients*, 2017, **9**, 774.
96. S. Bautista-Expósito, E. Peñas, J. M. Silván, J. Frias and C. Martínez-Villaluenga, *Food Chem.*, 2018, **248**, 262.
97. S. N. Jamdar, R. Deshpande and S. A. Marathe, *Food Biosci.*, 2017, **20**, 1.
98. J. M. Monk, D. Lepp, C. P. Zhang, W. Wu, L. Zarepoor, J. T. Lu and K. A. Power, *J. Nutr. Biochem.*, 2016, **28**, 129.
99. J. M. Monk, W. Wu, A. L. Hutchinson, K. P. Pauls, L. E. Robinson and K. A. Power, *J. Nutr. Biochem.*, 2018, **56**, 215.
100. J. M. Monk, D. Lepp, W. Wu, K. P. Pauls, L. E. Robinson and K. A. Power, *J. Nutr. Biochem.*, 2017, **49**, 89.
101. D. Guardado-Félix, S. O. Serna-Saldivar, E. O. Cuevas-Rodríguez, D. A. Jacobo-Velázquez and J. A. Gutierréz-Uribe, *Food Chem.*, 2017, **226**, 69.
102. D. Guajardo-Flores, S. O. Serna-Saldívar and J. A. Gutiérrez-Uribe, *Food Chem.*, 2013, **141**, 1497.
103. C. Patanè, V. Cavallaro and S. L. Cosentino, *Ind. Crops Prod.*, 2009, **30**, 1.
104. J. G. Xu, C. R. Tian, Q. P. Hu, J. Y. Luo, X. D. Wang and X. D. Tian, *J. Agric. Food Chem.*, 2009, **57**, 10392.
105. M. Mendoza-Sánchez, R. G. Guevara-González, E. Castaño-Tostado, E. M. Mercado-Silva, J. A. Acosta-Gallegos, N. E. Rocha-Guzmán and R. Reynoso-Camacho, *Food Chem.*, 2016, **212**, 128.
106. Y. Gao, Y. Yao, Y. Zhu and G. Ren, *J. Agric. Food Chem.*, 2015, **63**, 2701.
107. J. C. Abdul, R. Gopi, B. Sankar, P. Manivannan, A. Kishorekumar, R. Sridharan and R. Panneerselva, *S. Afr. J. Bot.*, 2007, **73**, 190.
108. H. Pedranzani, R. Sierra-de-Grado, A. Vigliocco, O. Miersch and G. Abda, *Plant Growth Regul.*, 2007, **52**, 111.
109. D. T. Khang, T. N. Dung, A. A. Elzaawely and T. D. Xuan, *Foods*, 2016, **5**, 27.
110. M. M. Win, A. Abdul-Hamid, B. S. Baharin, F. Anwar, M. C. Sabu and M. S. Pak-Dek, *Pak. J. Bot.*, 2011, **43**, 1635.
111. M. L. L. D. Francisco and A. V. A. Resurreccion, *J. Food Compos. Anal.*, 2009, **22**, 16.
112. Y. Aguilera, M. F. Díaz, T. Jiménez, V. Benítez, T. Herrera, C. Cuadrado, M. Martín-Pedrosa and M. A. Martín-Cabrejas, *J. Agric. Food Chem.*, 2013, **61**, 8120.
113. B. G. Tarzia, M. Gharachorlooa, M. Baharinia b and S. A. Mortazavic, *Iran J. Pharm. Res.*, 2012, **11**, 1137.

114. M. Gharachorloo, B. G. Tarzi, M. Baharinia and A. H. Hemaci, *J. Med. Plants Res.*, 2012, **6**, 4562.
115. A. Ahmed Fouad and F. M. Ali Rehab, *Acta Sci. Pol. Technol. Aliment.*, 2015, **14**, 233.
116. A. López, T. El-Naggar, M. Dueñas, T. Ortega, I. Estrella, T. Hernández, M. P. Gómez-Serranillos, O. M. Palomino and M. E. Carretero, *Food Chem.*, 2013, **138**, 547.
117. X. Huang, W. Cai and B. Xu, *Food Chem.*, 2014, **143**(2014), 268.
118. R. K. Mamilla and V. K. Mishra, *LWT -- Food Sci. Technol.*, 2017, **75**, 51.

CHAPTER 13

Legumes and Inflammation

M. A. RICARDO-ANTUNES AND D. GUAJARDO-FLORES*

Tecnológico de Monterrey, Centro de Biotecnologia-FEMSA., Av. Eugenio Garza Sada 2501 Sur, 64849, Monterrey, Mexico
*E-mail: danielgdo@itesm.mx

13.1 Inflammation

Inflammation is the first biological response of the immune system to noxious stimuli, trauma or infection and it is characterized by redness, warmth, swelling, pain and sometimes numbness.[1,2] The purpose of inflammation is to protect the host from potentially harmful stimuli, maintain homeostasis and the restore tissue integrity. The intensity of the inflammatory response depends on the type of noxious stimulus and its persistence.[3] Acute inflammation is a controlled response of the organism followed by restoration of the affected tissues to their normal structure and function, and it is also known as resoleomics a self-limiting process of inflammation controlled by the innate immune system.[4–7] However, inefficient regulation of the mechanisms involved in inflammation leads to chronic inflammation by over activation of the innate immune response.[8–10] When the state of inflammation persists for a prolonged period of time this becomes chronic, and it can be detrimental for the host.[11]

Chronic inflammation has been implicated in an increased risk of a wide range of chronic degenerative diseases such as rheumatoid arthritis, atherosclerosis, cardiovascular disease, Alzheimer's disease, asthma, cancer,

Food Chemistry, Function and Analysis No. 8
Legumes: Nutritional Quality, Processing and Potential Health Benefits
Edited by María Ángeles Martín-Cabrejas

Published by the Royal Society of Chemistry, www.rsc.org

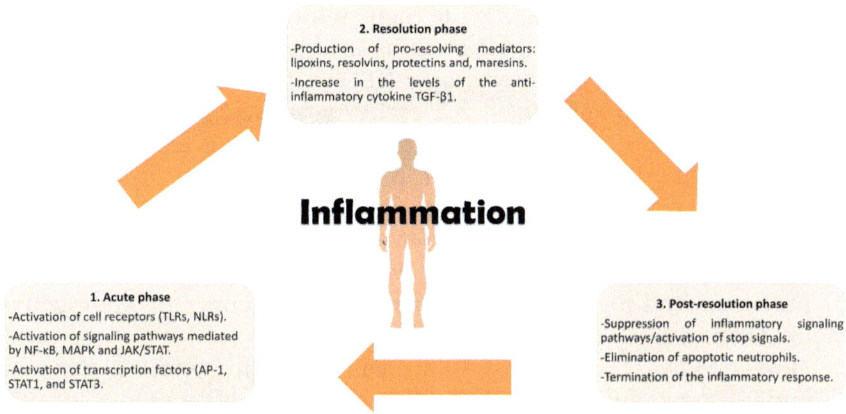

Figure 13.1 Phases of the inflammation response. TGF-β1: transforming growth factor-β1; TLRs: toll-like receptors; NLRs: Nod-like receptors; NF-κB: nuclear factor kappa-light-chain-enhancer of activated B cells; MAPK: mitogen activated protein kinases; JAK/STAT: janus tyrosine kinase/signal transducer and activator of transcription; AP-1: activator protein 1; STAT1: signal transducer and activator of transcription 1 STAT3: signal transducer and activator of transcription 3.

congestive heart failure, multiple sclerosis, diabetes, gout, obesity, inflammatory bowel disease, acquired immunodeficiency syndrome (AIDS), central nervous system depression and other neurodegenerative disease.[12–22] Recently studies reveled that chronic inflammation aggravates telomere dysfunction and cell senescence by increasing reactive oxygen species (ROS) generation and the activation of the nuclear factor kappa-light-chain-enhancer of activated B cells (NF-κB) signaling pathway, resulting in an increase in pro-inflammatory cytokines, especially interleukins IL-6 and IL-8, decreasing regenerative potential in multiple tissues and accelerating cell ageing.[23]

A typical inflammatory response involves four elements: inflammatory inducers, signaling molecules, downstream mediators, and the tissue affected,[3] and this process can be divided into three phases as summarized in Figure 13.1:

1. *Acute phase*: It this phase activation of inflammatory signaling pathways after a harmful stimulus occurs and this process is usually mediated by receptors of the innate immune system such as the Toll-like receptors (TLRs) and NOD (nucleotide-binding oligomerization-domain protein)-like receptors (NLRs).[24,25] These receptors induce an inflammatory response through the activation of the signaling pathways such as, NF-κB, mitogen-activated protein kinases (MAPK) and Janus kinase (JAK)/signal transducer and activator of transcription (STAT), which in turn, activate NF-κB, activator protein-1 (AP-1), STAT1, and STAT3 transcription factors. The NF-κB is an important transcriptional factor which plays critical roles in innate inflammatory, immune

anti-apoptotic responses.[26,27] It regulates the expression of genes of pro-inflammatory cytokines such as interleukin IL-1, tumor necrosis factor alpha (TNF-α), interferon-γ (IFN-γ), followed by the synthesis/activation of phospholipase A2, cyclooxygenase-2 (COX-2), and inducible nitric oxide synthase (iNOS).[28–30] This phase is also accompanied by high secretion of chemokines, and growth factors, and an increase in the migration of endothelial adhesion molecules. Likewise, polymorphonuclear neutrophils (PMNs) produce high levels of leukotrienes (LTX) and prostaglandins (PGs) using arachidonic acid (AA) as precursor and regulated by the activity of the enzymes lipoxygenases (5-LOX and 15-LOX), and COX-2.[31,32]

2. *Resolution phase*: This phase is also called the eicosanoid switch phase and it is governed by the production of pro-resolving mediators (SPM) that are produced by cells of the innate immune system to switch from pro- to anti-inflammatory eicosanoids production.[33] This group of specialized molecules or lipid mediator are biosynthesized from AA and essential polyunsaturated fatty acid (PUFA) precursors (EPA: eicosapentaenoic acid; DHA: docosahexaenoic acid) and include lipoxins, resolvins, protectins/neuroprotectins, and, maresins.[34] In this phase there is also an increase in anti-inflammatory cytokine transforming growth factor-β1 (TGF-β1) secretion from macrophages, which can suppress proinflammatory signaling from Toll-like receptors.[35,36]
3. *Post-resolution phase*: this phase is represented by the suppression of inflammatory signaling pathways mediated through activation of stop signals, and it plays an essential role in tissue reparation.[3] SPM reduce vascular permeability, retarding the entry of neutrophils to the inflammation site, and stimulate the elimination of apoptotic neutrophils by activated macrophages.[35,36] Accumulation of pro-inflammatory transcription factors as NF-κB, AP-1, STAT1, and STAT3 are also attenuated in this phase.[4] Finally, these actions lead to the termination of the inflammatory response and homeostasis is reestablished.

13.1.1 Cytokines in Inflammation

Cytokines are critical mediators involved in the regulation of the host response to a stimulus as in the case of an infection, immune response, inflammation, trauma and tumorigenesis.[37–40] Cytokines have been classified as interleukins (IL), interferons (IF), colony stimulating factors (CSF), TNF-α and growth factors. Unlike hormones, cytokines are produced in a specific tissue, but may be synthesized in any cell.[40]

According to their activity, cytokines can be classified as pro-inflammatory and anti-inflammatory. Under normal conditions, the levels of both types of cytokines remain balanced, but after a stimulus, the amount of pro-inflammatory cytokines tends to increase.[39,40] Anti-inflammatory cytokines are a number of immunoregulatory molecules that control the response of pro-inflammatory cytokines to a stimulus. The pro-inflammatory cytokines are those that have been shown to have a predominantly pro-inflammatory effect, and within this group are found TNF-α, IL-1, IL-6, IL-12, and IFN-γ. Some cytokines such as IL-6, IL-10 and TGF-β1 may exhibit both, pro- and anti-inflammatory effects depending on the physiological or physio-pathological condition.[41] Pro-inflammatory cytokines are not expressed simultaneously in the inflammatory process. Depending on the type of inflammation, acute or chronic, specific pro-inflammatory cytokines are produced. The cytokines IL-1, TNF-α, IL-6, IL-11, IL-8 play a key role in the mediation of acute inflammatory response. Moreover, cytokines that mediate the chronic inflammation can be divided into those involved in humoral inflammation (IL-3 to IL-10, IL-13 and TGF-β1) and cytokines that contribute to cellular inflammation (IL-1 to IL-4, IL-7, IL-9, IL-10, IL-12, interferons, INF-γ, TGF-β1, TNF-α and TNF-β).[42,43] TNF-α is the most studied pro-inflammatory cytokine, and it is synthesized as an inactive pro-protein which is activated by the action of the TNF-α converting enzyme (TACE).[44,45] It is synthesized primarily by monocytes and macrophages, but can also be expressed in other cell types, including neutrophils, natural killer cells (NK) and T cells in synovial membrane cells.[42,43] It possesses at least one receptor class in all cell types.[44] TNF-α acts as a lethal mediator of the acute or chronic immune response in inflammatory and infectious diseases, increasing the body temperature and producing sleep and fatigue by its action on the central nervous system.[45]

13.1.2 Cyclooxygenase in Inflammation

The cyclooxygenase (COX) enzymes, COX-1 and COX-2 are functional homodimers[46] which differ particularly in their active site. The action of COX-1 responds to hormonal stimulation and to diverse prostanoids synthesized by physiological processes involved in homeostasis.[46,47] It has been reported that mRNA from the COX-1 is detectable in normal physiological conditions, but its expression remains unchanged before an inflammatory reaction. On the other hand, COX-2 is not normally found in the tissue, but it is induced in inflammatory cells such as monocytes and macrophages upon stimulation by cytokines, mitogens and endotoxins.[48] The cyclooxygenases exert a double action on arachidonic acid: first they are responsible for its oxygenation and cyclization to form the prostaglandin G2 (PGG2), and then they reduce PGG2 to prostaglandin H2 (PGH2), which is considered the precursor of prostaglandins, important mediators of inflammation.[49]

The drugs available for clinical use are mainly nonsteroidal anti-inflammatory drugs (NSAIDs) that inhibit both COX-1 and COX-2. These drugs produce side effects by affecting the normal functioning of organs such as the stomach and kidney.[50,51] New generations of NSAID are directed to the selective inhibition of the inducible isoform of cyclooxygenase, COX-2, which decreases side effects on other organs. More recently dual inhibitors that block COX and LOX simultaneously have been developed. These drugs share the anti-inflammatory effect of COX-2 inhibitors, but also inhibit leukotriene synthesis by blocking the action gastrotoxic 5-LOX.[52,53] It has been suggested that dual inhibitors may be associated with fewer adverse effects than COX-1 or COX-2.[54,55]

13.1.3 Nitric Oxide in Inflammation

Nitric oxide is a wide-spread intracellular and intercellular signaling molecule in mammals. It is produced from L-arginine by three nitric oxide synthase (NOS) enzymes, endothelial NOS (eNOS), neuronal NOS (nNOS), and inducible NOS (iNOS).[56,57] nNOS and eNOS are considered calcium dependent and are constitutively expressed, whereas iNOS is considered calcium independent and it is only expressed after cellular activation in response to interferon-γ, lipopolysaccharide (LPS) and a variety of pro-inflammatory cytokines.[56] It has a broad spectrum of regulatory functions in the central nervous, cardiovascular, and immune systems, as well as in platelet inhibition, programmed cell death, and host responses to infection, among others.[56] Over expression of this mediator could lead to chronic disorders. The enzyme responsible for the prolonged production of nitric oxide is iNOS, and hence it is important in the process of chronic inflammation. Therefore, compounds that inhibit the expression or activity of this enzyme are potential anti-inflammatory agents.[58,59]

13.2 Inflammation Processes Regulated by Phytonutrients

Nutrition makes a substantial contribution to the etiology, prevention and progression of several diseases in which the anti-inflammatory mechanisms are involved. The most produced and consumed legumes are beans, peas, lentils and chickpeas that differ in their content of phytonutrients that are considered bioactive components. Epidemiological evidence often demonstrates that legume consumption lowers the risk of chronic diseases such as cardiovascular disease, diabetes, the metabolic syndrome, and certain cancers.[60] Several *in vivo* and *in vitro* studies have correlated these health benefits to the secondary metabolites in legumes, such as polyphenols, saponins, phytosterols, oligosaccharides, protein and peptides content (Figure 13.2). The nutritional and phytochemical components of pulses suggest a potential synergistic effect.

13.2.1 Consumption of Legume-based Diets

Legumes are an important component of various diets worldwide. Throughout recent years several *in vivo* studies have been performed in which legume-based diets have been investigated to determine the

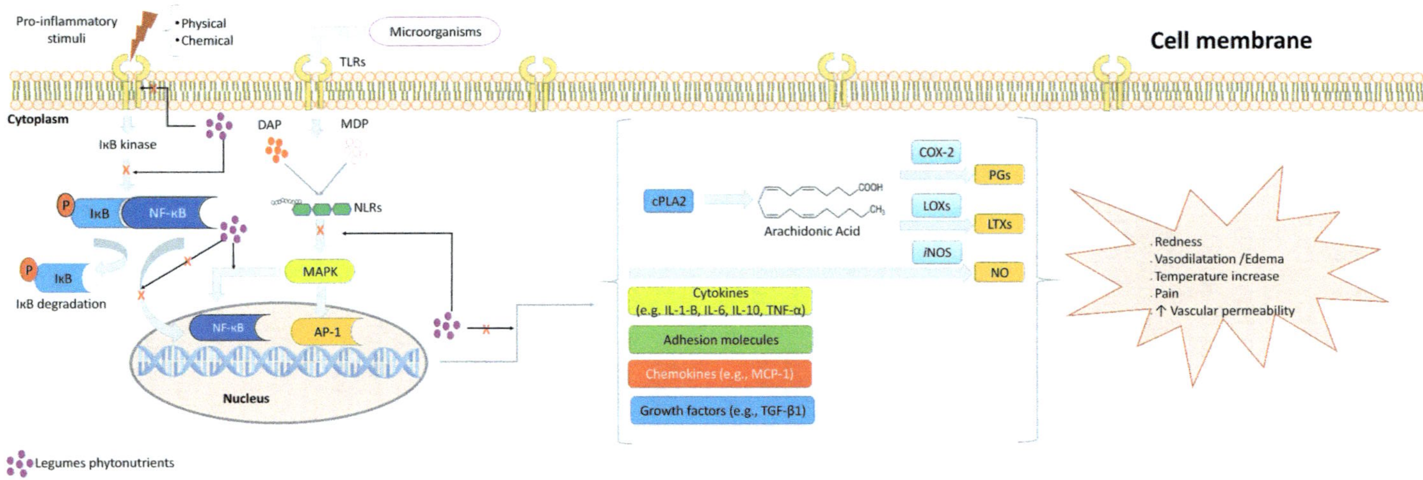

Figure 13.2 Anti-inflammatory mechanisms reported for phytonutrients contained in legumes. IκB: IκB kinase enzyme complex; NF-κB: nuclear factor kappa-light-chain-enhancer of activated B cells; TLRs: toll-like receptors; MDP: muramyl dipeptide found in Gram-positive bacteria; DAP: meso-diaminopimelic acid found in Gram-negative bacteria; NLRs: Nod-like receptors; MAPK: mitogen activated protein kinases; AP-1: activator protein 1; cPLA2: cytosolic phospholipase A2; COX-2: cyclooxygenase 2; PGs: prostaglandins; LOXs: lipoxygenases; LTXs: leukotrienes; *i*NOS: inducible nitric oxide synthase; NO: nitric oxide; IL: interleukin; TNF-α: tumor necrosis factor-alpha; MCP-1: monocyte chemoattractant protein-1; TGF-β1: transforming growth factor-β1.

potential antioxidant and anti-inflammatory effect of their consumption. Yook *et al.*[60] investigated the effects of black adzuki bean (*Vigna angularis*) on colon inflammation triggered by high-fat diet induced obesity in mice. Black adzuki bean consumption decreased the concentrations of circulating proinflammatory cytokines such as TNF-α, IL-1, and IL-6 in mice with induced obesity. Moreover, black adzuki bean inhibited induction of iNOS, COX-2, and activation of NF-κB in the colon. Furthermore, green pea supplementation of high-fat diet fed mice with induced colitis, reduced neutrophil infiltration, mRNA expression of monocyte chemoattractant protein-1 (MCP-1) and inflammatory markers IL-6, COX-2, IL-7, IFN-γ, and iNOS.[61] Further, *in vivo* studies in human-randomized, controlled, crossover trials demonstrated that 3 day intake of 100 g of black bean meal improved the arthritic condition by significantly reducing pain and inflammation.[62]

13.2.2 Legume Crude Extracts

There has been a growing interest to determine the phytonutrient composition of legumes and their potential health benefits. Legume extract studies were first carried out to determine the more bioactive fractions and identify the phytochemicals that directly affect specific mechanism of disease prevention. Ali *et al.*[63] evaluated the inhibition of inflammatory mediator NO *in vitro* and the inhibition of ear edema and reduction of response to pain stimulus *in vivo* induced by mung bean extracts. The results showed that germinated and fermented mung bean aqueous extract exhibited potent anti-inflammatory and antinociceptive activities in a dose dependent manner, by inhibiting the potent inflammatory mediator NO at both 2.5 and 5 mg mL^{-1}. Moreover, the *in vivo* studies showed that mung bean extract (1000 mg kg^{-1}) can significantly reduce ear edema in mice caused by arachidonic acid. Furthermore, the anti-inflammatory activity of mung bean extracts could be related to the presence of polyphenols gallic acid, vitexin, and isovitexin which are considered as potent antioxidant phytochemicals. It has been shown that mung bean extracts from different sources showed different inhibitory activities on IL-1β, IL-6 and COX-2 mRNA expression.[64] The study suggested that mung bean extracts significantly suppressed the expression of LPS-induced IL-1β mRNA, IL-6 and COX-2 mRNA. In this study, different extracts showed different inhibitory effect suggesting that commercial mung beans may differ in their bioactive components and health properties, and mung bean components may have selective anti-inflammatory effects. The results indicated that mung bean components (polyphenols) contributing to anti-oxidant activities may also play a role in anti-inflammation through suppressing IL-1β or IL-6 mRNA expression, but not COX-2 gene expression. Zhu *et al.*[65] explored the therapeutic potential of mung bean coat *in vitro* and *in vivo*. The results showed that mung bean extract attenuated LPS-induced release of HMGB1 and several chemokines in macrophage cultures, while oral administration of mung bean extracts significantly increased animal survival rates from 30% to 70%.

Furthermore, Oomah *et al.*[66] evaluated four bean cultivar extracts for antioxidant and anti-inflammatory activities in relation to their phenolic contents. Acetone hull extracts exhibited significantly higher antioxidant and stronger inhibitory effects on both cyclooxygenases COX-1 and COX-2, while bean hull water extracts were stronger inhibitors of lipoxygenase 15-LOX. Results suggested that the anti-inflammatory activity of bean hulls was dependent on their phenolic content, cultivar and extracting solvent. In another study, García-Lafuente *et al.*[67] determined the anti-inflammatory activities *in vitro* of white kidney beans and round purple beans extracts. Results showed that phenol rich extracts reduced the expression of iNOS, IL-1β, IL-6 and TNF-α *via* inactivation of the NF-κB pathway. Particularly, purple bean extracts exhibited a higher anti-inflammatory activity by decreasing the production of NO and expression of cytokine mRNA of LPS-stimulated macrophages.

Park *et al.*[68] used *in vitro* and *in vivo* experimental models to investigate the anti-inflammatory potential of the butanol fraction of red bean ethanol extract, which contains catechin-7β-D-glucopyranoside. This study showed that red bean extract inhibited nitric oxide production in LPS-stimulated macrophages through suppression of extracellular signal regulated kinase and inhibitory kappa B alpha (IκBα) activation. The *in vivo* murine sepsis model evaluated the oral administration of red bean extract which improved mouse survival rate by decreasing tumor-necrosis factor-α and interferon-γ and increasing interleukin-10. The survival rate of mice injected with LPS increased from 0 and 40% in ICR and BALB/c mice, respectively, to 100% in both mouse types with red bean extract. Yu *et al.*[69] studied an adzuki bean (*Phaseolus angularis*) ethanol extract which dose-dependently suppressed the release of prostaglandin E2 (PGE2) and NO in LPS-, Poly(I:C)-, and pam3CSK-activated macrophages. It also has an activity in the down-regulated LPS-induced mRNA expression of iNOS and COX-2. Furthermore, the adzuki bean extract markedly inhibited NF-κB, activator protein AP-1, and cAMP response element binding protein (CREB) activation and blocked the activation of the upstream signaling molecules spleen tyrosine kinase (Syk), p38, and transforming growth factor β-activated kinase 1 (TAK1). Also, Gabriele *et al.*[70] focused on an extract from common bean *Phaseolus vulgaris* flour, which was enriched in antioxidant during fermentation. The study assessed the protective effect on endothelial cells treated with oxidized-LDL (ox-LDL) *in vitro*. Results showed that the extract for fermented flour strongly reduced oxidative stress and significantly enhanced ICAM-1, IL-6 and ET-1 levels over basal levels. It was concluded that extract from fermented bean flour was able to inhibit the oxidation process, but that it was only marginally active against inflammation and ET-1 production in HMEC-1 exposed to ox-LDL.

Zia-Ul-Haq *et al.*[71] investigated crude methanolic extracts of selected legumes, namely black gram (*Vigna mungo* L.), green gram (*Vigna radiata*

(L.) R. Wilczek), soybean (*Glycine max* (L.) Merr.) and lentil (*Lens culinaris* Medik.) for anti-inflammatory activity using a COX-2 producing, PGE2 inhibitory *in vitro* assay. They observed 73.9%, 79.8%, 92.2%, and 74.5% inhibition for black gram, green gram, soybean, and lentil, respectively, at 20 mg mL^{-1} extract concentration, and the 100 μg mL^{-1} concentration showed an increase in the percentage inhibition, except for soybean.

13.2.3 Polyphenols

Nowadays, it is well known that colored legumes such as pinto beans and black beans, are a rich source of polyphenols with antioxidants properties. In recent years, polyphenols, such as flavonols, flavones, isoflavones, anthocyanins, tannins, and phenolic acids have been studied due their potential use as ingredients for functional food development or nutraceutical industry.[72a,b] In an animal study, 20% navy and black bean (*P. vulgaris* L.) flours, with different phenolic compound levels and profiles, were used to induce colitis in C57BL/6 mice.[73] Bean diets reduced the mRNA expression of colonic inflammatory cytokines (IL-1β, TNFα, IFNγ, IL-17A, and IL-9), increased anti-inflammatory IL-10, increased histological injury score and apoptosis, and alleviated symptoms of colitis and colon inflammation. Moreno-Jiménez *et al.*[74] studied the extracts of raw and processed common beans, which induced an overexpression of COX-1, especially due to the residual phenolics content. This study showed an inhibitory effect on the expression of TNF-α, NF-κB and IL-8 by rich phenolic extracts. Consumption of the common bean has anti-inflammatory effects, inhibiting the expression of COX-2, TNF-α and NF-κB, and inducing the expression of the anti-inflammatory cytokine IL-10. Pitura[75] measured the cellular anti-inflammatory activity of seed coat crude extracts of colored common beans (*P. vulgaris* L.) in LPS-stimulated murine macrophage RAW 264.7 cells. The black beans contained flavonol triglycosides (kaempferol, quercetin, and myricetin), which were strongly correlated to the anti-inflammatory activity, as indicated by decreasing TNF-α levels. Chao *et al.*[76] proved that red bean extract suppresses the inflammatory response at concentrations of 50–200 μg mL^{-1}. The mechanism involving the inflammatory response of red bean extracts occurs *via* the LPS-stimulated macrophages through the reduction of cellular NO and downregulation of the gene expressions of iNOS, COX-2, TNF-α, and IL-6 in a dose-dependent manner. In an animal study, white and dark kidney bean diets exerted a systemic anti-inflammatory effect during colitis by reducing serum levels of IL-17A, interferon IFNγ, TNFα, IL-1β, and IL-6.[77] Choi *et al.*[78] reported that red bean extracts decrease the levels of these inflammatory indicators by inhibiting acetic acid-mediated increases in mRNA levels of nitric oxide synthase, interleukin-1β, and tumor necrosis factor-α, which protected rats against intestinal colitis damage. The procyanidins found in common bean hulls can down regulate the transcription and secretion of proinflammatory cytokines, including interleukins IL-1β, IL-2, IL-6, TNF-α and interferon-γ,[79] which can be related to rheumatoid arthritis. Moreover, the phenolic acids, *p*-coumaric acid and ferulic acid, in cranberry extracts may possess an ability to attenuate the secretion of IL-8 as induced by TNF-α at low phenolic concentrations and short pre-treatment time; these results suggested that phenolic extracts are effective at managing oxidative stress related to inflammation.[80] Monk *et al.*[81] assessed the gut health priming effect of a 20% cranberry bean flour diet from two bean varieties with differing profiles of phenolic compounds. This study showed that during colitis, beans diets reduced the levels of colonic and circulating inflammatory cytokines (IL-1β, IL-6, IFNγ and TNFα), and increased gene expression of barrier function promoting genes, which resulted in a reduction of colonic histological damage.

The anti-inflammatory activity of ethanol hull extracts from faba beans inhibited the enzyme 15-LOX, which is involved in the molecular mechanisms of atherosclerosis, participating in oxidative modifications of low-density lipoproteins (LDL).[82] Boudjou *et al.*[83] studied faba bean (*Vicia faba* L.) and lentil seeds hulls and examined the correlation between antioxidant activity and their phenolic contents. The phenolic acid and procyanidin contents in extracts of lentil hulls exhibited high antioxidant and anti-inflammatory activities, preferentially inhibiting 15-LOX (IC_{50}, 55 μg mL^{-1}), with moderate COX-1 (IC_{50}, 66 μg mL^{-1}) and COX-2 (IC_{50}, 119 μg mL^{-1}) inhibitory effects on the COX pathway, whereas faba bean hull extracts rich in tannins exerted relatively mild LOX inhibitory activity. In addition, the antioxidant and anti-inflammatory activities were positively correlated with the total and individual phenolic contents of nondigested lentils.[84] This study showed dose-dependent anti-inflammatory activity of phenolics as seen in significant inhibition of the pro-inflammatory cytokines COX-2, IL-1β and IL-6 in TNF-α-induced inflammation in Caco-2 cells. Moreover, the flavonoids in the mung bean ethanol extracts, such as vitexin and isovitexin, significantly reduced the amounts of plasma IL6, TNF-α and MCP-1 that were detected in mung bean testa (MBT) groups.[85]

In a related study, Ojwang *et al.*[86] evaluated four colored cowpea types varying in phenol content and flavonoid profile for their anti-inflammatory effects on *in vitro* cells lines. Interestingly, flavonoid profile, but not phenolic content, down regulated pro-inflammatory cytokines (Il-8, TNF-α, VCAM-1), transcription factor NF-κB and modulated microRNA-126 (specific post-transcriptional regulator of VCAM-1). Agah *et al.*[87] studied the interactive effect of combining apigenin and quercetin, with flavone-dominated white sorghum and flavonol-dominated white cowpea extracts, against LPS-induced NF-κB and downstream cytokines (TNF-α, IL-6, IL-8) gene and protein expression. The extracts exhibited a synergistically downregulated LPS-induced NF-κB gene and protein expression in a dose-dependent manner. Nair *et al.*[88] studied the role of quercetin glycosides as an inhibitor of COX-2 activity and as an antioxidant. This study also showed the ability of anthocyanins to suppress the nuclear

NF-κB translocation and the protein expression of iNOS as well as NO production by inhibiting signaling factors, through a non-ROS suppression mechanism possibly downstream of ROS in macrophage cells.

13.2.4 Proteins and Peptides

Legumes are an important source of proteins from food, a serving of half a cup of cooked seeds contains 7–8 g of protein (20–38% dry weight).[89] Biological activities of proteins and peptides from legume seeds have been observed.[90] Lunasin, a peptide derived from soybean (*Glycine max*) that is commercialized as an ingredient of soy drinks and as a dietary supplement in capsules or powder, has been explored with regard to the health benefits of its anticancer, anti-inflammatory, antioxidant, and cholesterol lowering activities.[91] Lunasin is a 43-amino-acid-long peptide sequence that inhibits NF-κB activation, and thus pro-inflammatory cytokine and mediator production, in macrophages and significantly inhibited cell proliferation of cancerous cells.[92] Different studies investigated the mechanism by which lunasin potently inhibits lipopolysaccharide (LPS)-induced production of the pro-inflammatory mediators IL-6, TNF-α, and PGE2 in RAW 264.7 cells through suppression of NF-κB pathways.[93,94] Dia *et al.*[95] reported that lunasin exerts its anti-inflammatory activity through modulation of the COX-2/PGE2 and iNOS/NO pathways. Germination has proven to be an effective process to increase the inhibiting pro-inflammatory markers iNOS, PGE2, COX-2 and TNF-α in LPS-stimulated RAW 264.7 macrophages.[96] In similar studies hydrolysates of the common bean (*Phaseolus vulgaris*) inhibited inflammation in lipopolysaccharide-stimulated macrophages through the transactivation of NF-κB, the nuclear translocation of the NF-κB p65 subunit[97] and inhibition of NO production even after simulated gastrointestinal digestion of germinated extracts.[98]

13.2.5 Saponins

Soyasaponins are found in soybeans (*Glycine max*) and other legumes, such as common beans (*Phaseolus vulgaris*), adzuki bean (*Vigna angularis*), green peas (*Pisum sativum* L.) and lentils (*Lens culinaris*). Crude extracts of soyasaponins from soybean significantly inhibited the production of the pro-inflammatory cytokines (TNF-α and MCP-1, PGE2 and NO), the inflammatory enzymes (COX-2 and iNOS) and the degradation of IκB-α (an inhibitor of NF-κB, in LPS-stimulated macrophages).[99]

Isolated soyasaponins were confirmed to contribute to these anti-inflammatory properties. Lan *et al.*[100] investigated the anti-inflammatory activities of isolated soyasaponins by an assay of NO release using LPS-stimulated RAW264.7 cells. In a similar study, Zha *et al.*[101] investigated the inhibitory effects of soyasaponins on the induction of NO and iNOS in murine RAW 264.7 cells activated with LPS. The soyasaponins suppressed both the iNOS enzyme activity and down-regulated iNOS mRNA expression in a dose-dependent manner. Moreover, an *in vitro* study showed the ability of soyasaponin Ab to inhibit the production of PGE2 and NO and the expression of TNF-α and IL-1β. In addition, an animal study with oral administration of soyasaponin Ab prevented the binding of LPS and increased the expression of IL-10, an immunosuppressive and anti-inflammatory cytokine that inhibits NF-κB activation and consequently suppresses the production of pro-inflammatory cytokines.[102] In the same study, *in vivo* results showed soyasaponin Ab also inhibited the expression of TLR4 and the phosphorylation of IRAK1, the inhibitor of NF-κB kinase (IKK)-β, and the NF-κB subunit p65 in the colon of TNBS (2,4,6-trinitrobenzenesulfonic acid)-induced colitic mice. Zha *et al.*[101] investigated the *in vitro* effect of soyasaponins A1, A2, and I to inhibit iNOS enzyme activity and NF-κB transcriptional activity in LPS-stimulated murine RAW 264.7 macrophages. In the same study, soyasapogenols A and B exhibited no anti-inflammatory activities, indicating that the sugar chains in the structures of soyasaponins are critical. The oral administration of soyasaponins Ab and I also suppressed the release of the proinflammatory cytokine interleukin IL-1β in LPS-stimulated mouse peritoneal macrophages and potently ameliorated body weight reduction, colon shortening, macroscopic score, and myeloperoxidase activity in the colon of TNBS-induced colitic mice.[103] In addition, soyasapogenol I inhibited NO production by downregulating the expression levels of iNOS and COX-2 enzymes *via* the TNF-α and IL-1β cytokines.[104] However, another study showed that soyasaponin III had a stronger anticomplementary activity than did soyasaponin I, by inhibiting NF-κB activation and consequently diminishing its ability to scavenge the lipid peroxides produced by induced colitis.[105] Further studies demonstrated that soy bean and mung bean soyasaponin extracts were found to block the G1-to-S phase cell cycle transition and inhibit the proliferation of Th cells, leading to the decreased mRNA transcription and secretion of cytokine IL-2 and IFN-γ in Th cells, excessive activation of which occurs during severe inflammation.[106] Finally, Jiang *et al.*[107] reported that adzukisaponins III and VI from adzuki bean (*Vigna angularis*) exhibited inhibitory effects on NO production in LPS-activated RAW264.7 macrophages.

13.3 Future Trends in Legume Phytonutrients with Anti-inflammatory Activity

In general, dietary polyphenols display poor *in vivo* bioavailability, mainly due to their limited absorption and rapid metabolism in the gut. Most of the anti-inflammatory phytonutrients are absorbed in the gastrointestinal tract when the integrity of the intestinal mucosa and its barrier function are severely impaired, so that phytonutrients might more easily interact with attacking inflammatory agents. For this reason, advances in scientific technologies are currently being directed towards target-based delivery products. These technologies are used to target active molecules to specific sites of action, to initiate controlled release, and to increase the half-time and reduce the amount of active component in order to minimize their side effects. One of the major challenges in implementing the use of phytonutrients as co-therapies for different diseases is the accurate formulation of each active ingredient, due to their unique and complex chemical nature.

The encapsulation of plant extracts in nanoparticles (embedded, adsorbed or bonded) represents an excellent approach for controlled release of active molecules, thereby decreasing the number of doses and toxic effects that occur with conventional delivery. Nanoparticles provide one of the most attractive methods of active compounds delivery due to their wide range of applications for example with polymers, lipids and metal.[108] Encapsulation is an ingenious delivery system that enhances the phytochemicals poor solubility, low bioavailability, and poor stability, and masks any bitter taste and unpleasant odor. Furthermore, encapsulation could also decrease systemic toxicity, and protect unstable compounds from degradation in the gastrointestinal tract, where particle size and surface charge, factors that most influence their bioactivity, might be adversely affected. Badri *et al.*[109] describe different techniques that have been used for the encapsulation of anti-inflammatory active ingredients using polymer and lipid-based carries. Different methods such as spray drying, spray chilling, melt extrusion, melt injection, fluid bed coating, emulsification, liposome entrapment and inclusion complexation can be used to performed the encapsulation process.

Stability of the phytonutrients in the gastrointestinal tract (bioaccessibility) and the limited delivery of these into the systemic circulation due to the presence of epithelial barriers (bioavailability) are the principal challenges to overcome in the design of pharmaceutical or food formulations with incorporated phytonutrients in order to increase their half-life *in vivo* and improve the biological effect.[110] The encapsulation process allows the masking of undesirable flavors of phytonutrients, protects them from the gastrointestinal environment, controls their release, and improves their absorbance. Aditya *et al.*[111] demonstrated that the bioavailability of curcumin, a low soluble phenolic compound, can be increased (>75%) by its incorporation into lipid-based nanoencapsulation systems including nanostructured lipid carriers (NLCs), solid-lipid nanoparticles (SLNs), and lipid nanoemulsions (LNEs). Likewise, Aditya *et al.*[112] used lipid-based nanocarriers for encapsulation of quercetin, a bioactive flavonoid with many health benefits, but low water solubility. Results showed that the bioaccessibility of quercetin in intestinal medium was increased 50% in LNEs, compared with quercetin in the pure form.[113] On the other hand, the water solubility of quercetin was increased 20 000-fold when it was loaded into lipid nanoparticles formulated by polyvinylpyrrolidone (PVP).[114] Stability of (−)-epigallocatechin-3-gallate (EGCG) under simulated intestinal conditions was improved by the encapsulation of ECGC with nanoparticles of food grade chitosan (CS) in combination with caseinophosphopeptides (CPP). Additionally, EGCG-loaded

CS–CPP nanoparticles demonstrated a significant increase in *in vitro* permeability across a Caco2 cell monolayer compared with non-encapsulated EGCG, showing apparent permeation rate (Papp) values of 1.3×10^{-6} cm s^{-1} and 3.5×10^{-7} cm s^{-1}, respectively. With a Papp value > 1×10^{-6} cm s^{-1} EGCG-loaded CS–CPP nanoparticles can be described as having high intestinal permeation.[115] Likewise, Ydjedd *et al.*[116] showed that the microencapsulation of carob (*Ceratonia siliqua* L.) polyphenols protected them from the effect of pH changes and enzymatic activities during digestion, allowing a controlled release and targeted delivery of the encapsulated compound, which in turn, contributed to an increase in its bioaccessibility in the gut. The recovery of the flavonoids apigenin and quercetin-rhamnoside after the intestinal digestion of encapsulated carob extract, was increased by 80% and 95%, respectively compared with non-encapsulated extract. Moreover, the antioxidant potential exhibited by the encapsulated extract was significantly higher than in the non-encapsulated carob extracts. Ruiz-Rico *et al.*[117] studied the bioaccessibility of the folic acid during the simulated digestion of fruits, using mesoporous silica particles (MSPs) as an encapsulation system. The folic acid was protected at the salivary and gastric stages, and vitamin bioaccessibility was enhanced at the intestinal stage.

After overcoming the solubility challenge and resistance to the gastrointestinal tract environment, the bioavailability of the phenolic compounds is a critical step to ensure the expected biological action. Ji *et al.*[118] showed that SLNs developed with soya lecithin and glyceryl monostearate significantly improved the intestinal permeability of curcumin *in situ*, whereas in the *in vivo* pharmacokinetic study, the curcumin in SNLs showed an area under the curve ($AUC_{0\rightarrow t}$) 12.27-fold greater than a curcumin suspension and the effective permeability value was 57.1% higher than with a suspension formulation. Luteolin-loaded nanoliposomes showed about 10-fold higher bioavailability *in vivo* compared with luteolin in its free form.[119] Jia *et al.*[120] describe the encapsulation of phenolic compounds using animal proteins (gelatin, caseins and whey proteins) or plant-derived protein matrices (zein, karifin, soy protein, *etc*). The encapsulation efficiency of phenolic compounds with these proteins ranged from 50% to more than 90%, depending on the encapsulation method and matrix materials, the absorption of phenolic compounds in the intestine was improved and, in most cases, the biological activity of the phenolic compounds in the encapsulation systems was also improved.

Physicochemical properties of peptides and proteins, and therefore their bioactivity, can also be affected under gastrointestinal conditions. In this sense, the use of enzyme inhibitor(s), absorption enhancers and the modification of dosage form through the use of food-grade colloidal systems (*i.e.*, microemulsions, emulsions, nanoemulsions, solid-lipid nanoparticles, multiple emulsions, liposomes, and microgels) seems to be the most popular approach to protect them from enzymes, improve their physical stability, mask bitterness, and improve their bioavailability profiles.[121,122] Chen and Hong[123] describe various techniques used for the encapsulation of proteins (*i.e.*, interfacial polymerization, spray drying, and solvent evaporation) to improve target delivery, loading efficiencies and life span.[124] Additionally, proteins have the capacity to bind to other compounds through crosslinks, forming colloidal particles to encapsulate polyphenolics.[125] Moreover, it has been demonstrated that proteins extracted from soybean, pea, sunflower, rice and wheat seeds can be efficiently used as stabilizers and emulsion coatings by coacervation and encapsulation processes.[126] These findings constitute a very important advance in the development of more stable and efficient pharmaceutical or food formulations.

In addition to the improvement of bioavailability of nanoformulation of phytochemicals as bioactive ingredients, studies regarding the impact of these formulations on anti-inflammatory activity have been undertaken. Nanoformulation-based technologies of polyphenols, carbohydrates and saponins are designed to result in a better intracellular concentration, slow and sustained release of the compounds, as well as an improvement in bioactivity, which optimize the bioefficacy of these natural compounds. As previously stated polyphenols possess a strong free radical scavenging action that is involved in their anti-inflammatory action. However, the efficacy of these bioactive molecules depends on their ability to maintain their structural integrity. Nanodelivery systems protect these molecules inside the nanoparticulate core, thus preventing degradation and maintaining their bioactivity. Quercetin, a potent inflammatory has been produced in a nanocapsule loaded with polylactic-*co*-glycolide (PLGA) that has increased antioxidant action, resulting in a more effective anti-inflammatory therapy.[127] In another study, a synthesized quercetin-loaded nanoparticle with a compound loading of around 99% demonstrated an enhanced antioxidant activity compared to quercetin.[128] The potency of orally administrated quercetin in a rat model was also demonstrated, in which the nanoparticle prevented 90% of the induced inflammation as compared to 20% of the free quercetin.[129] A study with synthesized ellagic acid-loaded PLGA nanoparticles showed controlled release, improving the cytotoxic activity and *in situ* intestinal permeability of these nanoparticles.[130]

In the last decades, there has been considerable interest in curcuminoids, among which the most abundant is curcumin (90%), studied for its antioxidant, anti-inflammatory, anticancer, antiviral, and antifungal properties.[131] Recent *in vitro* studies have suggested that the anti-inflammatory action of curcumin may be linked to its ability to limit the production of inflammatory mediators.[132] The *in vivo* topical application of curcumin nanoparticle was efficacious in slowing osteoarthritis progression and in pain relief.[133] An increment in curcumin anti-inflammatory activity was demonstrated when formulated in oil in water (o/w) nanoemulsions compared to untreated curcumin.[134]

Many plant polysaccharides have a strong anti-inflammatory action, with promising activity shown in the skin of mice subjected to croton oil induced inflammation.[135] The most recent studies on nanodelivery systems to improve anti-inflammatory carbohydrates deal with extracts of *Aloe vera*.[136]

Liposomes containing *Aloe vera* gel having intrinsic anti-inflammatory action, have been prepared to enhance skin collagen synthesis and growth of skin cell lines.[137] Similarly, several studies demonstrated that terpenoid ingredients can suppress nuclear factor-κB (NF-κB) signaling, the major regulator in the pathogenesis of inflammatory diseases and cancer.[138] Nanoemulsions based on aqueous propolis and lycopene, were tested as protective agents against acute UV-A-induced inflammation on rat paw.[139] Nanoemulsions showed better therapeutic effect against inflammation compared to a standard suspension.

Furthermore, complexes of cyclodextrin compounds have been used to increase concentration in aqueous solution. A method to load lycopene into maltodextrin and cyclodextrin has been recently proposed.[140] The comparison of the anti-inflammatory effect of nanoencapsulated and free lycopene on macrophage cell lines demonstrated that nanoencapsulation of lycopene can further improve its anti-inflammatory effect during tissue-damaging inflammatory conditions. Encapsulation of *p*-cymene in cyclodextrins has been successfully proved to inhibit mice edema *via* oral administration, indicating that nanocarriers can improve *p*-cymene anti-inflammatory effects.[141]

Nowadays, phytochemicals have promising potential to prevent and/or treat inflammatory diseases. However, poor water solubility, stability and bioavailability, together with the side effects reported when used in therapy, have limited their clinical application so far. Nanodelivery system formulations have been employed in the pharmaceutical, cosmetic and food industries to enhance their absorption, and stability, protect them from premature enzymatic degradation or metabolism in the body, and prolong their circulation time, limiting their side effects. Nanotechnology represents a promising approach to providing significant improvements on previously mentioned drawbacks to bringing anti-inflammatory phytochemicals much closer to clinical applications. Additional research is needed to improve nanotoxicological awareness and sensibility, and the cost-effectiveness and long-term safety of nanoparticles delivery systems.

References

1. J. Calixto, M. Campos, M. Otuki and A. Santos, *Planta Med.*, 2004, **70**, 93.
2. J. Calixto, M. Otuki and A. Santos, *Planta Med.*, 2003, **69**, 973.
3. M. O. Freire and T. E. Van Dyke, *Periodontology*, 2013, **63**, 149.
4. M. M. Bosma-den Boer, M. L. van Wetten and L. Pruimboom, *Nutr. Metab.*, 2012, **9**, 32.
5. C. N. Serhan and N. Chiang, *Rheum. Dis. Clin. North Am.*, 2004, **30**, 69.
6. R. Medzhitov, *Nature*, 2008, **454**, 428.
7. C. Nathan and A. Ding, *Cell*, 2010, **140**, 871.
8. J. F. Navarro, C. Mora, M. Muros de Fuentes and J. García, *Nat. Rev. Nephrol.*, 2011, **7**, 327.
9. A. García-Lafuente, E. Guillamón, A. Villares, M. Rostagno and J. Martínez, *Inflammation Res.*, 2009, **58**, 537.

10. P. Haddad, G. Azar, S. Groom and M. Boivin, *Evidence-Based Complementary Altern. Med.*, 2005, **2**, 513.
11. Y. Lei, K. Wang, L. Deng, Y. Chen, E. C. Nice and C. Huang, *Med. Res. Rev.*, 2015, **35**, 306.
12. Y. Kim, J. B. Keogh and P. M. Clifton, *Nutrients*, 2017, **9**, 336.
13. G. Marelli, A. Sica, L. Vannucci and P. Allavena, *Curr. Opin. Pharmacol.*, 2017, **35**, 57.
14. I. B. M. Kolseth, T. M. Reine, K. Parker, A. Sudworth, B. J. Witczak, T. G. Jenssen and S. O. Kolset, *J. Diabetes Complicat.*, 2017, **31**, 245.
15. I. Hameed, S. R. Masoodi, S. A. Mir, M. Nabi, K. Ghazanfar and B. A. Ganai, *World J. Diabetes*, 2015, **6**, 598.
16. T. Kishimoto and T. Tanaka, in: *Encyclopedia of Inflammatory Diseases*, ed. M. J. Parnham, Springer, Basel, 2015, p. 1.
17. S. Skeoch and I. N. Bruce, *Nat. Rev. Rheumatol.*, 2015, **11**, 390.
18. R. Wisastra and F. J. Dekker, *Cancers*, 2014, **6**, 1500–1521.
19. K. Taniguchi and M. Karin, *Nat. Rev. Immunol.*, 2018, **18**, 309–324.
20. F. L. Heppner, R. M. Ransohoff and B. Becher, *Nat. Rev. Neurosci.*, 2015, **16**, 358.
21. K. Angelidou, P. W. Hunt, A. L. Landay, C. C. Wilson, B. Rodriguez, S. G. Deeks, R. J. Bosch and M. M. Lederman, *J. Infect. Dis.*, 2017, **218**, 239–248.
22. J. V. Fahy, *Nat. Rev. Immunol.*, 2015, **15**, 57.
23. D. Jurk, C. Wilson, J. F. Passos, F. Oakley, C. Correia-Melo, L. Greaves, G. Saretzki, C. Fox, C. Lawless, R. Anderson and G. Hewitt, *Nat. Commun.*, 2014, **5**, 4172.
24. D. De Nardo, *Cytokine*, 2015, **74**, 181.
25. S. Tartey and O. Takeuchi, *Int. Rev. Immunol.*, 2017, **36**, 57.
26. F. Armutcu, S. Akyol, S. Ustunsoy and F. F. Turan, *Exp. Ther. Med.*, 2015, **9**, 1582.
27. O. Takeuchi and S. Akira, *Cell*, 2010, **140**, 805.
28. Y. Liang, Y. Zhou and P. Shen, *Cell. Mol. Immunol.*, 2004, **1**, 343.
29. S. Vallabhapurapu and M. Karin, *Ann. Rev. Immunol.*, 2009, **27**, 693.
30. Q. Li and V. M. Verma, *Nat. Rev. Immunol.*, 2002, **2**, 725.
31. J. Dalli, *Mol. Aspects Med.*, 2017, **50**, 12.
32. O. Werz, J. Gerstmeier, S. Libreros, X. De la Rosa, M. Werner, P. C. Norris, N. Chiang and C. N. Serhan, *Nat. Commun.*, 2018, **9**, 59.
33. C. N. Serhan, N. Chiang and J. Dalli, *Semin. Immunol.*, 2015, **27**, 200.
34. C. N. Serhan and N. Chiang, *Curr. Opin. Pharmacol.*, 2013, **13**, 632.
35. C. N. Serhan and J. Savill, *Nat. Immunol.*, 2005, **6**, 1191.
36. C. N. Serha and N. A. Petasis, *Chem. Rev.*, 2011, **111**, 5922.
37. C. A. Dinarello, *Cancer Metastasis Rev.*, 2006, **25**, 307.
38. J. E. Stellar, N. John-Henderson, C. L. Anderson, A. M. Gordon, G. D. McNeil and D. Keltner, *Emotion*, 2015, **15**, 129.
39. S. M. Opal and V. A. Depalo, *Chest*, 2000, **117**, 1162.
40. C. A. Dinarello, *Chest*, 2000, **118**, 503.
41. S. T. Standiford, *Curr. Pharm. Des.*, 2000, **6**, 633.

42. T. Rodney, N. Osier and J. Gill, *Cytokine*, 2018, **110**, 248–256.
43. C. A. Feghali and T. M. Wright, *Front. Biosci.*, 1997, **2**, 12–26.
44. H. Blaser, C. Dostert, T. W. Mak and D. Brenner, *Trends Cell Biol.*, 2016, **26**, 249.
45. Q. Xu, S. Choksi and Z. Liu, *Mol. Cell Oncol.*, 2017, **5**, 1.
46. E. A. Dennis and P. C. Norris, *Nat. Rev. Immunol.*, 2015, **15**, 511.
47. P. C. Norris and E. A. Dennis, *Adv. Biol. Regul.*, 2014, **54**, 99.
48. A. Cesario, B. Rocca and S. Rutella, *Curr. Med. Chem.*, 2011, **18**, 2263.
49. C. H. Ruan, S. P. So and K. H. Ruanb, *Life Sci.*, 2011, **88**, 24.
50. K. Ng, J. A. Meyerhardt, A. T. Chan, K. Sato, J. A. Chan, D. Niedzwiecki and R. Whittom, *J. Natl. Cancer Inst.*, 2015, **107**, 345.
51. K. Brune and P. Patrignani, *J. Pain Res.*, 2015, **8**, 105.
52. P. Rao and E. E. Knaus, *World J. Pharm. Pharm. Sci.*, 2008, **11**, 81s.
53. G. de Gaetano, M. B. Donati and C. Cerletti, *Trends Pharmacol. Sci.*, 2003, **24**, 245.
54. R. C. Lino, D. P. B. da Silva, I. F. Florentino, D. M. da Silva, J. L. R. Martins, D. da Costa Batista, K. C. S. Leite, B. Villavicencio, G. A. Vasconcelos, A. L. P. Silva, R. I. de Ávila, H. Verlid, M. C. Valadares, E. S. Gil, B. G. Vaz, L. M. Lião, R. Menegatti and E. V. Costa, *Eur. J. Pharm. Sci.*, 2017, **106**, 231.
55. J. Martel-Pelletier, D. Lajeunesse, P. Reboul and J. P. Pelletier, *Ann. Rheum. Dis.*, 2003, **62**, 501.
56. J. R. Kingery, T. Hamid, R. K. Lewis, M. A. Ismahil, S. S. Bansal, G. Rokosh, T. M. Townes, S. T. Ildstad, S. P. Jones and S. D. Prabhu, *Basic Res. Cardiol.*, 2017, **112**, 19.
57. M. Hämäläinen, R. Nieminen, P. Vuorela, M. Heinonen and E. Moilanen, *Mediators Inflammation*, 2007, **2007**, 45673.
58. P. Cassini-Vieira, F. A. Araújo, F. L. da Costa Dias, R. C. Russo, S. P. Andrade, M. M. Teixeira and L. S. Barcelos, *Mediators Inflammation*, 2015, **2015**, 138461.
59. J. S. Torrisi, G. E. Hespe, D. A. Cuzzone, I. L. Savetsky, M. D. Nitti, J. C. Gardenier, G. D. García-Nores, D. Jowhar, R. P. Kataru and B. J. Mehrara, *Sci. Rep.*, 2016, **6**, 19817.
60. J. S. Yook, K. A. Kim, M. Kim and Y. S. Cha, *J. Med. Food*, 2017, **20**, 367.
61. S. Bibi, L. F. de Sousa Moraes, N. Lebow and M. J. Zhu, *Nutrients*, 2017, **9**, 509.
62. E. J. Reverri, J. M. Randolph, F. M. Steinberg, C. T. Kappagoda, I. Edirisinghe and B. M. Burton-Freeman, *Nutrients*, 2015, **7**, 6139.
63. N. M. Ali, H. Mohd Yusof, S. K. Yeap, W. Y. Ho, B. K. Beh, K. Long, S. P. Koh, M. P. Abdullah and N. B. Alitheen, *Evidence-Based Complementary Altern. Med.*, 2014, **2014**, 350507.
64. X. Zhang, P. Shang, F. Qin, Q. Zhou, B. Gao, H. Huang, H. Yang, H. Shi and L. L. Yu, *LWT-Food Sci. Technol.*, 2013, **54**, 171.
65. S. Zhu, W. Li, J. Li, A. Jundoria, A. E. Sama and H. Wang, *Evidence-Based Complementary Altern. Med.*, 2012, **2012**, 498467.
66. B. D. Oomah, A. Corbé and P. Balasubramanian, *J. Agric. Food Chem.*, 2010, **58**, 8225.

67. A. García-Lafuente, C. Moro, N. Manchón, A. Gonzalo-Ruiz, A. Villares, E. Guillamón, M. Rostagno and L. Mateo-Vivaracho, *Food Chem.*, 2014, **161**, 216.
68. S. Park, K. C. Choi, M. Fang, Y. C. Lim, Y. M. Jeon and J. C. Lee, *Food Sci. Biotechnol.*, 2011, **20**, 1125.
69. T. Yu, H. M. Ahn, T. Shen, K. Yoon, H. J. Jang, Y. J. Lee, H. M. Yang, J. H. Kim, C. Kim, M. H. Han and S. H. Cha, *J. Ethnopharmacol.*, 2011, **137**, 1197.
70. M. Gabriele, L. Pucci, M. La Marca, D. Lucchesi, C. M. Della Croce, V. Longo and V. Lubrano, *Cell Mol. Biol. Lett.*, 2016, **21**, 10.
71. M. Zia-Ul-Haq, P. Landa, Z. Kutil and M. Qayum, *J. Pharm. Sci. Exp. Pharmacol.*, 2013, **26**, 185.
72. (a) M. Utrilla, M. Peinado, R. Ruiz, A. Rodriguez-Nogales, F. Algieri, M. Rodriguez-Cabezas, A. Clemente, J. Galvez and L. A. Rubio, *Mol. Nutr. Food Res.*, 2015, **59**, 807; (b) S. L. Dwivedi, H. D. Upadhyaya, I. M. Chung, P. De Vita, S. García-Lara, D. Guajardo-Flores, J. A. Gutiérrez-Uribe, S. O. Serna-Saldívar, G. Rajakumar, K. L. Sahrawat and J. Kumar, *Front. Plant Sci.*, 2016, **7**, 763.
73. C. Zhang, J. M. Monk, J. T. Lu, L. Zarepoor, W. Wu, R. Liu, K. P. Pauls, G. A. Wood, L. Robinson, R. Tsao and K. A. Power, *Br. J. Nutr.*, 2014, **111**, 1549.
74. M. R. Moreno-Jiménez, V. Cervantes-Cardoza, J. A. Gallegos-Infante, R. F. González-Laredo, I. Estrella, T. D. J. García-Gasca, E. Herrera-Carrera, J. O. Díaz-Rivas and N. E. Rocha-Guzmán, *Food Res. Int.*, 2015, **76**, 79.
75. K. Pitura, Master Dissertation, University of Manitoba, 2011.
76. W. W. Chao, Y. C. Chung, I. P. Shih, H. Y. Wang, S. T. Chou and C. K. Hsu, *J. Med. Food*, 2015, **18**, 724.
77. J. M. Monk, C. P. Zhang, W. Wu, L. Zarepoor, J. T. Lu, R. Liu, K. P. Pauls, G. A. Wood, R. Tsao, L. E. Robinson and K. A. Power, *J. Nutr. Biochem.*, 2015, **26**, 752.
78. K. C. Choi, S. W. Cho and J. C. Lee, *Food Sci. Biotechnol.*, 2016, **25**, 349.
79. N. Martinez-Micaelo, N. González-Abuín, A. Ardevol, M. Pinen and M. T. Blay, *Biofactors*, 2012, **38**, 257.
80. P. X. Chen, H. Zhang, M. F. Marcone, K. P. Pauls, R. Liu, Y. Tang, B. Zhang, J. B. Renaud and R. Tsao, *J. Funct. Foods*, 2017, **38**, 675.
81. J. M. Monk, D. Lepp, C. P. Zhang, W. Wu, L. Zarepoor, J. T. Lu, K. P. Pauls, R. Tsao, G. A. Wood, L. E. Robinson and K. A. Power, *J. Nutr. Biochem.*, 2016, **28**, 129.
82. I. Turco, G. Ferretti and T. Bacchetti, *J. Food Nutr. Res.*, 2016, **55**, 283.
83. S. Boudjou, B. D. Oomah, F. Zaidi and F. Hosseinian, *Food Chem.*, 2013, **138**, 1543.
84. B. Zhang, Z. Deng, Y. Tang, P. X. Chen, R. Liu, D. D. Ramdath, Q. Liu, M. Hernandez and R. Tsao, *J. Funct. Foods*, 2017, **32**, 248.
85. I. Kang, S. Choi, T. J. Ha, M. Choi, H. R. Wi, B. W. Lee and M. Lee, *J. Med. Food*, 2015, **18**, 841.

86. L. O. Ojwang, N. Banerjee, G. D. Noratto, G. Angel-Morales, T. Hachibamba, J. M. Awika and S. U. Mertens-Talcott, *Food Funct.*, 2015, **6**, 145.
87. S. Agah, H. Kim, S. U. Mertens-Talcott and J. M. Awika, *Mol. Nutr. Food Res.*, 2017, **61**, 1600625.
88. V. Nair, W. Y. Bang, E. Schreckinger, N. Andarwulan and L. Cisneros-Zevallos, *J. Agric. Food Chem.*, 2015, **63**, 6355.
89. C. J. Rebello, F. L. Greenway and J. W. Finley, *J. Agric. Food Chem.*, 2014, **62**, 7029.
90. L. López-Barrios, J. A. Gutiérrez-Uribe and S. O. Serna-Saldívar, *J. Food Sci.*, 2014, **79**, R273.
91. B. Hernández-Ledesma, C. C. Hsieh and B. Ode Lumen, *Protein Pep. Lett.*, 2013, **20**, 424.
92. K. Majumder, Y. Mine and J. Wu, *J. Sci. Food Agric.*, 2016, **96**, 2303.
93. B. Hernández-Ledesma, C. C. Hsieh and O. Ben, *Biochem. Biophys. Res. Commun.*, 2009, **390**, 803.
94. A. M. S. Assreuy, M. D. D. Shibuya, G. J. Martins, M. L. P. De Souza, E. G. de Mejia and V. P. Dia, *Peptides*, 2009, **30**, 2388.
95. V. P. Dia, W. Wang, V. L. Oh, B. O. De Lumen and E. G. De Mejia, *Food Chem.*, 2009, **114**, 108.
96. M. G. Vernaza, V. P. Dia, E. G. de Mejia and Y. K. Chang, *Food Chem.*, 2012, **134**, 2217.
97. M. E. Oseguera-Toledo, E. G. de Mejia, V. P. Dia and S. L. Amaya-Llano, *Food Chem.*, 2011, **127**, 1175.
98. L. López-Barrios, M. Antunes-Ricardo and J. A. Gutiérrez-Uribe, *Food Chem.*, 2016, **203**, 417.
99. J. H. Kang, M. K. Sung, T. Kawada, H. Yoo, Y. K. Kim, J. S. Kim and R. Yu, *Cancer Lett.*, 2005, **230**, 219.
100. X. Lan, K. Deng, J. Zhao, Y. Chen, X. Xin, Y. Liu, A. I. Khan, S. Yang, T. Wang and Q. Xu, *J. Agric. Food Chem.*, 2017, **65**, 11065.
101. L. Y. Zha, L. M. Mao, X. C. Lu, H. Deng, J. F. Ye, X. W. Chu, S. Sun and H. Luo, *Bioorg. Med. Chem. Lett.*, 2011, **21**, 2415.
102. I. A. Lee, Y. J. Park, E. H. Joh and D. H. Kim, *J. Agric. Food Chem.*, 2011, **59**, 13165.
103. I. A. Lee, Y. J. Park, H. K. Yeo, M. J. Han and D. H. Kim, *J. Agric. Food Chem.*, 2010, **58**, 10929.
104. S. H. Yang, B. Le, V. P. Androutsopoulos, C. Tsukamoto, T. S. Shin, A. M. Tsatsakis and G. Chung, *Food Chem. Toxicol.*, 2018, **113**, 211.
105. S. R. Oh, J. Kinjo, Y. Shii, T. Ikeda, T. Nohara, K. S. Ahn, J. H. Kim and H. S. Lee, *Planta Med.*, 2000, **66**, 506.
106. S. J. Lee, J. Bae, S. Kim, S. Jeong, C. Y. Choi, S. P. Choi, H. S. Kim, W. W. Jung, J. Y. Imm, S. H. Kim and T. Chun, *Biotechnol. Lett.*, 2013, **35**, 165.
107. H. Jiang, Y. Tong, D. Yan, S. Jia, C. G. Ostenson and Z. Chen, *Sci. Rep.*, 2015, **5**, 15599.
108. B. Armendáriz-Barragán, N. Zafar, W. Badri, S. A. Galindo-Rodríguez, D. Kabbaj, H. Fessi and A. Elaissari, *Expert Opin. Drug Delivery*, 2016, **13**, 1165.

109. W. Badri, K. Miladi, Q. A. Nazari, H. Greige-Gerges, H. Fessi and A. Elaissari, *Int. J. Pharm.*, 2016, **515**, 757.
110. J. M. Carbonell-Capella, M. Buniowska, F. J. Barba, M. J. Esteve and A. Frígola, *Compr. Rev. Food Sci. Food Saf.*, 2014, **13**, 155.
111. N. Aditya, A. S. Macedo, S. Doktorovova, E. B. Souto, S. Kim, P. A. Chang and S. Ko, *LWT-Food Sci. Technol.*, 2014, **59**, 115.
112. N. Aditya, M. Shim, I. Lee, Y. Lee, M. H. Im and S. Ko, *J. Agric. Food Chem.*, 2013, **61**, 1878.
113. A. F. Esfanjani, E. Assadpour and S. M. Jafari, *Trends Food Sci. Technol.*, 2018, **76**, 56.
114. E. P. Porcu, M. Cossu, G. Rassu, P. Giunchedi, G. Cerri, J. Pourová, I. Najmanová, T. Migkos, V. Pilařová, L. Nováková and P. Mladěnka, *Int. J. Pharm.*, 2018, **541**, 224.
115. B. Hu, Y. Ting, X. Yang, W. Tang, X. Zeng and Q. Huang, *Chem. Commun.*, 2012, **48**, 2421.
116. S. Ydjedd, S. Bouriche, R. López-Nicolás, T. Sánchez-Moya, C. Frontela-Saseta, G. Ros-Berruezo, F. Rezgui, H. Louaileche and D. E. Kati, *J. Agric. Food Chem.*, 2017, **65**, 827.
117. M. Ruiz-Rico, É. Pérez-Esteve, M. J. Lerma-García, M. D. Marcos, R. Martínez-Máñez and J. M. Barat, *Food Chem.*, 2017, **218**, 471.
118. H. Ji, J. Tang, M. Li, J. Ren, N. Zheng and L. Wu, *Drug Delivery*, 2014, **23**, 459.
119. G. Wu, J. Li, J. Yue, S. Zhang and K. Yunusi, *Mol. Med. Rep.*, 2018, **17**, 2456.
120. Z. Jia, M. J. Dumont and V. Orsat, *Food Biosci.*, 2016, **15**, 87.
121. D. J. McClements, *Adv. Colloid Interface Sci.*, 2018, **253**, 1.
122. M. Segura-Campos, L. Chel-Guerrero, D. Betancur-Ancona and V. M. Hernandez-Escalante, *Food Rev. Int.*, 2011, **27**, 213.
123. Y. E. Chen and C. H. I. Hong, *Mater. Sci. Eng. C. Mater. Biol. Appl.*, 2017, **83**, 233.
124. A. Mohan, D. J. McClements and C. C. Udenigwe, *Food Chem.*, 2016, **213**, 143.
125. J. Xue, C. Tan, X. Zhang, B. Feng and S. Xia, *J. Agric. Food Chem.*, 2014, **62**, 4677.
126. J. Quintero, J. Rojas and G. Ciro, *Food Res.*, 2018, **2**, 208.
127. H. Pool, D. Quintanar, J. D. D. Figueroa, M. Marinho, C. Mano, J. E. H. Bechara, L. A. Godinez and S. Mendoza, *J. Nanomater.*, 2012, **12**, 86.
128. T. H. Wu, F. L. Yen, L. T. Lin, T. R. Tsai, C. C. Lin and T. M. Cham, *Int. J. Pharm.*, 2008, **346**, 160.
129. S. Chakrabort, S. Stalin, N. Das, S. T. Choudhury, S. Ghosh and S. Swarnakar, *Biomaterials*, 2012, **33**, 2991.
130. I. Bala, V. Bhardwaj, S. Hariharan, S. V. Kharade, N. Roy and M. N. Ravi Kumar, *J. Drug Targeting*, 2006, **14**, 27.
131. N. Chainani-Wu, *J. Altern. Complementary Med.*, 2003, **9**, 161.
132. M. Mathy-Hartert, I. Jacquemond-Collet, F. Priem, C. Sanchez, C. Lambert and Y. Henrotin, *Inflammation Res.*, 2009, **58**, 899.

133. Z. Zhang, D. J. Leong, L. Xu, Z. He, A. Wang, M. Navati, S. J. Kim, D. M. Hirsh, J. A. Hardin, N. J. Cobelli and J. M. Friedman, *Arthritis Res. Ther.*, 2016, **18**, 128.
134. X. Wang, Y. Jiang, Y. W. Wang, M. T. Huang, C. T. Ho and Q. Huang, *Food Chem.*, 2008, **108**, 4194.
135. S. V. Popov, R. G. Ovodova, G. Y. Popova, I. R. Nikitina and Y. S. Ovodov, *Biochemistry*, 2005, **70**, 10812.
136. A. Surjushe, R. Vasani and D. Saple, *Indian J. Dermatol. Venereol. Leprol.*, 2008, **53**, 163.
137. M. Takahashi, D. Kitamoto, Y. Asikin, K. Takara and K. Wada, *J. Oleo Sci.*, 2009, **58**, 643.
138. A. Salminen, M. Lehtonen, T. Suuronen, K. Kaarniranta and J. Huuskonen, *Cell. Mol. Life Sci.*, 2008, **65**, 2979.
139. M. V. Butnariu and C. V. Giuchici, *J. Nanobiotechnol.*, 2011, **9**, 3.
140. E. Y. Seo, M. H. Kim, W. K. Kim and M. J. Chang, *J. Nutr. Health*, 2015, **48**, 459.
141. L. J. Quintans-Júnior, R. S. S. Barreto, P. P. Menezes, J. R. G. S. Almeida, A. F. S. C. Viana, R. C. M. Oliveira, D. P. Gelain, W. L. Júnior and A. A. S. Araújo, *Basic Clin. Pharmacol. Toxicol.*, 2013, **113**, 167.

CHAPTER 14

Legumes and Obesity

M. F. ZERÓN-RUGERIO[a,b] AND M. IZQUIERDO-PULIDO*[a,b]

[a]Department of Nutrition, Food Science and Gastronomy, School of Pharmacy and Food Science, Barcelona, Spain; [b]Food Safety Research Institute (INSA-UB), University of Barcelona, Av. Prat de la Riba, 171, Edifici de Recerca (Gaudí), Santa Coloma de Gramenet, 08921, Spain
*E-mail: maria_izquierdo@ub.edu

14.1 Introduction

Obesity remains a continuing global health concern as it is has become a global epidemic. Moreover, obesity has been associated with increased risk of numerous non-communicable diseases, such as cardiovascular diseases (CVD), musculoskeletal disorders, type 2 diabetes (T2D) and some types of cancers (including endometrial, ovarian, breast, prostate, liver, gallbladder, kidney, and colon).[1–3] Overweight and obesity are defined as abnormal or excessive fat accumulation that may impair health.[1] Body Mass Index (BMI), the relationship between body weight (kg) and height in square meters (m^2), is the most widely used formula to define overweight (BMI ≥ 25–30) and obesity (BMI ≥ 30).[1,2] At an individual level, any given risk associated with raised BMI can be highly variable, secondary to a patient's gender, age, race or ethnicity, cardiovascular fitness and many other factors.[2]

In 2016, the World Health Organization (WHO) estimated that 39% of adults aged 18 years and over (40% women) were overweight and about 13%

Food Chemistry, Function and Analysis No. 8
Legumes: Nutritional Quality, Processing and Potential Health Benefits
Edited by María Ángeles Martín-Cabrejas

Published by the Royal Society of Chemistry, www.rsc.org

of the world's adult population were obese. In addition, the prevalence of childhood overweight (5–19 years) has risen from 4% in 1975 to over 18% in 2016, and from 1% to 6% respectively in the case of obesity.[1] During childhood and adolescence, obesity is an even more concerning problem, since it increases the chances of suffering from obesity during adulthood, premature death and disability.[1] Furthermore, obese children experience breathing difficulties, increased risk for fractures, hypertension and early markers for CVD, insulin resistance and psychological effects.[1] In the same year 2016, the United Nations (UN) declared the International Year of Pulses, highlighting pulses, edible non-oil seeds legumes (*e.g.* beans, peas, chickpeas and lentils), as a vital source of plant-based proteins and amino acids that should be eaten as part of a healthy diet to address obesity, as well as, to prevent and help manage chronic diseases such as diabetes, coronary conditions and cancer.[3] In addition to being rich in protein, pulses are low-energy-density foods, high in antioxidants and dietary fibre, which can also contribute to obesity management and reduction of its comorbidities.[4]

Systematic reviews and meta-analysis have demonstrated that consumption of legumes produce an overall but significant weight reduction (−0.34 kg) in diets that contained dietary pulses (132 g day^{-1} = 1 serving/day) compared with diets without a dietary pulse intervention.[5,6] Moreover, the consumption of four servings per week of legumes for 8 weeks combined with a low energy diet produced greater weight loss than an energy restricted diet that excluded legumes.[7]

In addition to an elevated BMI, obesity is mainly associated with an increase in white adipose tissue commonly found under the skin and around visceral organs.[8] The increase in size of white adipose cells in obesity leads to disruption of hormones and release of inflammatory cytokines and adipokines, which alters the normal energy homeostasis leading to a wide variety of non-communicable diseases.[8] Additionally, the increase in adipose tissue depots is associated with the magnitude of chronic disease risk; for that reason, attention is now being directed to measures of body composition, mainly fat percentage and distribution (android or gynoid), and body size, primarily waist circumference (when raised above 88 cm for women and 102 cm for men).[2] Regarding the effect of legume consumption and adiposity measures (body fat and waist circumference), it seems legumes also play a role in the shift from android to gynoid adiposity, which will be discussed later. In addition, a recent meta-analysis showed that there is a trend that diets including legumes seem to reduce waist circumference, even though further studies are warranted in order to clarify the legume effect on adiposity measures.[5,6]

Mechanistically, the imbalance between energy intake and expenditure is known to be the fundamental cause of the excessive fat accumulation that characterizes overweight and obesity.[1,2] Thus, it is recommended that obesity by treated by introducing changes in lifestyle such as improving diet quality and increasing physical activity, as well as contributing to both sides of the energy balance: (i) reducing energy intake and (ii) increasing energy expenditure. Furthermore, in order to achieve successful weight-loss and maintenance, it is highly recommended to manage appetite and reduce cravings, while maintaining a healthy diet and lifestyle behaviors.[2,9,10] Although these strategies are useful for short-term weight loss, long-term adherence to energy-reduced eating plans is hard to achieve. In fact, approximately 80% of weight loss interventions are unsuccessful, resulting in regain of weight lost during the trial. This may be due, in part, to hunger and food cravings that compromise the adherence to energy-reduced eating plans.[10] In this regard, studies reveal that participants tend to feel fewer cravings for sweet and fatty foods when consuming a meal containing legumes.[11]

According to obesity guidelines, a weight-loss rate of between 5 and 10% within 6 months is highly recommended in order to produce clinically relevant health improvements, including reductions in plasma triglycerides, blood glucose, and risk for developing T2D. Additionally, larger weight-loss reduces additional risk factors of CVD and decreases the need for medication to control CVD and T2D. Trials have generally found greater weight loss in dietary pulse interventions compared to controls in long-term randomized feeding trials. Thus, it is plausible that the greater effects seen for dietary pulse interventions in negative energy balance trials are due to lessened cravings and/or hunger and thus greater adherence compared to control groups.[10] Hence, according to evidence, legumes constitute a great diet alternative.

A growing body of scientific evidence shows that certain nutrients and bioactive compounds found in legumes (Figure 14.1) may play a key role in improving energy balance by lowering caloric intake and increasing energy expenditure. Plausible mechanisms involved in obesity management by pulses are discussed later in this chapter, the main goal of which it is to explain the role of legumes in the prevention and management of obesity.[3,7,9–12]

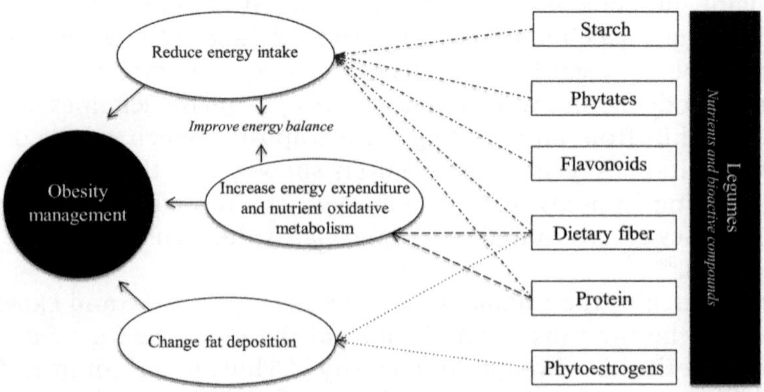

Figure 14.1 Mechanisms for obesity management and legume nutrients and the bioactive compounds involved.

14.2 Pulse Nutrients and Obesity Management

Legumes show high nutritional properties, which have been linked with weight control and obesity management (Figure 14.1). In this section, evidence will be provided regarding nutrients that are predominantly found in legumes (starch, fibre, and protein), their classification (if applicable) and their potential relationship with obesity management, while plausible mechanisms will be described in further sections of the chapter.

Carbohydrates are the main nutritional component of legumes, ranging from between 50 and 65% of their dry weight, starch being starch the main source of carbohydrates, providing approximately 22 to 45% of pulse grain weight. Starch is a polysaccharide that constitutes an important source of energy (glucose) and is also considered important for gastrointestinal function and colonic health.[4,7] Regarding legume starch profile, it is important to highlight two main components amylose and resistant starch (Table 14.1), both of which are key factors in obesity management by reducing energy intake (Figure 14.1).

Dietary fibre is another main component of legumes, providing around 8–30% of their dry weight.[7] Legumes contain a wide range of macromolecules that resist digestion by human endogenous enzymes. They are essentially composed of plant cell wall components, including

Table 14.1 Starch classification and their effects on obesity management.

Starch classification	Effect on obesity management.
Amylose starch	Legumes contain 30–40% of amylose starch, a highly branched polysaccharide slowly digested in the small intestine. Slowly digested carbohydrates take between 20 and 120 minutes to pass through the gastrointestinal tract, resulting in a slow and prolonged release of glucose into the blood stream, coupled to a low glycaemic response. Thus, through mechanisms explained in the text, amylose is known to aid weight loss by prolonging satiety and delaying glucose absorption (Section 14.2).[7]
Resistant starch	Resistant starch is a kind of starch that has the ability to retrograde or re-associate during the cooling process after cooking, making it resistant to digestion in the small intestine. Thus, resistant starch limits calorie intake and bioavailability (Figure 14.1). Moreover, because resistant starch is not digested or absorbed in the small intestine, it is now recognized as a valuable feedstock for the gut microbiota,[7,13] as it is highly fermentable and contributes to short-chain fatty acids (SCFAs) production in the colon, mainly propionate, butyrate and acetate.[7] These SCFAs are known to play a key role in satiety, thus reducing energy intake (Section 14.3.3), as well as in modulation of nutrient metabolism (Section 14.5) and changes in fat deposition (Section 14.6), all of which constitute important aids regarding obesity management.[13,14]

Table 14.2 Dietary fibre and its beneficial effect on obesity management.

Dietary fibre classification	Effect on obesity management
Soluble dietary fibre:[a] pectin, gums, inulin-type and fructans	Soluble dietary fibre shows the capability to form viscous solutions upon dissolution in water, thus they are known as soluble dietary fibre.[7] Due to this ability, this kind of fibre is known to prolong chewing time, allowing satiety sensations to the brain by promoting gastric distension, triggering vagal signals of fullness in the stomach, and slowing the rate of gastric emptying (described in Section 14.2). Additionally, soluble dietary fibre slows intestinal transit and the absorption rate of nutrients. Moreover, exposure of the intestinal mucosa to dietary fibre induces the release of appetite regulating peptides, thus inducing satiety and controlling postprandial glucose and lipid levels.[7,15] Soluble dietary fibre is also easily fermentable, contributing to the production of SCFA.
Insoluble dietary fibre: lignin, cellulose and some hemicelluloses	Insoluble dietary fibre contributes to faecal bulk, stool frequency, and reducing intestinal transit time, thus promoting laxation and decreasing nutrient absorption.[7,16] Additionally, insoluble dietary fibre seems to play a role in changes in fat deposition which will be described further ahead in Section 14.6.
Oligosaccharides: raffinose, verbascose and stachyose	Oligosaccharides are non-digestible and rapidly fermentable compounds, which are considered prebiotics as they can contribute to modulation of the colonic metabolome with increased production of SCFAs. Although oligosaccharides have been linked to flatulence and bloating, which limits legume consumption, evidence suggests that tolerance may be developed.[7]

[a]Resistant starch is also considered a component of dietary fibre.

oligosaccharides, and soluble and insoluble dietary fibre (described in Table 14.2). All of those components have a beneficial effect on weight management, as they are associated with lowering food caloric content and bioavailability as well as trigger satiety signals (Figure 14.1). Additionally, dietary fibre seems to modulate nutrient metabolism through short-chain fatty acids (SCFAs).

Beyond SCFAs production and their role in satiety and the modulation of nutrient metabolism, fermentation of soluble dietary fibre, oligosaccharides, and resistant starch has been shown to modulate the composition of the gut microbiota, decreasing the ratio of *Firmicutes*:*Bacteroidetes*. In this regard, it is well known that an increased ratio of *Firmicutes*:*Bacteroidetes* is associated with obesity. Nevertheless, the mechanisms implicated need to be elucidated. It seems that the increase of *Firmicutes* with respect to *Bacteroidetes* is associated with a higher presence of enzymes for complex carbohydrate degradation and fermentation.[16,17]

Finally, legumes are known as valuable sources of plant-based proteins, an average of 17 to 30% of their dry weight being protein.[3] Historically, legumes have been recommended to prevent or treat malnutrition, primarily under-nutrition. The consumption of legumes in combination with other plant-based proteins (*i.e.* cereals) and/or animal protein sources, provides a high-quality meal or diet.[4] Regarding obesity, legume protein content seems to have an impact on satiety (Section 14.3.3) and energy expenditure (Section 14.3). Plausible mechanisms are further described in the sections mentioned.

14.3 Pulse Consumption and Reduction of Energy Intake

As mentioned above, the role of energy intake reduction in weight management has been widely described. Different approaches have been proposed in this regard, including (i) limiting daily caloric intake (1200 to 1500 kcal per day for women and 1500 to 1800 kcal for men), (ii) restricting the amount of calories per day (500 to 750 kcal) and/or (iii) restricting certain foods (*i.e.* sugars or high fat foods).[9] It is estimated that this dietary approach, alone or with exercise, results in a mean weight loss of approximately 5 to 9% within the first 6 months, while at 24 months, weight loss mean ranges approximately from 2 to 5%. Additionally, very-low-energy diets (≤800 kcal per day) are sometimes recommended, although evidence suggests that while they result in a dramatic weight loss, they are usually followed by a substantial weight regain.[18] As mentioned earlier, weight regain is due to the lack of adherence over time. Thus it is important to take advantage of certain food components (nutrients or bioactive compounds) that have been proven to aid obesity management, as are found in legumes.

In this regard, evidence supports that legumes play a role in reducing energy intake as they have been linked to a lower caloric intake, reduced bioavailability of calories, and satiety induction (Figure 14.1).[5,7,20] These three aspects will be discussed in the following subsections. Firstly, evidence will be presented regarding the effect of pulse consumption on the caloric intake; secondly, we will address the possible mechanisms that can give a reduced bioavailability of calories; and finally, we will focus on the role that legumes play in satiety.

14.3.1 Pulse Consumption and Lower Caloric Intake

Lowering caloric intake is one of the most accepted strategies used for weight control; in this regard, pulse consumption is considered a great strategy for many reasons. First of all, legumes are low energy-dense foods (offering an

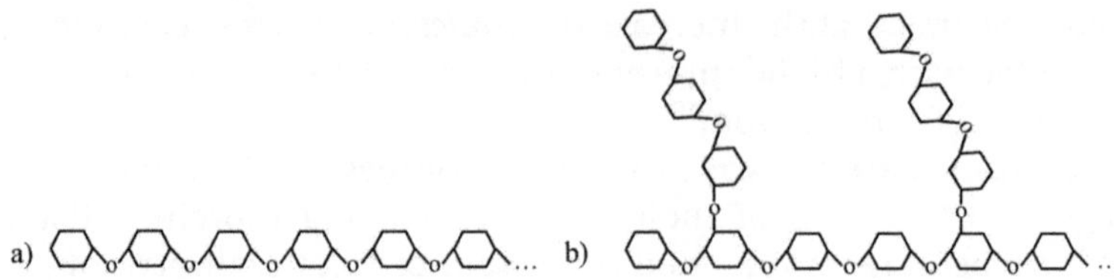

Figure 14.2 Starch chemical structure: amylose (a) and amylopectin (b).

average of 1.3 kcal g^{-1}), since they are high in dietary fibre, which contributes to lowering the energy density and reducing the glycaemic response. On the other hand, most beans are very low in fat, generally containing <5% of energy as fat, with the exception of chickpeas, lupin seeds and soybeans. Moreover, as mentioned above, legumes are also rich in carbohydrates (mainly starch) providing approximately 22–45% of the dry weight of pulses, depending on the source, cooking and processing.[4,7]

Starch is composed of amylose (30–40%) (Figure 14.2a), which is a linear glucan with few branches that tends to be more resistant to digestion, and amylopectin (Figure 14.2b), which is a larger, more digestible, and highly branched molecule. Pulse starches generally have a higher content of amylose compared with cereal and tuber starches. As far as digestibility is concerned, foods containing starches that are high in amylose tend to be more resistant to digestion,[7] which has an impact on the amount of calories provided. In addition to their higher amylose content, the high capacity for retrogradation of legume starch may also reduce the starch digestion rate, rendering them either slowly digestible and/or resistant to digestion.[10] It is well established that cooked legumes are prone to retrogradation or re-associate on cooling more quickly than cereals, making them resistant to digestion; starch that is not hydrolysed or digested is called resistant starch, as has been described above and extensively in Chapter 4. It has been pointed out that legumes are high in resistant starch.[7] Therefore, the carbohydrate profile of legumes is quite peculiar and new scientific evidence suggests that this can play an important role in weight management, as is discussed below.[7]

The rate of digestion determines: (i) how sustained the supply of nutrients will be as digestion continues, (ii) how prolonged its contribution to delaying the urge to eat again will be (satiety will be discussed further later), and (iii) the food caloric content.[7,13] Taking into account all of the above, the postulate that all carbohydrates, simple or complex, equally provide 4 kcal g^{-1} seems no longer to be accepted, since carbohydrates vary greatly in their digestibility and metabolic potential. It is now known that fully fermentable starch provides approximately 2 kcal g^{-1}, whereas glucose provides 4 kcal g^{-1}.[6,13] In conclusion, legumes are a potential aid in weight loss since both the higher proportion of amylose and the starch retrogradation result in legumes providing less energy than would usually be expected according to Atwater factors.[21]

On the other hand, legumes also constitute a great alternative in order to lower the caloric content of a food product while maintaining its rheological properties. Pulse flours are nowadays used as a substitute for meat products (*e.g.* burgers, sausages, chicken nuggets) in order to lower energy density (specially fat and saturated fat content) while providing important nutrients and maintaining an acceptable texture and appearance.[7,22,23] This is why, in recent years, there has been an increasing interest in developing innovative technologies to expand the use of legumes (see Chapter 11). Furthermore, the National Health And Nutrition Examination Survey (NHANES) 2009–2010 showed that pulse consumption by individuals in the United States has been associated with greater nutrient intake, lower body weight, and smaller waist circumference, compared with meat consumers.[23–25]

14.3.2 Pulse Consumption and Reduced Bioavailability of Calories

Besides their low energy density, legumes can also reduce energy bioavailability (Figure 14.1) constituting another important mechanism by which energy balance regulation can be achieved.[2,12,13,19] The reduced bioavailability of calories produced by pulses could be due to different mechanisms that are detailed below:

i. Legumes have intact cell plant walls that might hinder the digestive action of intestinal enzymes, which contributes to lower the energy bioavailability of legumes themselves.
ii. Legumes contain enzyme inhibitors as well as certain compounds that may interfere with the energy availability. Those enzymes and compounds are summarized below and extensively described in Chapter 7 of this book.[6,11]
   - *Several α-amylase inhibitors and other enzyme inhibitors* are found in pulses, which could reduce starch digestibility and thus energy availability. However, these substances may be partially inactivated through cooking or other processing methods.[11]
   - *Certain phenolic compounds* are found in pulses, with darker grains such as lentils, black beans, and red kidney beans generally having higher amounts.[11] Their potential role in weight management is unclear, although studies indicate that specific phenolic compounds could interfere with enterocyte glucose absorption through the blocking of glucose transporters.[11] Additionally, lentils have shown their potential to inhibit α-glucosidase and pancreatic lipases, which leads to decreased glucose and fat digestion and absorption in the intestine.[26]
   - *Phytates* have been shown to reduce the *in vitro* rate of starch digestion and delay postprandial glucose absorption in humans. This last effect could also contribute to satiety and delay the return of hunger, as discussed below.[11]

iii. Legumes are characterized for their high fibre content (8–30%) with soluble fibre in the range of 3.3–13.8% dry weight.[7] The viscosity of soluble fibre traps fat and simple sugars, reducing their absorption and retarding their passage through the gastrointestinal tract.[7,27]

14.3.3 Pulse Consumption and Satiety

The regulation of food intake (hunger and satiation) is a complex system controlled by the hypothalamus and relys on internal systems (homeostatic and hedonic), as well as external cues. The homeostatic regulation of food intake is a complex system (Figure 14.3), in which neural, nutrient, and hormonal signals allow communication between gut, pancreas, liver, adipose tissue, and hypothalamus. To this end, separate groups of chemically distinct neurotransmitters AGRP/NPY (Agouti-related Peptide/Neuropeptide Y) and POMC/CART (Proopiomelanocortin/Cocaine and Amphetamine Regulated Transcript) are sensitive to circulating metabolites and hormones signalling availability of energy, such as leptin, ghrelin, insulin and glucose.[2,9] The role of the hypothalamus is to detect different signals, including neurotransmitters, gut hormones and nutrients, external cues, *etc.*, and translate them into behaviour, in order to make the decision weather to eat or not (hunger/satiety).[2] Although, it is believed that cross-talk does occur between the internal systems, homeostatic and hedonic, little is known about this process.[9] Legumes seem to be linked to this homeostatic and hedonic regulation of food intake, leaning towards inducing satiation, rather than hunger, which is a favourable effect regarding obesity management. In a recent meta-analysis, it has been shown that consumption of dietary pulses might increase acute satiety (+31%) in healthy participants. The authors pointed out that even though this increase in satiety does not manifest in a detectable decrease in second meal food intake, it might explain some of the long-term weight loss benefits associated with dietary pulse consumption by supporting adherence to weight loss diets.[10] Increasing satiety is another strategy commonly used

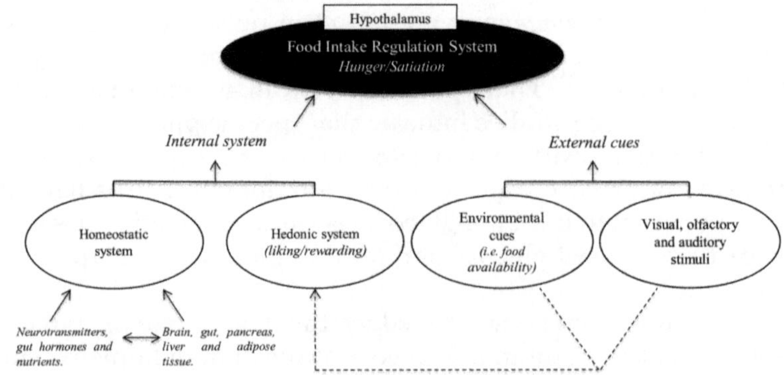

Figure 14.3 Regulation of food intake.

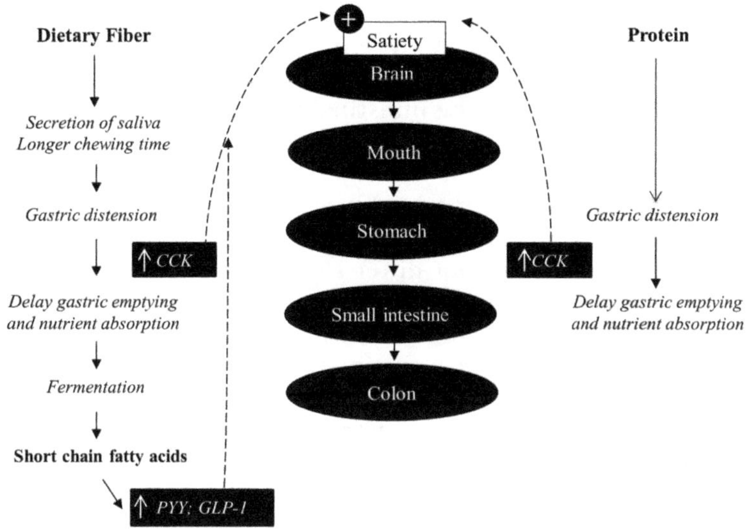

Figure 14.4 Mechanisms by which pulse consumption could modulate satiety.

to lower energy intake and reduce adiposity. Several studies have demonstrated that pulse grains modulate satiety.[20]

Concerning the hedonic system, it seems that the liking and the rewarding qualities of food strongly influence food intake. The "dessert effect" is a great example of the influence of the hedonic system on food intake, because in this kind of situations people continue eating even after reports of satiation.[2] In this regard, it is postulated that the greater effects seen in weight loss interventions where pulses were included are due to lessened cravings and/or hunger, compared to the control groups.[10]

Additionally, several mechanisms involving different nutrients support the potential role of legumes in satiety (Figure 14.4), as they improve the secretion of several gut hormones reflecting the nutritional status of the gut.[2] Nowadays, it is established that a meal containing legumes modulates satiety by increasing cholecystokinin (CCK) and peptide tyrosine tyrosine (PYY) secretion as well as other gut hormones (*i.e.* GLP-1, glucagon-like peptide-1) related to satiety. All of them play a key role in the satiety modulation by interacting with receptors on the vagus nerve, facilitating gastric distension and slowing the rate of gastric emptying.[11,20,28]

It has been shown that dietary fibre, besides limiting nutrient bioavailability, prolongs chewing time and limits caloric intake by stimulating the secretion of saliva and gut hormones that cause stomach distension, thereby promoting satiety (Figure 14.4), in addition viscous dietary fibre delays gastric emptying.[4,15,19] An intervention study carried out with healthy volunteers, which perfectly illustrates the effect of legume fibre on satiety, showed that the intake of lentils along with pasta and sauce reduced food intake, body weight, and waist circumference. The authors hypothesized that the synergistic combination of fibre and flavonoids present in lentils would enhance satiety and lower the amount of food intake.[26] Dietary fibre also retards nutrient passage through the gastrointestinal tract, enhancing the interaction between the intestinal wall and nutrients, to stimulate the release of appetite regulating hormones such as CCK (Figure 14.4).[7,29] A cross-over study, whose aim was to study acute insulin and CCK hormone levels and lipid responses, showed that CCK response was twice as high after a bean-containing meal than after a low dietary fibre meal, while insulin response did not differ significantly between test meals.[30] Furthermore, Reverri *et al.* (2017) evaluated the satiety effects of beans as an intrinsic dietary fibre source in a standardized breakfast meal compared with added fibre. Twelve participants with metabolic syndrome were involved in this study: insulin, gut hormones (CCK and PYY) and subjective satiety were measured 5 hours after the meal. Results showed that CCK and PYY response was higher and postprandial insulin was reduced, both significantly, in the meals containing beans in comparison to the meals with added dietary fibre; however, all meals induce fullness to a similar degree.[29]

The SCFAs derived from resistant starch and oligosaccharide fermentation have been also associated with satiety. Colonic SCFA (butyrate, acetate and propionate) produced by bacterial fermentation may trigger signalling cascades through acting on SCFA receptors, resulting in increased release of gut peptides such as GLP-1 and PYY,[31] two gut hormones mentioned earlier for their role in inducing satiety (Figure 14.4).[20] A randomized cross-over study showed that an evening meal of brown beans lowered blood glucose (−15%) and insulin (−16%) responses, increased satiety hormones (PYY + 51%), suppressed hunger hormones (ghrelin, −14%), and hunger sensations (−15%), and increased GLP-2 concentrations (+8.4%) at a subsequent standardised breakfast. Additionally, propionate (+16%) and isobutyrate (+18%) were significantly increased after brown bean ingestion, indicating a higher colonic fermentative activity after brown beans compared with white wheat.[27]

Furthermore, as shown in Figure 14.4, the satiating effect of legumes could also be attributed to its protein content for two reasons. (i) Proteins have more satiating effect than carbohydrates and lipids, and (ii) proteins promote both gastric distension and CCK secretion, apparently through the same mechanism as fibre.[11,20] Researchers also suggest that trypsin inhibitors found in beans may also stimulate CCK release directly.[20]

Finally, and although glycaemic index (GI) remains a controversial topic regarding obesity, a large number of studies have shown the efficacy of foods with a low glycaemic index in the dietary management and prevention of obesity and its comorbidities.[13] Foods can be classified as high GI, which includes refined-grain products (*e.g.*, white bread or potato), and low GI foods, including whole grain or high fibre products, such as legumes.[13] The GI of legumes ranges from 28 to 48 due to their content of slowly digested starches and protein, as well as their high fibre content.[4,13] Carbohydrates in low GI foods are slowly digested and do not cause large fluctuations in postprandial glycaemia. Therefore, circulating glucose levels are effectively reduced while glucose uptake in the small intestine continues,[13] which could delay hunger and subsequent energy intake, thus participating in the homeostatic regulation of food intake (Figure 14.3).[6] Furthermore, several studies have shown that after a high amylose meal, plasma glucose levels are sustained and insulin requirements are lower, therefore linking amylose to a better energy balance and satiety.[7,13,32] As mentioned above, legumes are richer in amylose than in amylopectin.

Moreover, data from clinical trials suggest that low GI diets based on high amounts of legumes, fruits, vegetables and whole grains, are better than conventional diets for weight loss. Participants have shown a delay in hunger and decreased energy intake, compared with high GI diets. Benefits related to body composition have also been demonstrated since participants seem to have lost less lean mass and showed a reduced decrease in resting metabolic rate.[33] As an added value, legumes are a rich source of potassium, which has also been associated with increased insulin secretion by pancreatic beta cells.[7]

14.4 Pulse Consumption Increases Energy Expenditure and Nutrient Oxidative Metabolism

As previously mentioned, increasing energy expenditure (Figure 14.1) is another strategy used to achieve a negative energy balance and consequently weight-loss. In this regard, the high protein content of pulse grains suggests that they could increase energy expenditure primarily through effects on diet-induced thermogenesis (Figure 14.1).[7,20] Compared to carbohydrates, protein produces the highest thermic effect of food, secondary to the energetic costs associated with dietary peptide catabolism, protein synthesis, and gluconeogenesis.[20] A randomized trial showed that human subjects, consuming legume-rich diets for 8 weeks within a hypocaloric diet exhibited increased mitochondrial oxidation, which could contribute to a greater weight loss compared with controls. In addition, the increases in mitochondrial oxidation were similar to those in individuals consuming diets high in animal-based protein.[33]

Besides the effect of protein on thermogenesis, increasing physical activity is another strategy used in order to increase energy expenditure, and achieve a negative energy balance necessary to lose weight. In this regard, the WHO actually recommends engaging in 150 minutes per week of physical activity for adults and 60 minutes a day for children.[1] Likewise, the American College of Sports Medicine (ACSM) recommends practicing 150 to 250 minutes per week of moderate physical activity, which would result in modest weight loss, corresponding to an increase in weekly energy expenditure from about 1200 to 2000 kcal. In addition, the ACSM states that exercise of 250 to 300 minutes per week can enhance energy expenditure by up to 2000 kcal per week.[34] Nevertheless, energy expenditure is embedded in a much more complex system regulated by the hypothalamus, and is tightly coupled to the control of food intake (Figure 14.3). It is postulated that any increase in energy expenditure will be compensated by an increased in energy intake (hunger),[2] which could represent another plausible mechanism by which weight regain occurs after the first 6 months of treatment. Therefore, it is important to consider other strategies to increase energy expenditure, like increasing mitochondrial biogenesis as well as modulating genes involved in this process. In this regard, it has been suggested that amino acid composition could modulate the increase in energy expenditure.[7] Arginine, a major amino acid in the protein component of pulses,[7] possess thermogenic properties which enhance carbohydrate and fat oxidation *via* increased mitochondrial biogenesis and modulation of genes that regulate energy expenditure.[7,20] Moreover, pulses contain substantial amounts of glutamine,[7] which could improve insulin-mediated glucose disposal during the early postprandial period. Thus, reduced levels of carbohydrate would be available during the later postprandial period and force cells to rely on fat as a source of energy.[20]

On the other hand, evidence is emerging to support that pulse grain can modulate the gastrointestinal microbiome and facilitate the production of SCFA, which alter energy expenditure and nutrient metabolism.[20] Regarding energy expenditure, SCFAs have been shown to increase the rates of oxygen consumption, enhance both adaptive thermogenesis and fat oxidation and increase mitochondrial function in rodents.[35] Furthermore, it appears that SCFA have indirect effects on hormones from pancreas and adipose tissue that are involved in the regulation of energy metabolism.[13]

Concerning the effect of SCFA on nutrient oxidative metabolism, it has been pointed out that:

i. *Butyrate* modulates hepatic carbohydrate metabolism, increasing glycogen storage, and decreasing glycolysis, thereby stimulating hepatic fat oxidation to meet cell energy demands.[7] Additionally, it has been shown *in vitro* that resistant starch predominantly increase the levels of butyric acid.[20] Consistently, it has been demonstrated that meals containing a moderate amount of RS caused an increase in fat oxidation.[20] In addition to an increased fat oxidation, butyrate has been shown to contribute to colonic health by providing energy to the colonic epithelial cells,[14] constituting an internal barrier against the invasion of pathogens, promoting anti-inflammatory effects on colonocytes, and also functioning as a signalling molecule.[7,31]
ii. *Propionate* is absorbed and metabolized aerobically in the liver and may play a role in blood glucose control by suppressing the release of plasma triglycerides, which contributes to insulin resistance.[14] Propionate also seems to increase leptin expression and secretion in an adipose cell culture; however, it failed to show any increase in

adiponectin, an adipokine released from adipose tissue having insulin sensitizing and anti-inflammatory effects.[40,41]

Finally, a recent meta-analysis showed that low doses of isoflavones exhibited a nearly significant effect on BMI. It has been reported that isoflavones, mainly found in soy, interact with oestrogen receptors (ER) in adrenal medullary cells, where they stimulate catecholamine synthesis, which is associated with higher systemic energy expenditure. However the catecholamine-stimulating effect of isoflavones is dose dependent, where low doses stimulate catecholamine secretion, and high doses inhibit it. Nevertheless, more studies need to be performed to confirm this effect.[36]

14.5 Pulse Consumption and Changes in Fat Deposition

Obesity is characterized by an excess of adipose tissue, an organ in charge of accumulating lipids in response to an organism's energy status. During periods of caloric excess, adipose tissue sequesters circulating lipids, which then accumulates mainly as triglycerides in adipocytes. Conversely, during periods of caloric need, adipose tissue mobilizes triglycerides into the circulation to act as an energy source for peripheral tissues.[37] Moreover, excess of waistline adipose deposition, particularly visceral/ectopic fat, constitutes an independent risk factor for CVD, diabetes, and metabolic syndrome. Thus, by shifting fat deposition away from visceral depots into subcutaneous adipose tissue, risks factors related to excess eight could decline before weight loss. Regional fat deposition is in part, regulated by sexual hormones and their affinity for appropriate receptors dispersed throughout visceral and subcutaneous adipose tissue. In this regard, pulse grain-derived phytoestrogens could also contribute to the modulation of adipose deposition *via* their interaction with oestrogen receptors.[20]

A cross-over study conducted on hypercolesterolemic overweight subjects associated legume consumption with an enhanced insulin response and changes in adiposity distribution (android/gynoid). Participants ingested three different types of muffins for 28 days: whole pea flour (WPF), fractioned pea flour (FPF) or white wheat flour (WWF) muffin, followed by a wash out period. Daily intakes of WPF and FPF complied with the United States Department of Agriculture's recommended level of intake of half a cup of pulses/day (approximately 50 g day^{-1}). Findings showed that WPF effectively reduced fasting insulin concentrations and insulin resistance in the study population. The authors hypothesized that the SCFA production secondary to dietary fibre microbial fermentation could be responsible for the increased insulin sensitivity observed.[29] Interestingly, in this same study, women consuming WPF or FPF had a shift in fat distribution from android to gynoid, in comparison with the those consuming only WWF. These results suggested that bioactive compounds, other than dietary fibre, are responsible for the effects observed. WPF contains various proteins and phytochemicals including secoisolariciresinol (13 μg/100 g). It has been postulated that pea derived secoisolariciresinol is converted into enterolactone by colonic microbiota and subsequently absorbed. It has been described that the enterolactone is able to induce oestrogen-dependent reporter gene expression, which is linked to the regulation of regional fat deposition. As a conclusion, the authors of the study concluded that secoisolariciresinol could be considered as a lignin-type phytoestrogen.[39]

Phytoestrogens constitute a group of compounds that, because of their similarity in structure to 17-β-estradiol, are able to bind to the phytoestrogen receptors ERα and ERβ, mimicking oestrogen activity. ERs are expressed in various cell types and organs including adipose tissue, where preadipocytes express ERα, while adipocytes express both isoforms (ERα and ERβ).[41] Phytoestrogens exert their anti-obesogenic effects by decreasing the expression of different transcription factors such as: (i) SREBP-1 (Sterol Regulatory Element-Binding Protein 1), which induces the expression of fatty acid synthesis genes;[41,42] (ii) PPARγ (Peroxisome proliferator-activated receptor γ) essential for the adipogenesis process participating in the maturation of pre-adipocytes to adipocytes, and (iii) C/EBPα (CCAAT/enhancer binding protein α), an early key regulator of adipogenesis, playing an important role in adipocyte differentiation (Figure 14.5).[8]

Soy isoflavones are also known to be phytoestrogens. A recent meta-analysis has highlighted soy isoflavones for their anti-obesity effects in postmenopausal women. Thus, because of the low amount of oestrogen as a result of menopause, it seems that this population benefits from this compounds. Besides the effect of soy isoflavones in adrenal medullary cells, where they stimulate catecholamine synthesis and enhance thermogenesis, they have

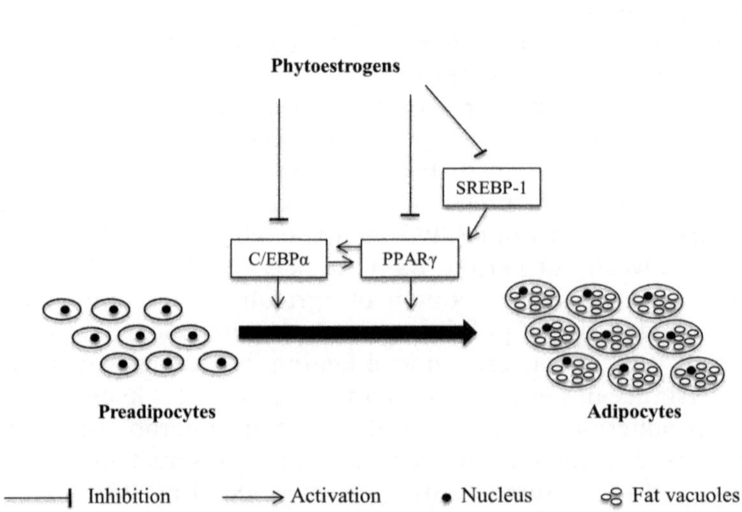

Figure 14.5 Differentiation of preadipocytes to adipocytes and molecular mechanisms.

been recognized for their anti-obesity potential, including the inhibition of adipogenesis and increased oxidation of fatty acids (explained above), leading to the reduction of fat depots. Furthermore, the effect of soy and its isoflavones on weight loss is known to be age and gender dependent. Likewise, soy protein has a gender and dose dependent effect on fat deposition. In trials conducted with women, mostly postmenopausal, reductions in waist circumference were reported after the consumption of <40 g of soy protein per day, whereas ≥40 g day^{-1} of soy protein per day increased weight.[36]

As mentioned above, reduced visceral adipose deposition was accompanied by a decrease in hyperinsulinemia and impaired glucose tolerance, as well as increased glucose disposal. It is important to note that homeostatic control of glucose and lipids is closely related to the balance between ER and the activation of PPARs by isoflavones.[41] This effect may play a role not only in obesity management, but also in the prevention of its comorbidities. As we all know, obesity pathogenesis is closely related to its comorbidities, such as T2D and CVD, especially since some alterations may have a pivotal role in the regulation of intermediate substrate metabolism as well as the regulation of food intake. Fat accumulation, with particular regard to visceral adiposity, independently sustains metabolic abnormalities through chronically altered, pro-inflammatory, and insulin-desensitizing signals. Impairments of insulin effects on target tissues (insulin resistance) may result in reduced glucose uptake and utilization by skeletal muscle, accompanied by alterations in protein turnover and lipid metabolic pathways in various tissues in obese insulin resistant-individuals.[2,38]

Finally insoluble dietary fibre could either directly or indirectly modulate visceral adiposity through the production of microbe-derived bioactive compounds. This type of dietary fibre has also been linked to fat oxidation, as it seems to activate the expression of two genes that facilitate hepatic fat oxidation: *FOXA2* (which encodes the transcription factor forkhead box protein A2) and *PPARGC1B* (which encodes PPARγ coactivator 1-β). In addition, compared to soluble fibre, mice consuming insoluble fibre demonstrated reduced adiposity despite no differences in dietary intake, digested energy or faecal output. Additionally, insoluble fibre attenuated the expression of *GPR40* (which encodes the protein G protein-coupled receptor 40), which is associated with obesity-induced hyperinsulinemia, impaired glucose and hepatic glucose output. The authors postulate that insoluble dietary fibre derived from pulse grains could modulate visceral adiposity *via* microbe-derived bioactives. However, more studies need to be carried out in order to clarify this effect and the plausible mechanisms involved.[20]

14.6 Legumes and the Evolution of Dietary Guidelines and Consumption

Based on NHANES, consuming approximately half a cup of beans results in higher intakes of fibre, protein, folate, zinc, iron, and magnesium with lower intakes of saturated fat and total fat, suggesting that plant-based protein sources improve dietary quality.[25] Additionally, a growing body of evidence demonstrates that legume intake is associated with reduced downstream chronic diseases, such as obesity, and better overall diet quality.[9,18] Owing to these benefits, legume consumption has been recommended as part of a healthy diet by the UN[3] and the WHO,[1] as well as by various international Dietary Guidelines including Canada's Food Guide, the Dietary Guidelines for Americans. Furthermore, in 2015 the Dietary Guidelines Advisory Committee from the Unites States (U.S.), concluded that a diet high in plant-based foods and lower in animal based foods promotes health and is associated with less environmental impact.[25]

In addition to the U.S. Dietary Guidelines, other healthy dietary patterns, such as the Mediterranean Diet and the DASH (Dietary Approach to Stop Hypertension) eating plan, recommend legume consumption as part of a healthy diet.[1,9,10,43] Both, the Mediterranean diet and the DASH eating plan, have been associated with the reduction of risk factors for the majority of non-communicable diseases (including obesity). Additionally, these guidelines also recommend the consumption of fruits, vegetables, nuts, lean meats and low-fat dairy,[9] suggesting that incorporating bioactive ingredients (such as fibre or flavonoids) into the diet, enhancing fatty acid content, or varying the macronutrient composition, are strategies that could help to promote satiety, decrease hunger, and improve insulin response and adherence to the meal plan, making weight loss and weight maintenance easier.[33]

Despite the health benefits of legumes, the UN estimated in 2016 that the per capita consumption of pulses had steadily declined in both developed and developing countries.[44] It was hypothesized that one of the reasons for the low consumption of legumes is the fact that they have been stigmatized as food for low-income groups in some countries, while meat has been associated with wealth and better class.[3,7] This was one of the reasons why the UN *via* the Food and Agriculture Organization had the initiative to call the 2016 the International Year of Pulses, not only because it promotes health and sustainability, but because it was an opportunity to celebrate foods and traditions that are often unnoticed.[3,44]

Besides classism, it has been postulated that legume consumption is low because people lack the abilities or knowledge required for legume preparation, mainly cooking. Thus it is proposed to increase nutritional education, focusing on developing culinary skills, such as food storage, food preparation, grocery shopping and meal planning, as an effective technique to engage people in legume consumption so they incorporate them into their everyday life, especially since they are associated with a lower caloric intake and bioavailability, satiety induction and many other benefits mentioned in this chapter.[4] Additionally, evidence suggests that an association exists between food preparation and a healthy diet in young adults,[45] concluding that those who prepared their own meals ate less fast-food and were more likely to meet dietary objectives for fat, calcium, fruit, vegetable and whole grain consumption.[46]

14.7 Conclusion

Usually, legumes are known for their high-caloric profile, as they are rich in starch and protein, which makes them ideal for undernourished individuals but not obesity management. However, we have provided sufficient information highlighting legumes as an important aid in obesity management, since they have been associated with weight-loss, satiety, lessened cravings and shifts in fat deposition. Legumes have a high content of starch (amylose and resistant starch) and dietary fibre, which reduces the legumes available nutrients and calories, and induces satiety. Additionally, proteins and bioactive compounds found in legumes have also been consistently associated with a decreased availability of calories, satiety, and higher energy expenditure, which would also have an impact on weight management and obesity. Finally, it is important to note that legume consumption should be encouraged for weight loss as part of a healthy diet.

Acknowledgement

MFZR was supported by a scholarship from the "*Consejo Nacional de Ciencia y Tecnología*" CONACYT from Mexico. MIP belongs to the "*Centro de Investigación Biomédica en Red Fisiopatología de la Obesidad y Nutrición; CIBEROBN*" (Spain).

References

1. World Health Organization, http://www.who.int/mediacentre/factsheets/fs311/en, (accessed February 2018).
2. K. M. Gadde, C. K. Martin, H. Berthoud and S. B. Heymsfield, *J. Am. Coll. Cardiol.*, 2018, **71**(1), 69.
3. Food and Agriculture Organization of the United Nations, http://www.fao.org/pulses-2016/about/en/, (accessed Februrary 2018).
4. S. Havemeier, J. Erickson and J. Slavin, *Ann. N. Y. Acad. Sci.*, 2017, **1392**(1), 58.
5. S. J. Kim, R. J. Souza, V. L. Choo, V. Ha, A. I. Cozma, L. Chiavaroli, A. Mirrahimi, S. Blanco, M. Di Buono, A. M. Bernstein, L. A. Leiter, P. M. Kris-Etherton, V. Vuksan, J. Beyene, C. W. Kendall, D. J. Jenkins and J. L. Sievenpiper, *Am. J. Clin. Nutr.*, 2016, **103**(5), 1213.
6. E. Viguiliouk, S. Blanco, C. W. Kendall and J. L. Sievenpiper, *Ann. N. Y. Acad. Sci.*, 2017, **1392**(1), 43.
7. C. J. Rebello, F. L. Greenway and J. W. Finley, *Obes. Rev.*, 2014, **15**(5), 392.
8. D. Moseti, A. Regassa and W. K. Kim, *Int. J. Mol. Sci.*, 2016, **17**(1), 124.
9. H. A. Raynor and C. M. Champagne, *J. Acad. Nutr. Diet.*, 2016, **116**(1), 129.
10. S. S. Li, C. W. Kendall, R. J. de Souza, V. H. Jayalath, A. I. Cozma, V. Ha, A. Mirrahimi, L. Chiavaroli, L. S. Augustin, S. Blanco, L. A. Leiter, J. Beyene, D. J. Jenkins and J. L. Sievenpiper, *Obesity*, 2014, **22**(8), 1773.

11. M. A. McCrory, B. R. Hamaker, J. C. Lovejoy and P. E. Eichelsdoerfer, *Adv. Nutr.*, 2010, **1**(1), 17.
12. J. C. E. Serrano, A. Cassanye and M. Portero-Otin, in *Scientific, Health and Social Aspects of the Food Industry*, ed. B. Valdez, InTech, London, 2012, vol. 14, p. 279.
13. A. Lovegrove, C. H. Edwards, I. de Noni, H. H. Patel, S. N. El, T. Grassby, C. Zielke, M. Ulmius, L. Nilsson, P. J. Butterworth, P. R. Ellis and P. R. Shewry, *Crit. Rev. Food Sci. Nutr.*, 2017, **57**(2), 237.
14. M. Miao, B. Jiang, S. W. Cui, T. Zhang and Z. Jin, *Crit. Rev. Food Sci. Nutr.*, 2015, **55**(12), 1642.
15. M. Kristensen and M. G. Jensen, *Appetite*, 2016, **56**(1), 65.
16. W. J. Dahl, N. C. Agro, A. M. Eliasson, K. L. Mialki, J. D. Olivera, C. T. Rusch and C. N. Young, *J. Am. Coll. Nutr.*, 2017, **36**(2), 127.
17. T. Requena, M. C. Martínez-Cuesta and C. Pelaez, *Food Funct.*, 2018, **9**, 688.
18. M. J. Franz, J. J. VanWormer, A. L. Crain, J. L. Boucher, T. Histon, W. Caplan, J. D. Bowman and N. P. Pronk, *J. Am. Diet. Assoc.*, 2007, **107**(10), 1755.
19. L. Trigueros, S. Peña, A. V. Ugidos, E. Sayas-Barbera, J. A. Perez-Alvarez and E. Sendra, *Crit. Rev. Food Sci. Nutr.*, 2013, **53**(9), 929.
20. C. Marinangeli and P. Jones, *Br. J. Nutr.*, 2012, **108**(S1), S46.
21. M. L. Zou, P. J. Moughan, A. Awati and G. Livesey, *Am. J. Clin. Nutr.*, 2007, **86**(6), 1649.
22. J. Boye, F. Zare and A. Pletch, *Food Res. Int.*, 2010, **43**(2), 414.
23. Y. Wang and M. A. Beydoun, *Int. J. Obes.*, 2009, **33**(6), 621.
24. Y. Papanikolaou and V. L. Fulgoni, *J. Am. Coll. Nutr.*, 2008, **27**(5), 569.
25. D. C. Mitchell, F. R. Lawrence, T. J. Hartman and J. M. Curran, *J. Am. Diet. Assoc.*, 2009, **109**(5), 909.
26. K. Ganesan and B. Xu, *Int. J. Mol. Sci.*, 2017, **18**(11), 2390.
27. J. M. Lattimer and M. D. Haub, *Nutrients*, 2010, **2**(12), 1266.
28. M. K. Sufian, T. Hira, K. Asano and H. Hara, *J. Agric. Food Chem.*, 2007, **55**(22), 8980.
29. E. J. Reverri, J. M. Randolph, C. T. Kappagoda, E. Park, I. Edirisinghe and B. M. Burton-Freeman, *Appetite*, 2017, **118**(1), 75.
30. I. Bourdon, B. Olson, R. Backus, B. D. Richter, P. A. Davis and B. O. Schneeman, *J. Nutr.*, 2001, **131**(5), 1485.
31. A. Nilsson, E. Johansson, L. Ekström and I. Björck, *PLoS One*, 2013, **8**(4), e59985.
32. G. H. Anderson, C. E. Cho, T. Akhavan, R. C. Mollard, B. L. Luhovyy and E. T. Finocchiaro, *Am. J. Clin. Nutr.*, 2010, **91**(4), 932.
33. I. Abete, D. Parra and J. A. Martinez, *J. Med. Food*, 2009, **12**(1), 100.
34. J. E. Donelly, S. N. Blair, J. M. Jakicic, M. M. Manore, J. W. Rankin and B. K. Smith, *Med. Sci. Sports Excercise*, 2009, **41**(2), 459.
35. C. S. Byrne, E. S. Chambers, D. J. Morrison and G. Frost, *Int. J. Obes.*, 2015, **39**(9), 1331.
36. M. Akhlaghi, M. Zare and F. Nouripour, *Adv. Nutr.*, 2017, **8**, 705.

37. P. Tandon, R. Wafer and J. E. N. Minchin, *J. Exp. Biol.*, 2018, **221**(S1), jeb164970.
38. R. Barazzoni, G. Gortan, M. Ragni and E. Nisoli, *Eating Weight Disord.*, 2018, **23**(2), 149.
39. C. P. Marinangeli and P. J. Jones, *Br. J. Nutr.*, 2011, **105**(1), 110.
40. T. Arora, R. Sharma and G. Frost, *Appetite*, 2011, **56**(2), 511.
41. S. Wang, Y. Wang, M. H. Pan and C. T. Ho, *Food Funct.*, 2017, **8**, 3831.
42. P. Ferre and F. Foufelle, *Horm. Res.*, 2007, **68**(2), 72.
43. L. Donini, L. Serra-Majem, M. Bullo, A. Gil and J. Salas-Salvado, *Br. J. Nutr.*, 2015, **113**(S2), S1.
44. Food and Agriculture Organization of the United Nations, http://www.fao.org/pulses-2016/news/news-detail/en/c/381491/, (accessed April 2018).
45. N. Larson, C. L. Perry, M. Story and D. Neumark-Sztainer, *J. Am. Diet. Assoc.*, 2006, **106**(12), 2001.
46. R. Polak, E. M. Phillips and A. Campbell, *Clin. Diabetes*, 2015, **33**(4), 198.

CHAPTER 15

Legumes and Cancer

M. GÓMEZ DE CEDRÓN*[a], A. RAMÍREZ DE MOLINA*[a] AND G. REGLERO[b,c]

[a]Precision Nutrition and Cancer Program, Molecular Oncology and Nutritional Genomics of Cancer Group, IMDEA Food Institute, CEI UAM + CSIC, Madrid, Spain; [b]Food Products for Precision Nutrition, IMDEA Food Institute, CEI UAM + CSIC, Madrid, Spain; [c]Instituto de Investigación en Ciencias de la Alimentación CIAL, CEI UAM + CSIC, Madrid, Spain
*E-mail: marta.gomezdecedron@imdea.org; ana.ramirez@imdea.org

15.1 Introduction

Cancer is the second leading cause of mortality being responsible for nearly one sixth of deaths worldwide. During 2015, there were 17.5 million cancer cases and 8.8 million patient's deaths[1] (data from The World Health Organization; WHO). In the course of this multistage (hyperplastic proliferative stage, dysplastic transformation stage and anaplastic metastatic stage) and multifactorial disease, normal cells acquire capabilities (hallmarks of cancer) enabling them to become tumorigenic and ultimately malignant. In addition, environmental factors play a critical role in cancer. The knowledge of the genetic alterations together with environmental factors that can foster cancer is required to develop integrative strategies to fight this disease. Herein, we first summarize the hallmarks of cancer as stated by Hanahan and Weinberg,[2] and then we highlight the relevance of unhealthy diets as modifying risk factors with impact in the progression and prognosis of cancer. Finally, we describe the nutritional and bioactive compounds found in

Food Chemistry, Function and Analysis No. 8
Legumes: Nutritional Quality, Processing and Potential Health Benefits
Edited by María Ángeles Martín-Cabrejas

Published by the Royal Society of Chemistry, www.rsc.org

legumes that can alleviate chronic diseases related to nutrition such as obesity and associated co-morbidities-type 2 diabetes mellitus (T2DM), metabolic syndrome (MS), cardiovascular discase (CVD), low grade inflammation, gut dysbiosis and cancer.

15.2 Cancer

Cancer is a multistage and multifactorial disease in which normal cells acquire capabilities enabling them to become tumorigenic and ultimately malignant.

15.2.1 Hallmarks of Cancer and Tumor Microenvironment

Development of human tumors is fueled by six hallmarks: sustained proliferative signaling, evasion of growth suppressors, resistance to cell death, replicative immortality, induction of angiogenesis and activation of invasion and metastasis. Underlying these hallmarks are genome instability, which generates the genetic diversity required for the acquisition of new capabilities, and inflammation that further fosters tumor progression. In the past decade, two emerging hallmarks have been added, *i.e.*, reprogramming of the energetic metabolism and evasion of destruction by the immune system. Importantly, cancer cells crosstalk with the extracellular matrix and cells in the adjacent microenvironment to create a permissive and supporting tumor microenvironment (TME).

1. *Sustained proliferative signaling*: different to normal cells, cancer cells present high rates of proliferation which can be achieved by increasing the production of growth factors, growth factor receptors or increased intracellular cell proliferation signaling pathways.
2. *Evasion of growth suppressors*: in response to the excess in cell proliferation, tumor suppressor genes activate intrinsic cellular defense mechanisms such as senescence and/or apoptosis. Thus, cancer cells need to overcome these negative-feedback pathways by the acquisition of new somatic mutations in tumor suppressor genes such as retinoblastoma-associated (RB) and TP53 proteins, which are the most mutated tumor suppressor genes in cancer. RB controls the entry into cell cycle division after integrating signals from diverse extracellular and intracellular sources, and TP53 triggers apoptotic cell death in response to stress and DNA damage.
3. *Resistance to cell death*: the increased replicative stress can lead to excessive DNA damage which can trigger apoptotic cell death. Cancer cells have developed several strategies to limit or circumvent apoptosis by increasing pro-survival Bcl2 family proteins (Bcl-xL, Bcl-w, Mcl-1) *versus* pro-apoptotic proteins (Bax and Bak), or *via* inactivating TP53 tumor suppressor.

4. *Replicative immortality*: most normal cells pass through a limited number of cell division cycles until they enter an irreversible viable but non-proliferative state, and finally a cell death crisis. This is attributed to the progressive shortening of telomeric DNA at the end of chromosomes which protects chromosomes from end to end fusions which could trigger senescence or apoptosis. Eventually, rare variants of cancer cells in crisis become immortalized due to their ability to maintain telomeric DNA length by expressing functionally significant levels of telomerase, the specialized DNA polymerase that adds telomere repeat segments.
5. *Angiogenesis*: tumors require constant flux of nutrients and oxygen to support cell proliferation, and for this the activation of angiogenesis is required. The tumor associated neovasculature is supported by the proliferation of endothelial and associated pericytes. Additionally, several types of bone marrow derived vascular progenitor cells have been found in certain cases.
6. *Activation of invasion and metastasis*: carcinomas arising from epithelial tissues can progress to higher pathological grades of malignancy, reflected in local invasion and distant metastasis. The epithelial–mesenchymal transition (EMT) is a developmental regulatory program by which transformed epithelial cells acquire the ability to invade, to resist apoptosis and to disseminate. In addition, disseminated cancer cells need to develop additional capabilities to colonize the new tissue microenvironment. Colonization does not only depend on cell-autonomous processes, instead, it requires the establishment of a permissive tumor microenvironment composed of critical stromal supporting cells.

The acquisition of all these capabilities depends on a succession of alterations in the genomes of neoplastic cells, which confers selective advantages on subclones of cells. This is made possible by the increased genome instability and mutational rates found in cancer cells. Consequently, defects in genome maintenance and repair are selectively advantageous and therefore instrumental for tumor progression. In addition, infiltration of inflammatory cells of the immune system fosters the development of incipient neoplasias into full-blown cancers.

The chronic and uncontrolled cell proliferation in cancer requires adjustments or reprogramming in the energetic metabolism to fuel cell growth and division. The Warburg effect, by which proliferating cancer cells upregulate glycolysis even in the presence of oxygen, is well known.[3,4] This glycolytic fueling has been shown to be associated with activated oncogenes (*e.g.*, RAS, MYC) and inactivation of tumor suppressors (*e.g.*, TP53).[5,6] Thus, glycolytic intermediates can easily be redirected to the synthesis of macromolecules including nucleotides, glycerol and aminoacids. In addition, the increase in the metabolism of glutamine is crucial to sustain the anaplerosis of intermediates of the tricarboxylic acid (TCA) cycle that can be further redirected to anabolism. In the last years, the importance of the altered lipid metabolism in cancer has been recognized. Lipids regulate very diverse cellular processes such as ATP synthesis, activation of essential cell-signaling pathways and plasmatic and organelles membrane organization and plasticity.[7] In addition, fibroblasts and adipocytes in the tumor microenvironment are a source of lipids for cancer cells influencing their malignancy.[8]

Cells and tissues are constantly monitored by the immune system, which recognizes and eliminates the majority of incipient cancer cells. Thus, solid tumors need to avoid the detection of the immune system. Highly immunogenic cancer cell clones are routinely eliminated in immunocompetent hosts, a process referred as "immunoediting", leaving behind only weakly immunogenic variants to grow and generate solid tumors. In other tumors, immunosuppressive cells, regulatory T cells (Tregs) and myeloid-derived suppressor cells (MDSCs), are recruited.

15.2.1.1 Tumor Microenvironment

Most solid tumors are constituted by an assemblage of distinct cell types that collectively enable tumor growth and progression. Increasing evidence in a variety of tumor types suggests that cells with properties of Cancer Stem Cells (CSCs) are more resistant to various commonly used chemotherapeutic treatments.[9] Prominent among the stromal constituents are the cells forming the tumor-associated vasculature. Inflammatory cells may promote tumors by secreting signaling molecules such as growth factors (EGF), pro-angiogenic growth factors (VEGF and FGF2), inflammatory cytokines or matrix-degrading enzymes (MMP-9, cysteine cathepsin proteases and heparanase). Fibroblasts may contribute to the desmoplastic stroma observed in many advanced carcinomas. Finally, immunosuppressive myeloid cells also support tumor progression and immunoevasion.

The intracellular signaling within the tumor microenvironment is not static but instead changes during tumor progression as a result of reciprocal signaling interactions between cancer cells of the parenchyma and stromal cells that convey the increasingly aggressive phenotypes that underlie growth, invasion, and metastatic dissemination.

15.2.2 Risk Factors in Cancer

In addition to genetic changes, environmental factors also contribute to cancer. Exposure to *carcinogenic* agents such as X-rays, UV light, viruses, pollutants, and many other chemicals increases the risk of development of several types of cancer as well as their prognosis. Additional factors related to life style patterns, such as tobacco smoking, alcohol consumption or high fat diets have been shown to contribute to cancer. The effect of environmental factors is not independent of cancer genes. Sunlight alters tumor suppressor genes in skin cells; cigarette smoke causes changes in lung cells, making them more sensitive to carcinogenic compounds in smoke. These factors probably act directly or indirectly on the genes that are already known to be involved in cancer. Moreover, individual genetic differences also affect the susceptibility of an individual to develop cancer.

Importantly, some cancers associated with environmental factors are preventable. In this sense, a change in lifestyle that minimizes exposure to environmental carcinogens is one effective manner of preventing cancer. Individuals who restrict their exposure to tobacco products, sunlight, and pollution can greatly decrease their risk of developing cancer.

15.2.2.1 Diet and Cancer

There is mounting evidence on the association of unhealthy diets and the increased risk of chronic diseases such as obesity, diabetes, CVD, gut dysbiosis and cancer.[10,11] Importantly, it is estimated that one third of the most common cancers are related to lifestyle and as such they are preventable.[12]

Obesity and the diseases arising from it (T2DM, CVD, MS, and cancer) constitute a major health challenge to the industrialized world. Being overweight or obese is associated with the increased risk of several cancers, including breast, endometrium, esophagus, renal, colon and rectum cancers It is considered that obesity increases the risk of metastasis although the underlying molecular mechanisms are not well known[13] and epidemiological studies link cholesterol levels with the risk of developing different tumors.[14,15] In 2016 the International Agency for Research on Cancer (IARC) committee reviewed more than 190 human meta-analyses together with animal studies regarding the association between obesity and cancer incidence. The committee concluded that excess of body weight was associated with at least 13 different types of cancers, including breast, colon and rectum, endometrial, esophageal, gallbladder, gastric, kidney liver, multiple myeloma, ovary, pancreas, and thyroid.[16] Although the specific mechanisms by which obesity increases the risk of cancer are not entirely clear, several potential mechanisms have been identified. These mechanisms include sex hormones systemic alterations, the insulin pathway and inflammation and immune system alterations.[17,18] All these factors can impact on carcinogenic pathways such as mTOR and PI3K and on the tumor microenvironment. Moreover, the importance of the local effects of fat tissue in tumor development are starting to be recognized. A link between obesity and hormone receptor positive (HR^+) breast cancer risk in postmenopausal women has been shown. The underlying mechanisms are complex but they are related to the increased levels of sex hormones (estradiol, testosterone, and DHEAs) and reduced levels of sex hormone-binding globulin (SHBG). Obesity leads to insulin resistance, where increased levels of insulin/insulin-like growth factors contribute to augmenting the risk of HR^+ breast cancer. Excess of sugars leads to elevated levels of insulin and insulin like growth, factors increasing the risk of colorectal and prostate cancer.[19,20] In addition, obesity has an impact on the levels of adipokines (increased leptin, decreased adiponectin) as well as proinflammatory mediators (TNFα, IL-1, IL-6, PGE2) affecting breast cancer-associated pathways and aromatase expression, which may explain in part the increased risk of HR^+ breast cancer and poorer outcomes in obese breast cancer survivors.

All in all, a proper control of obesity and associated metabolic alterations, such as diabetes, metabolic syndrome, and CDV, may impact on the risk of cancer development and/or prognosis. The benefits of adhering to healthy lifestyles have been quantified specifically in relation to cancer. In 2007, The World Cancer Research Fund (WCRF) and The American Institute for Cancer Research (AICR) reviewed all the relevant research regarding the association between food, nutrition and physical activity, and cancer prevention. Based upon evidence, ten recommendations to reduce the risk of developing cancer were produced,[21,22] among which, one of the most important recommendations was to maintain a healthy weight throughout life. The dietary recommendations are still valid and include consumption of a diet enriched in plant-based foods, and limitation of red and processed meat, salty foods, energy-dense foods and sugary drinks.[23,24]

High saturated fatty acids (FAs), such as those found in animal products and vegetable processed fat (trans double bonds, ω-6 polyunsaturated fatty acids (PUFAs) such as linoleic acid), have been shown to increase the carcinogenic process in preclinical models for breast and prostate cancers. Processed fats also increase the risk of colon cancer in humans by means of mutagenic secondary bile acid secretion. On the other hand, ω-3 PUFAs (α-linolenic acid, eicosapentaenoic acid (EPA) and docosahexaenoic acid (DHA)), inversely correlated with cancer risks.[25] Oleic acid monounsaturated fatty acid (MUFA) in olive oil has been shown to diminish carcinogenesis in breast cancer.[26]

In addition, many foods contain antioxidants and other nutrients that may help to prevent cancer. In this sense, The National Cancer Institute recommends a diet with large amounts of colorful fruits and vegetables. These foods supply ample amounts of vitamins A, C, and E, as well as phytochemicals and other antioxidants that help to prevent cancer. The anti-cancer effects of legumes have been explored extensively, and several studies have associated high intake of legumes with decreased risk of colorectal cancer (CRC).[27] In a case control study, including two small portions of legumes per week protected against different types of cancer, including CRC.[28] A meta-analysis of observational studies suggested an inverse association between legume intake levels and CRC risk.[29] Consumption of processed and red meat has been associated with various types of cancer such as colorectal cancer[30] (Figure 15.1).

15.3 Legumes and Cancer

Traditionally, legumes have been considered as foods with beneficial effects for human health.[31] Legumes are plants of the family *Fabaceae* whose seeds represent a basic pillar of human nutrition since ancient times. The

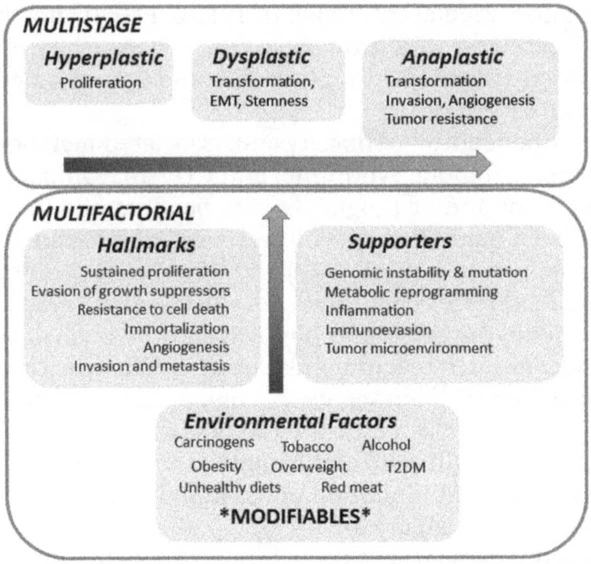

Figure 15.1 Hallmarks of cancer, tissue microenvironment and risk factors in cancer. Cancer is a multistage and multifactorial disease where normal cells acquire capabilities (hallmarks of cancer) enabling them to become tumorigenic and ultimately malignant. In addition, environmental factors play a critical role in cancer. A knowledge of the genetic alterations together with environmental factors that foster cancer is required to develop integrative strategies to fight this disease.

most consumed legumes are beans (*Phaseolus vulgaris*), chickpeas (*Cicer arietinum*), lentils (*Lens culinaris*), peas (*Pisum sativum*), beans (*Vicia faba*), peanuts (*Arachis hypogea*) and soy (*Glycine max*). Legumes present a very adequate nutrient content of carbohydrates, protein, fiber, minerals and vitamins, and thus legumes can be considered as functional foods against obesity and associated co-morbidities such as T2DM, MS, CVD, low grade inflammation, gut dysbiosis and cancer. In this sense, epidemiological studies have established an inverse relationship between legume consumption and the development of several neoplasms such as colorectal cancer, prostate cancer, breast cancer and endometrium cancer,[32–37] a lower incidence of colorectal cancer being found in the countries with the highest consumption of legumes.

In addition, legumes are a source for a variety of compounds considered non-nutrients among which are protease inhibitors (Bowman–Birk protease inhibitors, BBI), lectins, agglutinin, phenolic compounds, phytic acid, phytosterols and saponins, considered the bioactive compounds of these foods and for which antioxidant, hypoglycemic, hypolipidemic and anticarcinogenic properties have been described.[37] All in all, macronutrients, micronutrients and phytochemicals from legumes can contribute to cancer prevention and prognosis (Figure 15.2).

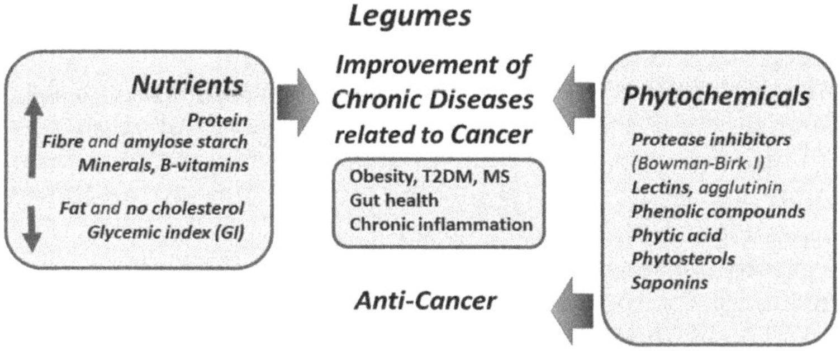

Figure 15.2 Nutritional benefits of legumes. Macronutrients, micronutrients and phytochemicals from legumes contribute to cancer prevention and prognosis.

15.3.1 Nutritional Properties of Legumes: Macronutrients and Vitamins

Legume seeds are recognized to be significant sources of protein in the diet. Protein content in legumes is variable, soybean content being comparable to that of meat. High consumption of proteins present in red meat has been positively associated with various forms of cancer such as breast,[38] ovary,[39] prostate[40] and colon[38] cancers, thus, protein from legumes provides a healthier alternative.

Legumes have low fat and cholesterol content (less than 5%), except soybeans and lupins (30%), making them functional foods in the management of obesity and/or hypercholesterolemic conditions. The predominant fatty acid in beans is linoleic acid (18 : 2) (20–50% of total fats), although they also contain small amounts of α-linolenic acid (3–20% of total fats). Soybeans are quite high in fats, contributing significantly to α-linolenic acid essential for the synthesis of ω-3 polyunsaturated fatty acids (PUFAs), such as EPA and DHA. Among pulses, chickpeas have the highest MUFA content (30%), and kidney beans the highest PUFA content (70%) followed by lentils (50%). Lupins are also high in monounsaturated fat with a lower saturated fat content.[41] Some types of legumes showed particularly interesting values for the ratio of PUFAs *n*-6/*n*-3, such as lentils where it ranged from 42.0% to 57.4%, while in adzuki beans it was 57.5%.[42] In general, although fat content is low, dried legumes contain FAs with health benefits that are favorable in reducing the risk of cancer.[43–46]

Legumes present a high content of dietary fiber (structural carbohydrates) and relatively high amylose starch (non-structural carbohydrates) which resist digestion by human endogenous enzymes. Dietary fiber which is described in detail in Chapter 4, includes water insoluble carbohydrates (cellulose, hemicelluloses, lignin) or water-soluble carbohydrates (pectins, gums, mucilages). Although fermentation of insoluble fiber is low, it contributes to augmenting the fecal bulk. On the other hand, soluble fiber is easily fermented by microflora in the large intestine forming viscous solutions slowing nutrient absorption that contributes to controlling postprandial glucose levels. In addition, legumes are rich in starch such as amylose and amylopectin, which are important sources of energy in human diets. End products of oligosaccharide fermentation are short chain fatty acids (SCFAs) such as acetate, butyrate and propionate. SCFAs have been shown to improve gastrointestinal health as they stimulate the growth of beneficial microbiota, exert anti-inflammatory properties, provide energy to colonic epithelial cells and improve the intestinal barrier against pathogens. Legumes are considered low glycemic index (GI) foods, increasing satiety and stabilizing blood sugar and insulin levels. As high insulin levels increases the risk of colorectal,[47] prostate,[48] and breast[49] cancer, legume consumption may impact on cancer risk by stabilizing insulin levels, and so is highly recommended for weight management and for people with diabetes. In addition, legumes contribute to reducing blood pressure and heart rate among diabetic individuals.[50] Moreover, randomized controlled trials (RCT) have shown that legume consumption lowers total cholesterol and LDL-C (LDL cholesterol) levels[51] which is related, in part, to the capacity of soluble fiber to bind bile acid, decreasing its reabsorption. SCFAs also contribute to decreasing hepatic cholesterol synthesis.[52,53]

Legumes are rich in minerals (iron, magnesium, potassium, phosphorus, zinc) which are essential micronutrients in human health. Iron is crucial for respiration and energy metabolism due to its function as an oxygen carrier and electron transporter. A deficiency of potassium, which is linked to sodium retention, has been associated with glucose intolerance and impaired insulin secretion and detrimental effects on blood pressure. Legumes are also a good source of B vitamins, namely riboflavin, thiamin, niacin, pyridoxine and folic acid, which play important roles in energy metabolism and fatty acid metabolic pathways.[27] Beans are an excellent source of folate. Low folate status has been associated with DNA hypomethylation which is found in many human diseases, including cancer.[54]

15.3.2 Legume Consumption as Part of a Healthy Diet

As indicated previously, compelling evidence from epidemiological and experimental studies supports the relevance of obesity and dietary patterns in cancer development and prognosis. Despite all the advances in cancer diagnosis and in the development of new targeted therapies, still many tumors continue to be untreatable. Importantly, overweight and obesity are largely preventable. Traditional dietary patterns enriched in legumes may represent an important part of the solution for epidemics of obesity, diabetes, and cardiovascular disease. Dietary legumes result in a clinical improvement in blood sugar control, increased satiety and weight loss, as dietary fiber,

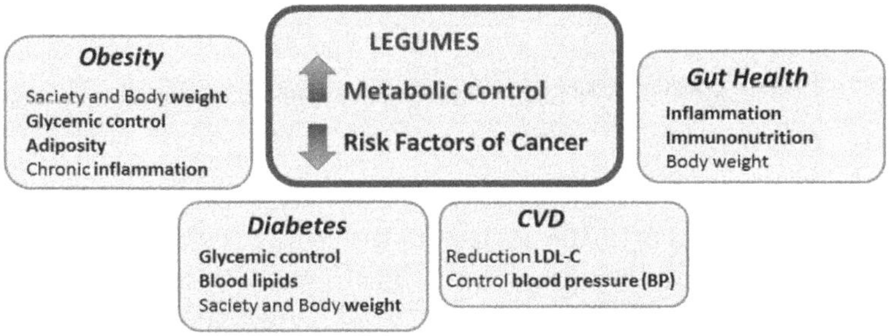

Figure 15.3 Legume consumption as part of a healthy diet. Legume consumption contributes to the management of chronic diseases resulting from unhealthy diets and to reducing the associated risk of cancer development and prognosis.

trypsin inhibitors and lectins contribute to reducing food intake by facilitating and prolonging cholecystokinin secretion. Dietary fiber and resistant starch alter energy expenditure, substrate trafficking and fat oxidation as well as visceral adipose deposition, meanwhile high amounts of insoluble fiber such as those found in pulses improve colon health helping to prevent colorectal cancer (Figure 15.3).

15.3.2.1 Diabetes

Globally, the prevalence of diabetes is increasing due to the increase of body weight and obesity resulting from unhealthy diets and reduction of physical activity. The International Diabetes Federation estimates that the number of people with diabetes in 2035 will rise to 592 M from the 382 M in 2014 (Diabetes Atlas. Sixth Edition. Update 2014. www.diabetesatlas.org). A major goal in the management of diabetes is to achieve optimal glycemic control, and dietary legumes, with high dietary fiber and low glycemic index carbohydrates, decrease the risk for developing type 2 diabetes.[50,55]

A review from several studies analyzing pulse consumption and diabetes outcome, provided evidence for improvements in glycemic control, reduction of blood lipids and body weight. In acute feeding trials with a single pulse serving (1 cup), postprandial blood glucose response is attenuated. Consumption of 5 cups per week improves glycemic control in the long term. Moreover, pulse consumption reduces the risk for CVD associated with hyperlipidemia in persons with type 2 diabetes. Satiety is improved contributing to reduced food intake and body weight, with benefits in diabetes management.[56,57] Researchers from the Unit of Human Nutrition at the Universitat Rovira i Virgili in Tarragona, Spain, together with other investigators from the PREDIMED study, revealed that those individuals with a higher intake of legumes were 35% less likely to develop T2DM than their counterparts who consumed smaller amounts of legumes. Of all the legumes studied, lentils had the strongest association with a low risk of T2DM.[58]

15.3.2.2 CVD

Cardiovascular disease is a growing problem closely associated with diabetes in the developing world.[59] A series of systematic reviews and meta-analyses indicate that dietary legumes significantly reduce LDL-C levels which is a risk factor of CVD.[59] High blood concentration of lipids and associated risk of CVD is a modifiable factor with diet. In a large observational study, consumption of dietary pulses (beans, chickpeas, lentils and peas) correlated with a reduction in CVD.[60] Several clinical trials have shown improvements in the levels of LDL-C.[61] Additional cardiometabolic risk factors, such as body weight, blood pressure and glucose levels are improved in diets with high legume content.[62] In the American College of Cardiology and the American Heart Association guidelines to reduce cardiovascular risk, the benefits of the consumption of legumes, fruits, whole grains, low-fat dairy products, poultry, fish and nuts in the diet have been highlighted.[63]

15.3.2.3 Obesity

The prevalence of obesity worldwide doubled in the last ten years, with an estimated number of overweight and obese adults of 2 billion by 2030. The increase in caloric intake together with a decrease in physical activity are the bases of obesity, which is characterized by hypertrophy and hyperplasia of adipose tissue. Individuals with a body mass index (BMI) between 25 kg m^{-2} and 30 kg m^{-2} are considered overweight and those with BMI >30 kg m^{-2} are considered obese. Nowadays, The WHO indicates that 15% of the overall adult population worldwide are obese.[64] Obesity is often accompanied by dyslipidemia, hypertension and impaired glucose homeostasis, known as metabolic syndrome. In addition, obesity increases the risk of chronic diseases such as CVD, type-2 diabetes[65] and of several types of cancer.[66] In addition, excess visceral fat increases the malignancy of tumors that develop in a microenvironment rich in adipocytes, like breast, ovary, or colon tumors.[67] A large volume of epidemiological studies as well as meta-analyses have demonstrated that total levels of insulin and insulin growth factors together with the systemic inflammation associated with adipose tissue correlate to several types of cancer.[68,69] Importantly, both overweight and obesity are largely preventable by means of diet and educational programs for health promotion and consumption of legumes, and, more specifically, dietary pulses contribute to the control of satiety and food intake, which are the bases of the energetic imbalance in obesity.[70] Dietary pulses have been shown to improved markers of longer term glycemic control in people with and without diabetes, which may contribute to preventing and treating obesity as part of a dietary strategy.[70–72]

15.3.2.4 Gut Health

The host and its associated gut microbiota have co-evolved together resulting in bidirectional crosstalk. Unhealthy diets disrupt the harmonic relationship between them, leading to dysbiosis, which has been shown to impact on obesity, metabolic disease, inflammatory bowel, allergies and cancer.[73–75] The host provides substrates to gut bacteria, resistant starch, dietary fibre, and amino acids, to produce SCFAs. Intestinal bacteria in this process are members of both the *Firmicutes* and the *Bacteroidetes Phyla*. In humans, a higher ratio *Firmicutes*:*Bacteroidetes* is found in obese individuals compared to lean ones,[76] suggesting diet and/or metabolites secreted into the gut in response to these diets (*e.g.*, bile acids) may be implicated. Intestinal microbiota affects host physiology in many ways such as interacting with the immune system, supporting nutrients available to the host, modulating the expression of nutrient transporters at enterocytes and modifying lipid and bile acids metabolism to regulate nutrient absorption by the host.[77,78]

Dietary fibre decreases the energy density of food and limits absorption of amino acids and small peptides in the upper digestive tract.[53] The beneficial effects of resistant starch are mediated by bacterial degradation of non-digestible dietary polysaccharides to monosaccharides and their fermentation to SCFAs. These SCFAs stimulate intestinal gluconeogenesis, which is sensed by the neural system adjacent to the portal vein which signals the brain, modulating the energy and glucose metabolism in the host.[79–81] In addition, SCFAs (acetate, propionate and butyrate) activate free fatty acid receptors (FFAR) 2 and 3 expressed in enterocytes, adipocytes and immune cells.[82] Stimulation of FFARs in adipocytes augments leptin secretion I adipocytes and peptide YY in enteroendocrine cells, resulting in appetite reduction. Moreover, in primary murine colonic cells, acetate and propionate enhance the secretion of glucagon-like peptide 1 (GLP-1) which contributes to improving insulin sensitivity and satiety.[83] Microbiota from obese and diabetic individuals has been shown to induce low-grade inflammation by mean of pro-inflammatory molecules such as LPS and/or peptidoglycans. Glycoprotein cluster of differentiation 14 (CD14) binds to LPS triggering an inflammatory response *via* the Toll-like receptor 4 (TLR4)/myeloid differentiation factor 2 (MD-2) receptor complex.[84,85] In the developed world, poor gut health is associated with increased risk of colon cancer and inflammatory bowel disease. Thus, diets enriched in legumes are a good alternative to stimulate gut health. For example, the non-digestible portion of common beans down-regulates signaling pathways leading to inflammation in a mouse model of colon cancer. When fermented by normal gut flora, the non-digestible fraction from common beans produces SCFAs that inhibit colon cancer HT-29 cell growth. The non-digestible fraction of common beans also protects against chemically induced crypt inflammation in rat colons.[35,86,87] In a clinical trial in humans, brown bean consumption exerted anti-inflammatory effects as shown by a decrease in the levels of cytokines IL-6 and IL-18 (ClinicalTrials.gov NCT01706042). Other studies have shown pulses to decrease markers of inflammation in colorectal cancer and cardiovascular disease.[88–90]

15.3.3 Bioactive Compounds in Legumes

The discovery of new bioactive compounds and the development of functional ingredients is a matter of interest for the prevention and treatment of cancer.[91–93] The anticancer effect of legumes can be attributed to several non-nutrient components considered to be the bioactive compounds such as protease inhibitors (BBI), lectins, phenolic compounds, phytosterols, phytic acid and saponins. The majority of these anti-nutritional compounds are thought to be a protective mechanism against predation from parasites, insects and herbivores. These phytoconstituents have been shown to exhibit antioxidant, hypoglycemic, hypolipidemic and anticarcinogenic properties[94–96] (Figure 15.4).

The protective effect of soy isoflavones on the risk of mammary tumors is known.[97,98] Some, such as genistein, also inhibit tumor progression in other neoplasms such as colon, pancreas[99,100] and prostate.[101] Protein inhibitors are among the protein components of legumes that inhibit the action of proteases such as trypsin, reducing the availability of essential amino acids for tumor cells. They can also act on the proteasome and on proteases

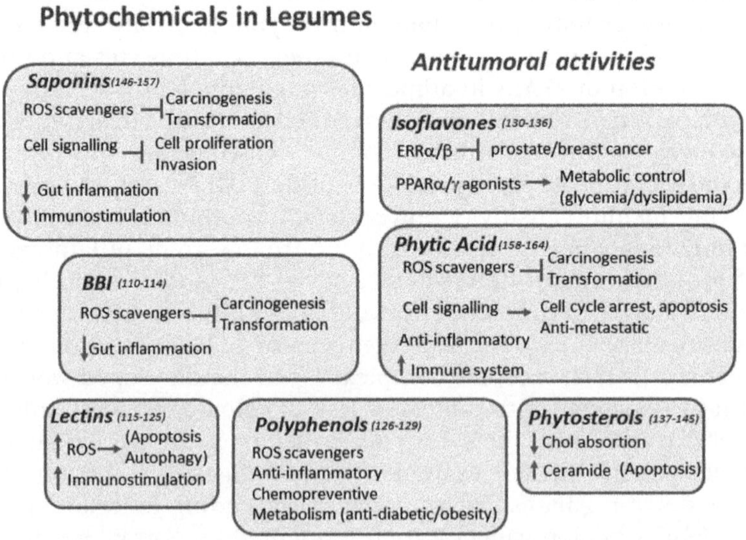

Figure 15.4 Summary of phytochemicals in legumes and associated benefits in cancer treatment. Non-nutrient components of legumes are considered the bioactive compounds, among which are highlighted protease inhibitors (BBI), lectins, phenolic compounds, phytosterols, phytic acid and saponins. These phytoconstituents have been shown to exert antioxidant, hypoglycemic, hypolipidemic and anticarcinogenic properties.

responsible for the degradation of the extracellular matrix, inhibiting the cell cycle and the formation of metastasis.[102–104] Saponins exert antiproliferative and chemo-sensitizing effects on cancer cells.[105,106] Moreover saponins also reduce blood cholesterol levels which could potentiate their antitumoral effects.[107,108] Another compound with an antioxidant effect is the hormone melatonin, whose presence has been detected in legumes such as beans, lentils and soybeans in appreciable quantities, especially in germinated sprouts, and which has also been proposed as an oncostatic agent, especially in mammary tumors.[34,109]

The wide variety of anti-cancer compounds present in legumes, together with synergism among them, make it difficult to describe the specific mechanisms of action. For this reason, the availability of extracts enriched in the different families of components would be of great help for a correct analysis of the potential and the identification of the molecular targets of these vegetables. At present, there are already extracts of legumes approved as safe by the Food and Drug Administration as is the case for white beans.[27]

15.3.3.1 Protease Inhibitors (Bowman–Birk Protease Inhibitors)

Protein hydrolyzation from legumes may produce bioactive peptides such as digestive enzyme inhibitors. Soybean, peas, lentils, beans and chickpeas are particularly rich in protease inhibitors of the BBI family, which are canonical serine protease inhibitors (Chapter 7). Several studies have demonstrated that BBIs are functionally and structurally resistant to the extreme conditions within the GIT (gastrointestinal tract). BBIs from pinto beans have been described to diminish cancer cell proliferation and to reduce the activity of matrix metalloproteases, diminishing tumor dissemination.[110] The therapeutic targets of BBI require further investigation but recent data suggest that BBIs diminish the levels of ROS, which is implicated in malignant transformation associated with DNA damage and lipid peroxidation of cellular membranes. Dysregulation of the epithelial intestinal barrier function plays a central role in intestinal inflammation and cancer development. *In vitro* and *in vivo* studies have demonstrated that BBI exert protective effects on inflammatory processes within the GIT. In this regard, BBI from different legumes have been shown to suppress chemical carcinogen-induced transformation *in vitro* and to inhibit the carcinogenic and inflammatory processes in the GIT in preclinical models: BBI from soybean diminished the frequency of colorectal tumors in rats treated with dimethylhydrazine (DMH). Similar results were reported with DMH-treated mice where BBI administration diminished colon carcinogenesis.[111–114]

15.3.3.2 Lectins (Agglutinins)

Lately, there is an increased interest in lectins as potential cancer therapeutic agents. Although lectins are considered to be toxic as well as anti-nutritional as shown in Chapter 7, many are non-toxic such as those found in beans, lentils and soya. Ingestion of active lectins can potentially cause red blood cell agglutination which may lead to complications such as hemolysis.[115] In recent years, scientific data have demonstrated that lectins can activate the innate immune system, which is the first line of defense against infection and also mediates the immune tolerance to normal bacterial flora. As obesity is characterized by low levels of intestinal barrier breach, stimulation of the innate immune system may contribute to stimulation of barrier function and thus may contribute, at least partially, to controlling obesity induced chronic inflammation.[116–118] Lectins are widely used for diagnostic and therapeutic purposes in cancer. The glycosylation profile of malignant cells is different to that of their normal counterparts[119] and these differences can be specifically detected by lectins. Several studies have associated branched and sialylated *N*-oligosaccharides to the acquisition of metastatic potential by tumor cells. Thus, due to their ability to specifically bind to carbohydrates at the cell surface, lectins can be used for cancer diagnosis and detection of metastasis. In this regard, *Phaseolus vulgaris* agglutinin (PHA) has been shown to discriminate hepatocellular carcinoma from benign liver disease.[120] Much scientific data has demonstrated *in vivo* and *in vitro* antitumor activities of lectins,[121,122] and some of them such as PHA are currently in different phases of clinical trials. Lentil lectins have a strong effect on reducing the onset of human hepatoma (H3B).[123] Lentil agglutinins are also known to have considerable effect on human Merkel skin carcinomas. *Pisum sativum* agglutinin (PSA) has high affinity for tumor cells, which may in turn reduce the tumorigenicity of these cells.

The mode of action of lectins on tumor cells depends on the type of tumor, the source of lectin and its biological activity. Some data suggest that lectins can inhibit cell division and induce both apoptosis and autophagy in cancer cell lines. This is the case for soybean lectins which have been shown to induce apoptosis and autophagy in HeLa cells by upregulating ROS levels.[124] The antitumoral properties of lectins have been related to their ability to modulate the immune system by mean of the regulation of the expression of interleukins and some protein kinases that modulate immune system responses. Lectins from various sources such as PSA are active immunomodulatory agents to stimulate lymphocyte proliferation, natural killer cell activity, antibody synthesis and cytokine regulation.[116] Lectins can improve the effect of chemotherapeutic drugs by inducing the activity of mononuclear cells.[125] More research is needed to characterize the effects of lectins in humans, as well as to determine the underlying molecular mechanisms supporting their therapeutic applications.

15.3.3.3 Phenolic Compounds

Legumes possess a high content of phenolic compounds, with lentils containing the highest amount compared to green peas and chickpeas.[126] Polyphenols are classified based on the number of phenolic rings and their structural elements. Two main groups can be identified based on the aromatic rings linked to the heterocyclic rings: flavonoids (flavones, flavanones, flavanonols, flavanols, catechins and anthocyanins) and the non-flavonoid groups (simple phenols, phenolic acids, hydroxybenzoic acids, tannins, phenylacetic acids; hydroxycinnamic acids, coumarins and stilbenes) as have been described in detail in Chapters 2 and 3. Polyphenols have been extensively described for their health benefits including antioxidant, antibacterial, anti-fungal, antiviral, cardioprotective, anti-inflammatory, antidiabetic, anti-cancer, anti-obesity, hypolipidemic and chemopreventive activities.[127] Many different studies have analyzed the anticancer and chemopreventive effects of polyphenol-rich legumes. In a study conducted with methanolic and aqueous extracts from 11 varieties of Leguminosae family plants, it was found that free radical scavenging activity underlay their chemoprevention activities.[128] In another study, the association between flavonol intake and breast cancer risk was evaluated in 90 630 premenopausal women.[129] Using multivariate relative risks and 95% confidence intervals, among the major food sources of flavonols, a significant inverse association with intake of beans or lentils was found, suggesting a protective role in breast cancer.[129]

Flavonoids represent about 60% of dietary polyphenols and they are classified into seven groups: flavones, flavonols, flavanones, isoflavones, catechins, anthocyanins, and chalcones. Soybeans and other legume seeds (mainly lentils, beans, peas) are the richest sources of isoflavones, including genistein and daidzein. Consumption of soybean, the principal dietary source of isoflavones, has been associated with a lower incidence of prostate and breast cancers in Asian countries.[130,131] The relationship between isoflavone intake and breast cancer has been investigated for many years. Epidemiological data indicates that soy intake is associated with decreased breast cancer risk, and maximum protection is associated with early consumption in life. Nevertheless, there is concern about isoflavone intake in patients with breast cancer based on results from rodent studies, where genistein, the main isoflavone of soybean, increased the growth of human breast cancer cells in a postmenopausal animal model with low plasma estradiol concentrations.[132] However, the clinical and epidemiological data suggest that isoflavone intake does not present any risk to patients with breast cancer.[133] Genistein possesses weak estrogenic activity compared to estradiol, and it can be considered a phytoestrogen. Studies of genistein have focused mainly on hormone-related cancers, such as prostate and breast cancer, but inhibitory effects on other tumor types have also been observed. Despite the variability of their chemical structure, phytoestrogens are structurally similar to natural and synthetic estrogens, including 17-β-estradiol. Phytoestrogens are biologically active compounds playing an important role in prevention of osteoporosis, relief of postmenopausal problems, decrease of body mass and fat tissue and neoplasms.[134,135] Genistein and daidzein can act as PPARα and PPARγ activators which are prescribed for dyslipidemia management and to restore glycemic balance. Genistein has been extensively studied in neoplastic cells where it inhibits cell proliferation, induces apoptosis and displays synergistic effect *in vitro* when combined with chemotherapeutic drugs, such as cisplatin, doxorubicin, bleomycin, docetaxel, and gemcitabine.[131] Moreover, genistein inhibits angiogenesis, limiting tumor growth *in vivo*.[136]

15.3.3.4 Phytosterols

Phytosterols are structurally similar to mammalian cholesterol but with some modifications involving the side chain and a double bond and/or methyl or ethyl group. The most common phytosterols are β-sitosterol, campesterol and stigmasterol. Among legumes, lupins (53.6 mg/100 g) and chickpeas (38.5 mg/100 g) have the highest content of β-sitosterol.[137,138] β-Sitosterol protected from chemically induced methylnitrosourea colon cancer in rats[139] and reduced colonocyte proliferation in rats treated with cholic acid, indicating its protective role against colon hyperplasia.[140] *In vitro* studies have shown antiproliferative effects of β-sitosterol on colon and prostate cancer cell lines. Moreover, in a xenograft mice model, β-sitosterol exerted an inhibitory effect on tumor growth.[141] Phytosterols have been described to diminish the expression of the plasma membrane enzymes, α-reductase and aromatase, modulating testosterone metabolism.[137,141] In the intestine, gut microbiota converts cholesterol and primary bile acids into secondary bile acids and copronastol, which has been suggested to play a role in the development of colon cancer.[137,139] Mechanistically, dietary phytosterols compete with cholesterol in the formation of micelles which diminish its absorption, and β-sitosterol affects the expression of intestinal genes implicated in cholesterol metabolism and transport to the liver. Due to its chemical similarity to cholesterol, most studies have focused on its role in CVD, finding that higher concentrations in plasma are inversely related with coronary heart disease,[142,143] but β-sitosterol exerts an antiproliferative role against several cancer types as well. β-sitosterol inhibits tumor cell proliferation and induces apoptosis in leukemia and colon, breast and prostate which has been associated with the activation of *de novo* ceramide synthesis.[144,145]

15.3.3.5 Saponins

Recently, saponins have attracted considerable interest due to their diverse range of biological and health promoting activities. Consumption of some saponins at high levels has been described to be deleterious to health because of their hemolytic activity and gastric irritant activities, nevertheless, only a few have been shown to be toxic. Saponins are found in lentils, chickpeas, soybeans, beans and peas. Soybeans are particularly high in saponins even after extensive cooking. The chemical structure of saponins is based on a triterpene or steroid aglycone coupled with polar sugar molecules. Saponin activity increases with the increase in lipophilicity.[146] Among the health benefits associated with saponins are a decrease of cholesterol and glucose levels in blood and protection against the risk of cancer development.[147] Moreover, anti-inflammatory, hypocholesterolemic and immune-stimulatory activities of saponins have also been reported.[41] However, more epidemiological and clinical studies are required for proper validation of these health promoting activities.

In vitro studies have indicated a strong cytotoxic effect of saponins against several cancer cell lines such as colon and lung carcinoma cells and leukemia cells.[148–150] Growth inhibition, G2/M cell cycle arrest and apoptotic cell death have been demonstrated in colon cancer cell lines treated with saponins. In addition, saponins diminish colon cancer cell invasion associated with a reduction in proangiogenic and metastatic factors (VEGF, bFGF, metalloproteinases 2 and 9). Saponins from *Astragalus* synergized with chemotherapeutic vinblastine to inhibit tumor growth in colon cancer tumor xenograft.[151] The ability of saponins to inhibit proliferation of cancer cells is mediated by their capacity to interact with free sterols and cholesterol in cell membranes, modulating cell permeability and intracellular oncogenic cell signaling pathways.[152,153] Saponins induce cell differentiation by inhibiting oncogenic protein kinase C signaling.[154] Colonic microflora also play an important role in saponin metabolism. Gut microbiota hydrolyze saponins to sapogenols which are more bioactive and cytotoxic against colon cancer cells.[146] In addition, saponins display antioxidant activities *in vitro* which prevents ROS-mediated cell membrane oxidative damage as well as DNA mutagenesis.[155–157]

15.3.3.6 Phytic Acid

Phytic acid (IP6) is a natural antioxidant present in legumes, cereals and natural oils as shown in Chapter 7. IP6 is considered an antinutritional component due to its ability to chelate with minerals (iron, zinc, and calcium), diminishing their absorption in the GIT. Nevertheless, IP6 has no significant effect on the bioavailability of these minerals in subjects eating balanced diets.[158] Cooking, germination and fermentation of legumes activates phytases which also helps in their breakdown. Nevertheless, total removal of phytates is not desirable as some studies suggest health beneficial effects: IP6 is a source of inositol which diminishes complications in diabetes, inhibits platelet aggregation,[159] controls level of serum lipids[160] and reduces the risk of developing bowel cancer associated with its antioxidant potential.[161] Studies have shown that IP6 decreases the expression of tumor necrosis factor-α (TNF-α), induces cell cycle arrest in G0/G1, activates apoptotic cell death and inhibits the activation of mitogen-activated protein kinases (MAPKs).[162,163] Importantly, IP6 modulates the functions of the immune system by increasing the activity of natural killer cells and decreasing the expression of pro-inflammatory cytokines. IP6 and inositol have been used in a few case studies and small preliminary clinical trials as an adjuvant where an enhanced antitumor activity without compromising the patient's quality of life was demonstrated in patients with advanced colorectal cancer.[164] Moreover, when combined with chemotherapy, IP6 and inositol diminished secondary effects of chemotherapy, such as a decrease in leukocytes and platelets. Additional pilot studies have shown an improvement in long-term survival and quality of life in patients with cancer, encouraging future controlled randomized clinical trials.[164]

15.4 Concluding Remarks

The WHO estimates that up to 80% of heart disease, type 2 diabetes and over a third of cancers could be prevented by eliminating risk factors, such as unhealthy diets. Legumes are part of a healthy and balanced diet with an important role in preventing chronic diseases as evidenced by epidemiological and animal studies. Consumption of one cup (60 grams) of pulses has been recommended by the World Food Program. Both nutrient and non-nutrient components of legumes are implicated in the reduction of cancer risk. There are still challenges to be addressed for the use of bioactive compounds in legumes as a complement to conventional cancer therapy to improve its effectiveness, among them: (1) delving into the molecular mechanisms underlying the biological effects of the non-nutrient components and/or identifying their molecular targets to develop personalized nutrition for patients with cancer; (2) improving the efficient extraction and fractionation of the bioactive components, and (3) developing new formulations, such as those including bioactive lipids, to increase the bioavailability of the bioactive components and to enhance, in a synergistic manner, the antitumoral and beneficial effects of pulses. Multiple chronic diseases are associated with the elderly and it is expected that including legumes as part of a healthy diet may improve expectancy and quality of life.

Acknowledgements

This work has been supported by Ministerio de Economía y Competitividad del Gobierno de España (AGL2016-76736-C3-3-R), Gobierno regional de la Comunidad de Madrid (P2013/ABI-2728, ALIBIRD-CM).

References

1. C. Fitzmaurice, *JAMA Oncol.*, 2017, **3**, 524.
2. D. Hanahan and R. A. Weinberg, *Cell*, 2011, **144**, 646.
3. O. Warburg, *Science*, 1956, **124**, 269.
4. O. Warburg, *Science*, 1956, **123**, 309.
5. R. J. DeBerardinis, J. J. Lum, G. Hatzivassiliou and C. B. Thompson, *Cell Metab.*, 2008, **7**, 11.
6. J. M. Friedman, P. A. Jones and G. Liang, *Cell Cycle*, 2009, **8**, 2313.
7. Z. T. Schug and E. Gottlieb, *Biochim. Biophys. Acta*, 2009, **1788**, 2022.
8. R. Ventura, K. Mordec, J. Waszczuk, Z. Wang, J. Lai, M. Fridlib, D. Buckley, G. Kemble and T. S. Heuer, *EBioMedicine*, 2015, **2**, 808.
9. A. Singh and J. Settleman, *Oncogene*, 2010, **29**, 4741.
10. T. Kurth, S. C. Moore, J. M. Gaziano, C. S. Kase, M. J. Stampfer, K. Berger and J. E. Buring, *Arch. Intern. Med.*, 2006, **166**, 1403.

11. H. Kirkegaard, N. F. Johnsen, J. Christensen, K. Frederiksen, K. Overvad and A. Tjonneland, *BMJ*, 2010, **341**, c5504.
12. M. Wiseman, *Proc. Nutr. Soc.*, 2008, **67**, 253.
13. J. Hu, C. La Vecchia, M. de Groh, E. Negri, H. Morrison and L. Mery, *Ann. Oncol. Off. J. Eur. Soc. Med. Oncol.*, 2012, **23**, 491.
14. C. M. Kitahara, A. Berrington de Gonzalez, N. D. Freedman, R. Huxley, Y. Mok, S. H. Jee and J. M. Samet, *J. Clin. Oncol. Off. J. Am. Soc. Clin. Oncol.*, 2011, **29**, 1592.
15. C. P. Marinangeli and P. J. Jones, *Br. J. Nutr.*, 2012, **108**(Suppl 1), S46.
16. M. Kyrgiou, I. Kalliala, G. Markozannes, M. J. Gunter, E. Paraskevaidis, H. Gabra, P. Martin-Hirsch and K. K. Tsilidis, *BMJ*, 2017, **356**, j477.
17. M. J. Khandekar, P. Cohen and B. M. Spiegelman, *Nat. Rev. Cancer*, 2011, **11**, 886.
18. M. Pollak, *Nat. Rev. Cancer*, 2012, **12**, 159.
19. E. Giovannucci, *J. Nutr.*, 2001, **131**, 3109S.
20. M. Pollak, *Epidemiol. Rev.*, 2001, **23**, 59.
21. R. Thompson, *J. Fam. Health Care*, 2010, **20**, 100.
22. B. Lauby-Secretan, C. Scoccianti, D. Loomis, Y. Grosse, F. Bianchini and K. Straif, *N. Engl. J. Med.*, 2016, **375**, 794.
23. Y. Choi, E. Giovannucci and J. E. Lee, *Br. J. Nutr.*, 2012, **108**, 1934.
24. C. A. Thomson, M. L. McCullough, B. C. Wertheim, R. T. Chlebowski, M. E. Martinez, M. L. Stefanick, T. E. Rohan, J. E. Manson, H. A. Tindle, J. Ockene, M. Z. Vitolins, J. Wactawski-Wende, G. E. Sarto, D. S. Lane and M. L. Neuhouser, *Cancer Prev. Res.*, 2014, **7**, 42–53, CAPR-13-0258.
25. H. Bartsch, J. Nair and R. W. Owen, *Carcinogenesis*, 1999, **20**, 2209.
26. R. W. Owen, R. Haubner, G. Wurtele, E. Hull, B. Spiegelhalder and H. Bartsch, *Eur. J. Cancer Prev.*, 2004, **13**, 319.
27. J. Curran, *Br. J. Nutr.*, 2012, **108**(Suppl 1), S1.
28. D. Aune, E. De Stefani, A. Ronco, P. Boffetta, H. Deneo-Pellegrini, G. Acosta and M. Mendilaharsu, *Cancer Causes Control*, 2009, **20**, 1605.
29. Y. Wang, Z. Wang, L. Fu, Y. Chen and J. Fang, *PLoS One*, 2013, **8**, e67335.
30. A. Tavani, C. La Vecchia, S. Gallus, P. Lagiou, D. Trichopoulos, F. Levi and E. Negri, *Int. J. Cancer*, 2000, **86**, 425.
31. Z. Madar and A. H. Stark, *Br. J. Nutr.*, 2002, **88**(Suppl 3), S287.
32. G. Bobe, K. G. Barrett, R. A. Mentor-Marcel, U. Saffiotti, M. R. Young, N. H. Colburn, P. S. Albert, M. R. Bennink and E. Lanza, *Nutr. Cancer*, 2008, **60**, 373.
33. G. L.-P. B. Rocio Campos-Vega, *Food Res. Int.*, 2009, **43**, 461.
34. M. A. Faris, H. R. Takruri, M. S. Shomaf and Y. K. Bustanji, *Nutr. Res.*, 2009, **29**, 355.
35. V. C. Hayde, G. G. Ramon, G. O. Lorenzo, O. B. Dave, R. C. Rosalia, W. Paul and L. P. Guadalupe, *Br. J. Nutr.*, 2012, **108**(Suppl 1), S145.
36. J. C. Mathers, *Br. J. Nutr.*, 2002, **88**(Suppl 3), S273.
37. X. Sanchez-Chino, C. Jimenez-Martinez, G. Davila-Ortiz, I. Alvarez-Gonzalez and E. Madrigal-Bujaidar, *Nutr. Cancer*, 2015, **67**, 401.
38. J. J. Anderson, N. D. M. Darwis, D. F. Mackay, C. A. Celis-Morales, D. M. Lyall, N. Sattar, J. M. R. Gill and J. P. Pell, *Eur. J. Cancer*, 2017, **90**, 73.

39. T. E. Crane, B. R. Khulpateea, D. S. Alberts, K. Basen-Engquist and C. A. Thomson, *Cancer Epidemiol., Biomarkers Prev. Publ. Am. Assoc. Cancer Res., Cosponsored Am. Soc. Prev. Oncol.*, 2014, **23**, 255.
40. K. M. Wilson, L. A. Mucci, B. F. Drake, M. A. Preston, M. J. Stampfer, E. Giovannucci and A. S. Kibel, *Prev. Res.*, 2016, **9**, 933.
41. A. N. Mudryj, N. Yu and H. M. Aukema, *Appl. Physiol., Nutr., Metab. Physiol. Appl., Nutr. Metab.*, 2014, **39**, 1197.
42. G. Caprioli, F. Giusti, R. Ballini, G. Sagratini, P. Vila-Donat, S. Vittori and D. Fiorini, *Food Chem.*, 2016, **192**, 965.
43. A. P. Simopoulos, *Forum Nutr.*, 2003, **56**, 67.
44. J. X. Kang, *World Rev. Nutr. Diet.*, 2005, **95**, 93.
45. S. H. Gottlieb, *Diabetes Forecast*, 2004, **57**, 28.
46. V. Messina, *Am. J. Clin. Nutr.*, 2014, **100**(Suppl 1), 437S.
47. A. Vasconcelos-Dos-Santos, H. F. Loponte, N. R. Mantuano, I. A. Oliveira, I. A. de Paula, L. K. Teixeira, J. C. de-Freitas-Junior, K. C. Gondim, N. Heise, R. Mohana-Borges, J. A. Morgado-Diaz, W. B. Dias and A. R. Todeschini, *Oncogenesis*, 2017, **6**, e306.
48. E. Giovannucci and D. Michaud, *Gastroenterology*, 2007, **132**, 2208.
49. X. F. Sun, Y. B. Shao, M. G. Liu, Q. Chen, Z. J. Liu, B. Xu, S. X. Luo and H. Liu, *Oncol. Lett.*, 2017, **13**, 2989.
50. D. J. Jenkins, C. W. Kendall, L. S. Augustin, S. Mitchell, S. Sahye-Pudaruth, S. Blanco Mejia, L. Chiavaroli, A. Mirrahimi, C. Ireland, B. Bashyam, E. Vidgen, R. J. de Souza, J. L. Sievenpiper, J. Coveney, L. A. Leiter and R. G. Josse, *Arch. Intern. Med.*, 2012, **172**, 1653.
51. H. V. Ho, J. L. Sievenpiper, A. Zurbau, S. Blanco Mejia, E. Jovanovski, F. Au-Yeung, A. L. Jenkins and V. Vuksan, *Eur. J. Clin. Nutr.*, 2016, **70**, 1340.
52. L. A. Bazzano, *Curr. Atheroscler. Rep.*, 2008, **10**, 473.
53. T. Jakob, A. J. Nordmann, S. Schandelmaier, I. Ferreira-Gonzalez and M. Briel, *Cochrane Database Syst. Rev.*, 2016, **11**, CD009753.
54. M. Kulis and M. Esteller, *Adv. Genet.*, 2010, **70**, 27.
55. S. H. Ley, O. Hamdy, V. Mohan and F. B. Hu, *Lancet*, 2014, **383**, 1999.
56. D. Ramdath, S. Renwick and A. M. Duncan, *Can. J. Diabetes*, 2016, **40**, 355.
57. J. L. Sievenpiper and P. D. Dworatzek, *Can. J. Diabetes*, 2013, **37**, 51.
58. N. Becerra-Tomas, A. Diaz-Lopez, N. Rosique-Esteban, E. Ros, P. Buil-Cosiales, D. Corella, R. Estruch, M. Fito, L. Serra-Majem, F. Aros, R. M. Lamuela-Raventos, M. Fiol, J. M. Santos-Lozano, J. M. Diez-Espino, O. Portoles and J. Salas-Salvado, *Clin. Nutr.*, 2018, **37**(3), 906.
59. V. Ha, J. L. Sievenpiper, R. J. de Souza, V. H. Jayalath, A. Mirrahimi, A. Agarwal, L. Chiavaroli, S. B. Mejia, F. M. Sacks, M. Di Buono, A. M. Bernstein, L. A. Leiter, P. M. Kris-Etherton, V. Vuksan, R. P. Bazinet, R. G. Josse, J. Beyene, C. W. Kendall and D. J. Jenkins, *Can. Med. Assoc. J. J. Assoc. Med. Can.*, 2014, **186**, E252.
60. L. A. Bazzano, A. M. Thompson, M. T. Tees, C. H. Nguyen and D. M. Winham, *Nutr., Metab. Cardiovasc. Dis.*, 2011, **21**, 94.

61. H. H. Hermsdorff, M. A. Zulet, I. Abete and J. A. Martinez, *Eur. J. Nutr.*, 2011, **50**, 61.
62. E. Viguiliouk, S. Blanco Mejia, C. W. Kendall and J. L. Sievenpiper, *Ann. N. Y. Acad. Sci.*, 2017, **1392**, 43.
63. R. H. Eckel, J. M. Jakicic, J. D. Ard, J. M. de Jesus, N. Houston Miller, V. S. Hubbard, I. M. Lee, A. H. Lichtenstein, C. M. Loria, B. E. Millen, C. A. Nonas, F. M. Sacks, S. C. Smith, L. P. Svetkey, T. A. Wadden and S. Z. Yanovski, *J. Am. Coll. Cardiol.*, 2014, **63**, 2960.
64. P. H. Lahmann, *et al.*, *Int. J. Cancer*, 2004, **111**, 762.
65. I. Martin-Timon, C. Sevillano-Collantes, A. Segura-Galindo and F. J. Del Canizo-Gomez, *World J. Diabetes*, 2014, **5**, 444.
66. M. Dobbins, K. Decorby and B. C. Choi, *ISRN Prev. Med.*, 2013, 680536.
67. B. Niemann, R. Pan, M. Teschner, A. Boening, R. E. Silber and S. Rohrbach, *Exp. Gerontol.*, 2013, **48**, 55.
68. T. H. Oh, J. S. Byeon, S. J. Myung, S. K. Yang, K. S. Choi, J. W. Chung, B. Kim, D. Lee, J. H. Byun, S. J. Jang and J. H. Kim, *J. Gastroenterol. Hepatol.*, 2008, **23**, 411.
69. S. C. Larsson and A. Wolk, *Br. J. Cancer*, 2007, **97**, 1005.
70. S. S. Li, *Obesity*, 2014, **22**, 1773.
71. S. J. Kim, R. J. de Souza, V. L. Choo, V. Ha, A. I. Cozma, L. Chiavaroli, A. Mirrahimi, S. Blanco Mejia, M. Di Buono, A. M. Bernstein, L. A. Leiter, P. M. Kris-Etherton, V. Vuksan, J. Beyene, C. W. Kendall, D. J. Jenkins and J. L. Sievenpiper, *Am. J. Clin. Nutr.*, 2016, **103**, 1213.
72. J. L. Sievenpiper, C. W. Kendall, A. Esfahani, J. M. Wong, A. J. Carleton, H. Y. Jiang, R. P. Bazinet, E. Vidgen and D. J. Jenkins, *Diabetologia*, 2009, **52**, 1479.
73. L. A. David, C. F. Maurice, R. N. Carmody, D. B. Gootenberg, J. E. Button, J. E. Wolfe, A. V. Ling, A. S. Devlin, Y. Varma, M. A. Fischbach, S. B. Biddinger, R. J. Dutton and P. J. Turnbaugh, *Nature*, 2014, **505**, 559.
74. R. N. Carmody, G. K. Gerber, J. M. Luevano, D. M. Gatti, L. Somes, S. K. Svenson and P. J. Turnbaugh, *Cell Host Microbe*, 2015, **17**, 72.
75. R. E. Ley, P. J. Turnbaugh, S. Klein and J. I. Gordon, *Nature*, 2006, **444**, 1022.
76. A. Schwiertz, D. Taras, K. Schafer, S. Beijer, N. A. Bos, C. Donus and P. D. Hardt, *Obesity*, 2010, **18**, 190.
77. S. Devkota, Y. Wang, M. W. Musch, V. Leone, H. Fehlner-Peach, A. Nadimpalli, D. A. Antonopoulos, B. Jabri and E. B. Chang, *Nature*, 2012, **487**, 104.
78. S. I. Sayin, A. Wahlstrom, J. Felin, S. Jantti, H. U. Marschall, K. Bamberg, B. Angelin, T. Hyotylainen, M. Oresic and F. Backhed, *Cell Metab.*, 2013, **17**, 225.
79. G. Mithieux, *Neuroendocrinology*, 2014, **100**, 89.
80. G. Mithieux, *Best Pract. Res. Clin. Endocrinol. Metab.*, 2014, **28**, 739.
81. G. Mithieux and A. Gautier-Stein, *Diabetes Res. Clin. Pract.*, 2014, **105**, 295.

82. F. Delaere, A. Duchampt, L. Mounien, P. Seyer, C. Duraffourd, C. Zitoun, B. Thorens and G. Mithieux, *Mol. Metab.*, 2012, **2**, 47.
83. G. Tolhurst, H. Heffron, Y. S. Lam, H. E. Parker, A. M. Habib, E. Diakogiannaki, J. Cameron, J. Grosse, F. Reimann and F. M. Gribble, *Diabetes*, 2012, **61**, 364.
84. L. G. Hersoug, P. Moller and S. Loft, *Nutr. Res. Rev.*, 2018, 1.
85. L. G. Hersoug, P. Moller and S. Loft, *Obesity Rev. Off. J. Int. Assoc. Study Obesity*, 2016, **17**, 297.
86. A. Nilsson, E. Johansson, L. Ekstrom and I. Bjorck, *PLoS One*, 2013, **8**, e59985.
87. H. A. Vergara-Castaneda, R. G. Guevara-Gonzalez, M. Ramos-Gomez, R. Reynoso-Camacho, H. Guzman-Maldonado, A. A. Feregrino-Perez, B. D. Ooma and G. Loarca-Pina, *Food Funct.*, 2010, **1**, 294.
88. A. Esmaillzadeh and L. Azadbakht, *J. Nutr.*, 2012, **142**, 334.
89. Z. Zhang, E. Lanza, P. M. Kris-Etherton, N. H. Colburn, D. Bagshaw, M. J. Rovine, J. S. Ulbrecht, G. Bobe, R. S. Chapkin and T. J. Hartman, *Lipids*, 2010, **45**, 765.
90. Z. Zhang, E. Lanza, A. C. Ross, P. S. Albert, N. H. Colburn, M. J. Rovine, D. Bagshaw, J. S. Ulbrecht and T. L. Hartman, *Eur. J. Clin. Nutr.*, 2011, **65**, 415.
91. M. Gonzalez-Vallinas, S. Molina, G. Vicente, A. de la Cueva, T. Vargas, S. Santoyo, M. R. Garcia-Risco, T. Fornari, G. Reglero and A. Ramirez de Molina, *Pharmacol. Res.*, 2013, **72**, 61.
92. L. Raj, T. Ide, A. U. Gurkar, M. Foley, M. Schenone, X. Li, N. J. Tolliday, T. R. Golub, S. A. Carr, A. F. Shamji, A. M. Stern, A. Mandinova, S. L. Schreiber and S. W. Lee, *Nature*, 2011, **475**, 231.
93. A. Ramirez de Molina, T. Vargas, S. Molina, J. Sanchez, J. Martinez-Romero, M. Gonzalez-Vallinas, R. Martin-Hernandez, R. Sanchez-Martinez, M. Gomez de Cedron, A. Davalos, L. Calani, D. Del Rio, A. Gonzalez-Sarrias, J. C. Espin, F. A. Tomas-Barberan and G. Reglero, *J. Pharmacol. Exp. Ther.*, 2015, **353**, 433.
94. D. L. Alekel, A. S. Germain, C. T. Peterson, K. B. Hanson, J. W. Stewart and T. Toda, *Am. J. Clin. Nutr.*, 2000, **72**, 844.
95. T. Wallerath, G. Deckert, T. Ternes, H. Anderson, H. Li, K. Witte and U. Forstermann, *Circulation*, 2002, **106**, 1652.
96. C. E. Smith, *Curr. Nutr. Rep.*, 2012, **1**, 161.
97. M. Messina, W. McCaskill-Stevens and J. W. Lampe, *J. Natl. Cancer Inst.*, 2006, **98**(18), 1275.
98. B. J. Trock, L. Hilakivi-Clarke and R. Clarke, *J. Natl. Cancer Inst.*, 2006, **98**, 459.
99. J. M. Pavese, R. L. Farmer and R. C. Bergan, *Cancer Metastasis Rev.*, 2010, **29**, 465.
100. Z. Zhang, C. Z. Wang, G. J. Du, L. W. Qi, T. Calway, T. C. He, W. Du and C. S. Yuan, *Int. J. Oncol.*, 2013, **43**, 289.
101. A. Clemente and C. Arques, *World J. Gastroenterol.*, 2014, **20**, 10305.

102. A. Clemente and C. Domoney, *Recent Prog. Med. Plants*, 2008, **20**, 397.
103. G. Hocman, *Int. J. Biochem.*, 1992, **24**, 1365.
104. F. Li, P. P. Fernandez, P. Rajendran, K. M. Hui and G. Sethi, *Cancer Lett.*, 2010, **292**, 197.
105. S. O. Lee, A. L. Simons, P. A. Murphy and S. Hendrich, *Exp. Biol. Med.*, 2005, **230**, 472.
106. D. Oakenfull and G. S. Sindhu, *Eur. J. Clin. Nutr.*, 1990, **44**, 79.
107. A. V. Rao and M. K. Sung, *J. Nutr.*, 1995, **125**, 717S.
108. Y. Aguilera, M. Rebollo-Hernanz, T. Herrera, L. T. Cayuelas, P. Rodriguez-Rodriguez, A. L. de Pablo, S. M. Arribas and M. A. Martin-Cabrejas, *J. Agric. Food Chem.*, 2014, **62**, 10736.
109. K. M. Nieman, I. L. Romero, B. Van Houten and E. Lengyel, *Biochim. Biophys. Acta*, 2013, **1831**, 1533.
110. Y. S. Chan, Y. Zhang, S. C. Sze and T. B. Ng, *J. Enzyme Inhib. Med. Chem.*, 2014, **29**, 485.
111. A. Clemente, M. C. Marin-Manzano, E. Jimenez, M. C. Arques and C. Domoney, *Br. J. Nutr.*, 2012, **108**(Suppl 1), S135.
112. A. Clemente, G. Sonnante and C. Domoney, *Curr. Protein Pept. Sci.*, 2011, **12**, 358.
113. A. Clemente, F. J. Moreno, M. C. Marin-Manzano, E. Jimenez and C. Domoney, *Mol. Nutr. Food Res.*, 2010, **54**, 396.
114. A. R. Kennedy, P. C. Billings, X. S. Wan and P. M. Newberne, *Nutr. Cancer*, 2002, **43**, 174.
115. I. E. Liener, *Carbohydr. Res.*, 1991, **213**, 1.
116. R. Hartmann and H. Meisel, *Curr. Opin. Biotechnol.*, 2007, **18**, 163.
117. J. E. Lima, A. L. Sampaio, M. G. Henriques and C. Barja-Fidalgo, *Immunopharmacology*, 1999, **41**, 147.
118. A. Pusztai, G. Grant, W. C. Buchan, S. Bardocz, S. de Carvalho and S. W. Ewen, *Br. J. Nutr.*, 1998, **79**, 213.
119. B. Blomme, C. Van Steenkiste, N. Callewaert and H. Van Vlierberghe, *J. Hepatol.*, 2009, **50**, 592.
120. L. L. Fu, C. C. Zhou, S. Yao, J. Y. Yu, B. Liu and J. K. Bao, *Int. J. Biochem. Cell Biol.*, 2011, **43**, 1442.
121. K. Sames, U. Schumacher, Z. Halata, E. J. Van Damme, W. J. Peumans, B. Asmus, R. Moll and I. Moll, *Exp. Dermatol.*, 2001, **10**, 100.
122. E. G. De Mejia and V. I. Prisecaru, *Crit. Rev. Food Sci. Nutr.*, 2005, **45**, 425.
123. H. Wang, T. B. Ng, V. E. Ooi and W. K. Liu, *Int. J. Biochem. Cell Biol.*, 2000, **32**, 365.
124. P. K. Panda, S. Mukhopadhyay, B. Behera, C. S. Bhol, S. Dey, D. N. Das, N. Sinha, A. Bissoyi, K. Pramanik, T. K. Maiti and S. K. Bhutia, *Life Sci.*, 2014, **111**, 27.
125. I. Pryme, S. Bardocz, A. Pusztai and S. W. Ewen, *Histol. Histopathol.*, 2006, **21**, 285.
126. B. Xu and S. K. Chang, *J. Agric. Food Chem.*, 2010, **58**, 1509.

127. S. N. Nichenametla, T. G. Taruscio, D. L. Barney and J. H. Exon, *Crit. Rev. Food Sci. Nutr.*, 2006, **46**, 161.
128. C. Spanou, D. Stagos, L. Tousias, A. Angelis, N. Aligiannis, A. L. Skaltsounis and D. Kouretas, *Anticancer Res.*, 2007, **27**, 3403.
129. C. A. Adebamowo, E. Cho, L. Sampson, M. B. Katan, D. Spiegelman, W. C. Willett and M. D. Holmes, *Int. J. Cancer*, 2005, **114**, 628.
130. P. L. Horn-Ross, E. M. John, A. Canchola, S. L. Stewart and M. M. Lee, *J. Natl. Cancer Inst.*, 2003, **95**, 1158.
131. S. Banerjee, Y. Li, Z. Wang and F. H. Sarkar, *Cancer Lett.*, 2008, **269**, 226.
132. Y. H. Ju, K. F. Allred, C. D. Allred and W. G. Helferich, *Carcinogenesis*, 2006, **27**, 1292.
133. J. J. Anderson, M. S. Anthony, J. M. Cline, S. A. Washburn and S. C. Garner, *Public Health Nutr.*, 1999, **2**, 489.
134. S. I. Mishra, V. Dickerson and W. Najm, *Am. J. Obstet. Gynecol.*, 2003, **188**, S66.
135. G. Sikand, P. Kris-Etherton and N. M. Boulos, *Curr. Cardiol. Rep.*, 2015, **17**, 39.
136. L. Varinska, P. Gal, G. Mojzisova, L. Mirossay and J. Mojzis, *Int. J. Mol. Sci.*, 2015, **16**, 11728.
137. T. A. Woyengo, V. R. Ramprasath and P. J. Jones, *Eur. J. Clin. Nutr.*, 2009, **63**, 813.
138. P. J. Jones and S. S. AbuMweis, *Curr. Opin. Clin. Nutr. Metab. Care*, 2009, **12**, 147.
139. R. F. Raicht, B. I. Cohen, E. P. Fazzini, A. N. Sarwal and M. Takahashi, *Cancer Res.*, 1980, **40**, 403.
140. A. B. Awad, A. Hernandez, C. S. Fink and S. L. Mendel, *Nutr. Cancer*, 1997, **27**, 210.
141. Y. Wang, J. G. Corr, H. T. Thaler, Y. Tao, W. R. Fair and W. D. Heston, *J. Natl. Cancer Inst.*, 1995, **87**, 1456.
142. V. Escurriol, M. Cofan, C. Moreno-Iribas, N. Larranaga, C. Martinez, C. Navarro, L. Rodriguez, C. A. Gonzalez, D. Corella and E. Ros, *J. Lipid Res.*, 2010, **51**, 618.
143. E. Windler, B. C. Zyriax, F. Kuipers, J. Linseisen and H. Boeing, *Atherosclerosis*, 2009, **203**, 284.
144. D. O. Moon, M. O. Kim, Y. H. Choi and G. Y. Kim, *Cancer Lett.*, 2008, **264**, 181.
145. R. L. von Holtz, C. S. Fink and A. B. Awad, *Nutr. Cancer*, 1998, **32**, 8.
146. D. M. Gurfinkel and A. V. Rao, *Nutr. Cancer*, 2003, **47**, 24.
147. B. Singh, J. P. Singh, K. Shevkani, N. Singh and A. Kaur, *J. Food Sci. Technol.*, 2017, **54**, 858.
148. P. Leterme, *Br. J. Nutr.*, 2002, **88**(Suppl 3), S239.
149. S. Jiang, J. Fan, Q. Wang, D. Ju, M. Feng, J. Li, Z. B. Guan, D. An, X. Wang and L. Ye, *Phytomed. Int. J. Phytother. Phytopharmacol.*, 2016, **23**, 243.
150. J. Shi, K. Arunasalam, D. Yeung, Y. Kakuda, G. Mittal and Y. Jiang, *J. Med. Food*, 2004, **7**, 67.

151. K. K. Auyeung, P. C. Law and J. K. Ko, *Nutr. Cancer*, 2014, **66**, 662.
152. A. V. Rao and M. K. Sung, *J. Nutr.*, 1995, **125**, 717S.
153. M. K. Sung, C. W. Kendall, M. M. Koo and A. V. Rao, *Nutr. Cancer*, 1995, **23**, 259.
154. Y. J. Oh and M. K. Sung, *Nutr. Cancer*, 2001, **39**, 132.
155. M. A. Berhow, C. L. Cantrell, S. M. Duval, T. A. Dobbins, J. Maynes and S. F. Vaughn, *Phytochem. Anal.*, 2002, **13**, 343.
156. M. A. Lazo-Velez, D. Guajardo-Flores, D. Mata-Ramirez, J. A. Gutierrez-Uribe and S. O. Serna-Saldivar, *Food Sci.*, 2016, **81**, C19.
157. D. Guajardo-Flores, M. Garcia-Patino, D. Serna-Guerrero, J. Gutierrez-Uribe and S. O. Serna-Saldivar, *Food Chem.*, 2012, **134**, 1312.
158. A. S. Sandberg, *Br. J. Nutr.*, 2002, **88**(Suppl 3), S281.
159. M. Friend, I. Vucenik and M. Miller, *BMJ*, 2003, **326**, 82.
160. S. Onomi, Y. Okazaki and T. Katayama, *Biosci. Biotechnol. Biochem.*, 2004, **68**, 1379.
161. E. Graf and J. W. Eaton, *Nutr. Cancer*, 1993, **19**, 11.
162. I. Vucenik and A. M. Shamsuddin, *Nutr. Cancer*, 2006, **55**, 109.
163. K. Cholewa, B. Parfiniewicz, I. Bednarek, L. Swiatkowska, E. Jezienicka, J. Kierot and L. Weglarz, *Acta Pol. Pharm.*, 2008, **65**, 75.
164. I. Bacic, N. Druzijanic, R. Karlo, I. Skific and S. Jagic, *J. Exp. Clin. Cancer Res.*, 2010, **29**, 12.

Subject Index